Elementary Education
North Carolina State University
Campus Box 7801
Raleigh, NC 27695-7801

D1521660

Children's Fractional Knowledge

Leslie P. Steffe · John Olive

Children's Fractional Knowledge

Leslie P. Steffe
University of Georgia
Athens, GA
USA
lsteffe@uga.edu

John Olive
University of Georgia
Athens, GA
USA
jolive@uga.edu

ISBN 978-1-4419-0590-1   e-ISBN 978-1-4419-0591-8
DOI 10.1007/978-1-4419-0591-8
Springer New York Dordrecht Heidelberg London

Library of Congress Control Number: 2009932112

© Springer Science+Business Media, LLC 2010
All rights reserved. This work may not be translated or copied in whole or in part without the written permission of the publisher (Springer Science+Business Media, LLC, 233 Spring Street, New York, NY 10013, USA), except for brief excerpts in connection with reviews or scholarly analysis. Use in connection with any form of information storage and retrieval, electronic adaptation, computer software, or by similar or dissimilar methodology now known or hereafter developed is forbidden.
The use in this publication of trade names, trademarks, service marks, and similar terms, even if they are not identified as such, is not to be taken as an expression of opinion as to whether or not they are subject to proprietary rights.

Printed on acid-free paper

Springer is part of Springer Science+Business Media (www.springer.com)

*We dedicate this book to the memory
of Arthur Middleton who was taken from
us at the age of 16 by a senseless act
of violence. The contributions that Arthur
made to our understanding of children's
powerful ways of constructing mathematical
knowledge will live on in these pages.
The world is a poorer place for the lack of
what we know he could have accomplished.
Our lives and our research are so much
richer for having known and worked
with him during our teaching experiment.*

Preface

The basic hypothesis that guides our work is that children's fractional knowing can emerge as accommodations in their natural number knowing. This hypothesis is referred to as *the reorganization hypothesis* because if a new way of knowing is constructed using a previous way of knowing in a novel way, the new way of knowing can be regarded as a reorganization of the previous way of knowing. In contrast to the reorganization hypothesis, there is a widespread and accepted belief that natural number knowing *interferes* with fractional knowing. Within this belief, children are observed using their ways and means of operating with natural numbers while working with fractions and the former are thought to interfere with the latter. Children are also observed dealing with fractions in the same manner as with natural numbers, and it is thought that we must focus on forming a powerful concept of fractions that is resistant to natural number distractions.

In our work, we focus on what we are able to constitute as mathematics of children rather than solely use our own mathematical constructs to interpret and organize our experience of children's mathematics. This is a major distinction and it enables us to not act as if children have already constructed fractional ways of knowing with which natural number knowing interferes. Rather, we focus on the assimilative activity of children and, on that basis, infer the concepts and operations that children use in that activity. Focusing on assimilative activity opens the way for studying reorganizations we might induce in the assimilative concepts and operations and, hence, it opens the way for studying how children might use their natural number concepts and operations in the construction of fractional concepts and operations.

The question concerning whether fractional knowing necessarily emerges independently of natural number knowing is based on the assumption that the operations involved in fractional knowing have their origin in continuous quantity and only minimally involve discrete quantitative operations. In a developmental analysis of the operations that produce discrete quantity and continuous quantity, we show that the operations that produce each type of quantity are quite similar and can be regarded as unifying quantitative operations. The presence of such unifying operations is essential and serves as a basic rationale for the reorganization hypothesis.

We investigated the reorganization hypothesis in a 3-year teaching experiment with children who were third graders at the beginning of the experiment. We selected the children on the basis of the stages in their construction of their number sequences. Our research hypothesis was that the children would use their number sequences in the construction of their fractional concepts and operations, that the nature and quality of the fractional knowledge the children constructed within the stages would be quite similar, and that the nature and quality of the fractional knowledge the children constructed across the stages would be quite distinct. We did not begin the teaching experiment with foreknowledge of how the children would use their number sequences in their construction of fractional knowledge, nor the nature and quality of the knowledge they might construct. This book provides detailed accounts of how we tested our research hypothesis as well as detailed accounts of the fractional knowledge the children did construct in the context of working with us in the teaching experiment and of how we engendered the children's constructive activity. We do not report on the children who began the teaching experiment in the initial stage of the number sequence because of the serious constraints we experienced when teaching them.

Our overall goal is to establish images of how the mathematics of children might be used in establishing a school mathematics that explicitly includes children's mathematical thinking and learning. Toward that goal, we provide accounts of how the reorganization hypothesis was realized in the constructive activity of the participating children as well as how their number sequences both enabled and constrained their constructive activity. Further, we provide models of children's fractional knowing that we refer to as children's fraction schemes and explain how these fraction schemes are based on partitioning schemes. We found that partitioning is not a singular construct and broke new ground in explaining six partitioning schemes that are inextricably intertwined with children's number sequences and the numerical schemes that follow on from the number sequences. Explaining fraction schemes in terms of the partitioning schemes provides a way of thinking about fractions in terms of children's fraction schemes rather than in terms of the rational numbers. We used our understanding of the rational numbers throughout the teaching experiment as orienting us in our various activities, but we make a distinction between our concepts of rational numbers and our concepts of children's fraction schemes. The former are a part of our first-order mathematical knowledge and the latter are a part of our second-order mathematical knowledge.

Athens, Georgia, USA										Leslie P. Steffe
														John Olive

Acknowledgments

The genesis of this work took place in the project entitled *Children's Construction of the Rational Numbers of Arithmetic*, codirected by Leslie P. Steffe and John Olive, with support from the National Science Foundation (project No. RED-8954678) and from the Department of Mathematics Education at the University of Georgia. Several doctoral students in the department took substantial roles in the Teaching Experiment that provided the data for our analyses. We would like to acknowledge the particular contributions of Dr. Barry Biddlecomb, Dr. Azita Manouchehri, and Dr. Ron Tzur who acted as teachers for several of the teaching episodes contained in the data presented in this book. Dr. Barry Biddlecomb also helped with the programming of the original TIMA software. We would also like to thank Dr. Heide Wiegel for the help she provided in organizing, cataloging, and analyzing much of the data. The extensive retrospective analysis of the data was also supported by a second grant from the National Science Foundation, project no. REC-9814853. We are deeply indebted to the NSF and to the Department of Mathematics Education for their extensive support.

This work would not exist, of course, without the collaboration of the school district personnel, school principal, teachers, and students of the anonymous school in which we conducted the Teaching Experiment. We are also grateful to the students' parents and guardians who gave us permission to work with their children over the 3-year period and trusted that we would do our best to help their children move forward in their mathematical thinking.

Finally, we would like to thank our wives, Marilyn Steffe and Debra Brenner, for bearing with us and providing encouragement during our construction of the ideas contained in this book and our writing of it.

Athens, Georgia, USA   Leslie P. Steffe
                                                                                     John Olive

Foreword

It is a rare experience in the life of an academic to stand in awe of a body of work. I confess to having had that feeling in the midst of reading Steffe's and Olive's (SO's) account of children's development of fraction knowledge from numerical counting schemes. Their enterprise is especially important, for several reasons – some having to do with fractions and others having to do with science. I'll first say something about the latter and then speak to the former.

Science

George Johnson (2008) tells the stories of ten experiments that emanated from people questioning accepted wisdom about the physical world and the way it works. His stories are not of individual genius. Rather, the stories are about the scientific method of postulating invisible forces and mechanisms behind observable phenomena, perturbing materials to see if they respond the way your model predicts, and, most importantly, revising your model in light of the specific ways your predictions failed. The stories, above all, are a quest for *understanding*.

Whether research in mathematics education is scientific has been under heavy debate recently. Psychologists, especially experimental psychologists, tend to think it has been unscientific because of its lack of randomized sampling, experimental controls, and statistical analyses. I would argue, however, that it is those who confuse method with inquiry who are too often unscientific. Science is not about *what works*. Science is about *the way things work*. Johnson's stories repeatedly reveal that scientific advances happen when new conceptualizations of phenomena lead to greater coherence among disparate facts and theories. Lavoisier's investigations into the nature of phlogiston, the "stuff" whose release from a substance produces flames, eventually led him to the conclusion that there is no such stuff as phlogiston! After Lavoisier, no one saw combustion as entailing the hidden forces and mechanisms that everyone saw 10 years prior. Were modern psychologists dominant in 1790, they would have criticized Lavoisier for his lack of experimental control. But he had a *strong* experimental control – an initial model of how combustion is supposed to work and of the materials involved in those processes, and it was his model of how combustion works that he investigated.

It is in this spirit that you must read this book – that SO's enterprise is to start with, test, and refine their models of children's fractional thinking. They also take seriously the constraints of employing a constructivist framework for their models, predictions, and explanations: Children's mathematical knowledge does not appear from nothing. It comes from what children know in interaction with situations that they construe as being somehow problematic. To be scientific in this investigation, SO take great pains to give precise model-based accounts of the ways of thinking that children bring to the settings that SO design for them.

Which brings up another point. The significance of children's behaviors can only be judged in the context of the tasks with which they engage and as they construe them. In fact, how a child construes a task often gives insight into the ways of thinking the child has. I urge readers to read SO's tasks carefully and to understand the computing environment that gave context to them. The computing environment (TIMA) afforded actions to children that are not possible with physical sticks, and hence children were able to express anticipations of acting that would not have been possible outside that environment. Read the tasks slowly so as to imagine what cognitive issues might be at play in responding to them. I urge you also to read teaching-experiment excerpts slowly. SO's models afford very precise predictions of children's behavior and very precise explanations of their thinking, so the smallest nuance in a child's behavior can have profound implications for the theoretical discussion of that behavior. For example, according to SO's models of number sequences, if a child partitions a segment into 10 equal-sized parts, but has to physically iterate one part to see how long 10 of them will be, this has tremendous implications for the fraction knowledge we can attribute to him or her. The contribution of SO's work is that their theoretical framework not only supports such nuanced distinctions, but also allows us to understand what might appear to some as uneven fraction knowledge instead as a coherent system of thinking that has evolved to a particular state (and will evolve further to states of greater coherence).

Finally, it is imperative in reading this book that you understand that *SO employ teaching as an experimental method.* Understanding this, however, requires an expanded meaning for experiment and a nonstandard meaning for teaching. The idea of a scientific experiment is to poke nature to see how it responds. That is, we start with an idea of how nature works in some area of interest, perturb nature to see whether it responds in the way our understandings would suggest, nature responds according to its own structures, and then we revise our understandings accordingly. It is in this respect – teaching as a designed provocation – that it can be used as an experimental method in understanding children's thinking. To be used effectively as an experimental method, though, you cannot think of teaching as a means for transmitting information to children. Rather, you must think of it as an interaction with children that is guided by your models of children's thinking and by what you discern of their thinking by listening closely to what they say and do. Of course, all this is with the backdrop that children are participating according to their ways of thinking and with the intent of understanding your, the teacher's, actions.

Fractions

SO's basic thesis is that children's fraction knowledge can emerge by way of a reorganization of their numerical counting schemes. This might, at first blush, seem like a weak hypothesis, as in "you can devise super special methods and invest super human effort to have students create fractions from their counting schemes." I propose a different interpretation:

> *If allowed*, children can, and in most cases will, use their counting schemes to create ways of understanding numerical and quantitative relationships that we recognize as powerful fractional reasoning.

The phrase "if allowed" is highly loaded. It does not mean that children should be turned loose, with no adult intervention, to create their own mathematics. We know that little of consequence will result. Rather, it means that the instructional and material environments must be shaped so that they are amenable to children using natural ways of reasoning to create more powerful ways of reasoning – they are designed to respect children's thinking and build from it.

There are three important aspects to SO's argument for the reorganization hypothesis. The first is that they did not start with it. Rather, it emerged from their interactions with children. In a sense, the children forced the reorganization hypothesis upon SO. Children whose number sequences did not progress to higher levels of organization simply were unable to progress in their fraction knowledge despite SO's best attempts to move them along. Children whose number sequences were limited, developmentally speaking, to early forms simply could not see fraction tasks in the ways that children with the generalized number sequence could.

Second, the reorganization hypothesis entails the claim that children's number sequences are very much at play as they develop spatial operations with continuous quantities. It is through their number sequences that children impose segmentations on continuous quantities and reassemble them as measured quantities.

Third, SO's reorganization hypothesis removes any need to think that the operation of splitting, as described by Confrey, appears independently of counting. In a very real sense, SO's explication of the reorganization hypothesis gives Confrey's work a developmental foundation. But it does more. As noted by Norton and Hackenberg (Chap. 11), the splitting operation described by Confrey is not sufficient for children to generate the highest level of fraction reasoning described by SO. More is required, and SO give a compelling argument for what that is.

Next Steps

Norton and Hackenberg (Chap. 11) give a highly useful analysis of potential connections between SO's research on fractions with other research programs in the development of algebraic and quantitative reasoning. My hope is that SO's research develops another set of connections – with pedagogy and curriculum. What sense might teachers make of the reorganization hypothesis? What reorganizations must they make to understand it and to use it? What professional development structures could help them understand and use it? How could the reorganization hypothesis inform the development of curriculum that in turn would support teachers as they attempt to actualize the reorganization hypothesis? I look forward to SO and protégés giving us insight into these questions.

Tempe, Arizona, USA   Patrick W. Thompson

Reference

Johnson G (2008) The ten most beautiful experiments. Knopf, New York

Contents

1	**A New Hypothesis Concerning Children's Fractional Knowledge**	1
	The Interference Hypothesis	2
	The Separation Hypothesis	5
	A Sense of Simultaneity and Sequentiality	6
	Establishing Two as Dual	7
	Establishing Two as Unity	8
	Recursion and Splitting	9
	Distribution and Simultaneity	10
	Splitting as a Recursive Operation	11
	Next Steps	12
2	**Perspectives on Children's Fraction Knowledge**	13
	On Opening the Trap	14
	Invention or Construction?	15
	First-Order and Second-Order Mathematical Knowledge	16
	Mathematics of Children	16
	Mathematics for Children	17
	Fractions as Schemes	18
	The Parts of a Scheme	20
	Learning as Accommodation	21
	The Sucking Scheme	21
	The Structure of a Scheme	22
	Seriation and Anticipatory Schemes	24
	Mathematics of Living Rather Than Being	25
3	**Operations That Produce Numerical Counting Schemes**	27
	Complexes of Discrete Units	27
	Recognition Templates of Perceptual Counting Schemes	29
	Collections of Perceptual Items	29
	Perceptual Lots	30
	Recognition Templates of Figurative Counting Schemes	32
	Numerical Patterns and the Initial Number Sequence	35

The Tacitly Nested Number Sequence	38
The Explicitly Nested Number Sequence	41
An Awareness of Numerosity: A Quantitative Property	42
The Generalized Number Sequence	43
An Overview of the Principal Operations of the Numerical Counting Schemes	45
The Initial Number Sequence	45
The Tacitly Nested and the Explicitly Nested Number Sequences	45
Final Comments	47

4 Articulation of the Reorganization Hypothesis ... 49

Perceptual and Figurative Length	50
Piaget's Gross, Intensive, and Extensive Quantity	51
Gross Quantitative Comparisons	52
Intensive Quantitative Comparisons	52
An Awareness of Figurative Plurality in Comparisons	53
Extensive Quantitative Comparisons	55
Composite Structures as Templates for Fragmenting	57
Experiential Basis for Fragmenting	58
Using Specific Attentional Patterns in Fragmenting	59
Number Sequences and Subdividing a Line	64
Partitioning and Iterating	67
Levels of Fragmenting	68
Final Comments	70
Operational Subdivision and Partitioning	72
Partitioning and Splitting	73

5 The Partitive and the Part-Whole Schemes ... 75

The Equipartitioning Scheme	75
Breaking a Stick into Two Equal Parts	75
Composite Units as Templates for Partitioning	76
Segmenting to Produce a Connected Number	78
Equisegmenting vs. Equipartitioning	78
The Dual Emergence of Quantitative Operations	80
Making a Connected Number Sequence	80
An Attempt to Use Multiplying Schemes in the Construction of Composite Unit Fractions	83
Provoking the Children's use of Units-Coordinating Schemes	83
An Attempt to Engender the Construction of Composite Unit Fractions	86
Conflating Units When Finding Fractional Parts of a 24-Stick	87
Operating on Three Levels of Units	89
Necessary Errors	90

Laura's Simultaneous Partitioning Scheme	92
An Attempt to Bring Forth Laura's Use of Iteration to Find Fractional Parts	95
Jason's Partitive and Laura's Part-Whole Fraction Schemes	98
Lack of the Splitting Operation	98
Jason's Partitive Unit Fraction Scheme	100
Laura's Independent Use of Parts	102
Laura's Part-Whole Fraction Scheme	107
Establishing Fractional Meaning for Multiple Parts of a Stick	110
A Recurring Internal Constraint in the Construction of Fraction Operations	112
Continued Absence of Fractional Numbers	113
An Attempt to Use Units-Coordinating to Produce Improper Fractions	114
A Test of the Iterative Fraction Scheme	116
Discussion of the Case Study	118
The Construction of Connected Numbers and the Connected Number Sequence	118
On the Construction of the Part-Whole and Partitive Fraction Schemes	119
The Splitting Operation	121

6 The Unit Composition and the Commensurate Schemes ... 123

The Unit Fraction Composition Scheme	124
Jason's Unit Fraction Composition Scheme	125
Corroboration of Jason's Unit Fraction Composition Scheme	126
Laura's Apparent Recursive Partitioning	128
Producing Composite Unit Fractions	129
Laura's Reliance on Social Interaction When Explaining Commensurate Fractions	133
Further Investigation into the Children's Explanations and Productions	136
Producing Fractions Commensurate with One-Half	138
Producing Fractions Commensurate with One-Third	142
Producing Fractions Commensurate with Two-Thirds	147
An Attempt to Engage Laura in the Construction of the Unit Fraction Composition Scheme	148
The Emergence of Recursive Partitioning for Laura	151
Laura's Apparent Construction of a Unit Fraction Composition Scheme	153
Progress in Partitioning the Results of a Prior Partition	157
Discussion of the Case Study	161
The Unit Fraction Composition Scheme and the Splitting Operation	162
Independent Mathematical Activity and the Splitting Operation	163

       Independent Mathematical Activity and the Commensurate
       Fraction Scheme 163
       An Analysis of Laura's Construction of the Unit Fraction
       Composition Scheme 164
       Laura's Apparent Construction of Recursive Partitioning
       and the Unit Fraction Composition Scheme 169

7 The Partitive, the Iterative, and the Unit Composition Schemes 171

    Joe's Attempts to Construct Composite Unit Fractions 172
    Attempts to Construct a Unit Fraction of a Connected Number 174
    Partitioning and Disembedding Operations 176
    Joe's Construction of a Partitive Fraction Scheme 180
    Joe's Production of an Improper Fraction 185
    Patricia's Recursive Partitioning Operations 188
       The Splitting Operation: Corroboration in Joe and Contraindication
       in Patricia 188
    A Lack of Distributive Reasoning 191
    Emergence of the Splitting Operation in Patricia 193
    Emergence of Joe's Unit Fraction Composition Scheme 195
    Joe's Reversible Partitive Fraction Scheme 197
       Fractions Beyond the Fractional Whole: Joe's Dilemma
       and Patricia's Construction 199
    Joe's Construction of the Iterative Fraction Scheme 204
    A Constraint in the Children's Unit Fraction Composition Scheme 208
    Fractional Connected Number Sequences 211
    Establishing Commensurate Fractions 214
    Discussion of the Case Study 217
       Composite Unit Fractions: Joe 217
       Joe's Partitive Fraction Scheme 218
       Emergence of the Splitting Operation and the Iterative
       Fraction Scheme: Joe 219
       Emergence of Recursive Partitioning and Splitting Operations:
       Patricia 220
       The Construction of the Iterative Fraction Scheme 221
       Stages in the Construction of Fraction Schemes 222

8 Equipartitioning Operations for Connected Numbers:
    Their Use and Interiorization 225

    Melissa's Initial Fraction Schemes 225
       Contraindication of Recursive Partitioning in Melissa 227
       Reversibility of Joe's Unit Fraction Composition Scheme 228
    A Reorganization in Melissa's Units-Coordinating Scheme 231
    Melissa's Construction of a Fractional Connected Number Sequence 236
    Testing the Hypothesis that Melissa Could Construct
    a Commensurate Fraction Scheme 241

Melissa's Use of the Operations that Produce Three Levels
of Units in Re-presentation 247
    Repeatedly Making Fractions of Fractional Parts
    of a Rectangular Bar 247
    Melissa Enacting a Prior Partitioning by Making a Drawing 251
    A Test of Accommodation in Melissa's Partitioning Operations 254
    A Further Accommodation in Melissa's Recursive
    Partitioning Operations 256
A Child-Generated Fraction Adding Scheme 260
An Attempt to Bring Forth a Unit Fraction Adding Scheme 263
Discussion of the Case Study 266
The Iterative Fraction Scheme 268
    Melissa's Interiorization of Operations that Produce
    Three Levels of Units 269
    On the Possible Construction of a Scheme of Recursive
    Partitioning Operations 271
    The Children's Meaning of Fraction Multiplication 273
    A Child-Generated vs. a Procedural Scheme
    for Adding Fractions 275

**9 The Construction of Fraction Schemes Using the Generalized
Number Sequence** 277
The Case of Nathan During His Third Grade 277
    Nathan's Generalized Number Sequence 278
    Developing a Language of Fractions 279
    Reasoning Numerically to Name Commensurate Fractions 284
    Corroboration of the Splitting Operation for Connected
    Numbers 286
    Renaming Fractions: An Accommodation of the IFS: CN 288
    Construction of a Common Partitioning Scheme 289
    Constructing Strategies for Adding Unit Fractions with Unlike
    Denominators 291
Multiplication of Fractions and Nested Fractions 295
Equal Fractions 298
    Generating a Plurality of Fractions 299
    Working on a Symbolic Level 301
Construction of a Fraction Composition Scheme 303
    Constraining How Arthur Shared Four-Ninths of a Pizza
    Among Five People 304
    Testing the Hypothesis Using TIMA: Bars 307
Discussion of the Case Study 310
    The Reversible Partitive Fraction Scheme 310
    The Common Partitioning Scheme and Finding
    the Sum of Two Fractions 311
    The Fractional Composition Scheme 313

10 The Partitioning and Fraction Schemes ... 315
The Partitioning Schemes ... 315
    The Equipartitioning Scheme ... 315
    The Simultaneous Partitioning Scheme ... 316
    The Splitting Scheme ... 317
    The Equipartitioning Scheme for Connected Numbers ... 319
    The Splitting Scheme for Connected Numbers ... 320
    The Distributive Partitioning Scheme ... 321
The Fraction Schemes ... 322
    The Part-Whole Fraction Scheme ... 322
    The Partitive Fraction Scheme ... 323
    The Unit Fraction Composition Scheme ... 328
    The Fraction Composition Scheme ... 330
    The Iterative Fraction Scheme ... 333
    The Unit Commensurate Fraction Scheme ... 335
    The Equal Fraction Scheme ... 336
School Mathematics vs. "School Mathematics" ... 337

11 Continuing Research on Students' Fraction Schemes ... 341
Research on Part-Whole Conceptions of Fractions ... 342
Transcending Part-Whole Conceptions ... 344
The Splitting Operation ... 345
Students' Development Toward Algebraic Reasoning ... 348

References ... 353

Index ... 359

List of Figures

Fig. 2.1.	A sharing situation	19
Fig. 2.2.	A diagram for the structure of a scheme	23
Fig. 3.1.	An attentional pattern: Sensory-motor item	28
Fig. 3.2.	The attentional pattern of a perceptual unit item	31
Fig. 3.3.	The attentional pattern of a perceptual lot	31
Fig. 3.4.	The attentional pattern of a figurative unit item	33
Fig. 3.5.	The attentional structure of a figurative lot	34
Fig. 3.6.	The attentional structure of a numerical pattern	36
Fig. 3.7.	The attentional structure of the initial number sequence	38
Fig. 3.8.	The attentional structure of the numerical composite "six"	38
Fig. 3.9.	Attentional structure of a composite unit of numerosity nine	40
Fig. 3.10.	The attentional structure of a composite unit containing an iterable unit	42
Fig. 4.1.	Two rows of blocks: Endpoints not coincident	52
Fig. 4.2.	Two rows of blocks: Endpoints coincident	52
Fig. 4.3.	The first subdivision of a line task	64
Fig. 4.4.	The second subdivision of a line task	64
Fig. 5.1.	Cutting a stick into two equal parts using visual estimation	76
Fig. 5.2.	Jason testing if one piece is one of the four equal pieces	77
Fig. 5.3.	Jason's completed test	77
Fig. 5.4.	The results of Laura marking a stick into seven parts	97
Fig. 5.5.	Jason's attempt to make a stick so that a given stick is five times longer than the new stick	99
Fig. 5.6.	Laura's attempt to make two equal shares of a stick	104
Fig. 5.7.	Laura's attempt to mark a stick into two equal shares	104
Fig. 5.8.	Using PARTS in partitioning	105

Fig. 6.1.	Jason's unit fraction composition scheme	127
Fig. 6.2.	The situation of protocol XVII	157
Fig. 6.3.	Laura's introduction of vertical PARTS	160
Fig. 6.4.	The result of partitioning one-ninth into thirds	160
Fig. 7.1.	A set of number-sticks in TIMA: Sticks	175
Fig. 7.2.	Making estimates for one-fourth of a 27-stick	177
Fig. 7.3.	Joe's estimates for one-fourth of a 27-stick	178
Fig. 7.4.	Finding one-seventh of a mystery stick	179
Fig. 7.5.	Find a stick that is one-fifth of the long blue bottom stick	180
Fig. 7.6.	Estimating a stick that is three times as long as a 1/3-stick	182
Fig. 7.7.	Joe's mark for four-fifths of a stick	185
Fig. 7.8.	Making six-fifths of a stick	186
Fig. 8.1.	Making one-eighth by partitioning one-fourth	249
Fig. 8.2.	Making one-thirty-second by partitioning one-sixteenth	249
Fig. 8.3.	Making one-one-hundred twenty-eighth by partitioning one-sixty-fourths	249
Fig. 8.4.	Melissa's partition of the first part of a nine-part stick into three parts	253
Fig. 8.5.	Melissa's drawing of a partition a partition of a partition	253
Fig. 8.6.	Joe partitioning one-sixth of the bar into three parts	255
Fig. 8.7.	Melissa's partition of one-eighteenth of a bar into two parts	255
Fig. 8.8.	Joe's partition of a one-twelfth of a bar into four pieces	257
Fig. 8.9.	Melissa coordinating her drawing and her notational system	257
Fig. 8.10.	Melissa filling one-eighth of the bar	259
Fig. 8.11.	Melissa's partition of one-eighteenth of a bar into two parts	262
Fig. 9.1.	Sharing two bars among three mats	281
Fig. 9.2.	Nathan makes five copies of two parts of a 6-part bar to complete eight-eighths from three-eighths	282
Fig. 9.3.	Nathan makes a bar that is two and a half times as much as a unit bar	283
Fig. 9.4.	Nathan gives two-thirds of a 15-part bar to the big mat	286
Fig. 9.5.	The results of partitioning three-tenths of a bar into four parts	295
Fig. 9.6.	Nathan compares one-thirteenth to one-fourth of three-tenths of a unit bar	296
Fig. 9.7.	Nathan makes one-fourth of nine-sevenths of a unit bar	297

Fig. 9.8.	Nathan makes one-third of a 7/7-bar	298
Fig. 9.9.	The fraction labeler from TIMA: Sticks	302
Fig. 9.10.	Four people share three-sevenths of a pizza stick	303
Fig. 9.11.	Nine iterations of three-fourths of one-seventh of a pizza stick	304
Fig. 9.12.	Sharing four-ninths of a pizza stick among five people	305
Fig. 9.13.	Two-thirds of one-seventh of a stick	306
Fig. 9.14.	Iterating three twenty-firsts to make a whole stick	306
Fig. 9.15.	Filling one-seventh of four-ninths of a bar	309
Fig. 11.1.	Task response providing indication of a splitting operation	345

Chapter 1
A New Hypothesis Concerning Children's Fractional Knowledge

Leslie P. Steffe

The basic hypothesis that guides our work is that *children's fraction schemes can emerge as accommodations in their numerical counting* schemes. This hypothesis is referred to as the reorganization hypothesis because if a new scheme is constructed by using another scheme in a novel way, the new scheme can be regarded as a reorganization of the prior scheme. There are two basic ways of understanding the reorganization of a prior scheme. The first is that the child constructs the new scheme by operating on the preceding scheme using operations that can be, but may not be, a part of the operations of that scheme. In this case, the new scheme is of the same type as the preceding scheme. But it solves problems and serves purposes that the preceding scheme did not solve or did not serve. It also solves all of the problems the preceding scheme solved, but it solves them better. It is in this sense that the new scheme supersedes the preceding, more primitive, scheme.

The first type of reorganization was important in my work on children's construction of numerical counting schemes (Steffe 1994a). For example, a child might only be able to count items in its perceptual field by coordinating the utterance of a number word in her number word sequence with pointing at each item. If the child abstracts its pointing acts from the acts of counting and uses the pointing acts as countable items, the pointing acts can stand in for perceptual items that are hidden from view. The child can still solve all the old counting problems, but now by counting pointing motions. And the child can solve new problems, such as counting the number of cookies where five cookies are showing and three more cookies are shown to the child and then hidden.

The second way of understanding the reorganization hypothesis is that the child constructs the new schemes by operating on novel material in situations that are not a part of the situations of the preceding schemes. The child uses operations of the preceding schemes in ways that are novel with respect to the situations of the schemes as well as operations that may not be a part of the operations of the preceding schemes. The new schemes that are produced solve situations, which the preceding schemes did not solve, and they also serve purposes, which the preceding schemes did not serve. But the new schemes do not supersede the preceding schemes because they do not solve all of the situations, which the preceding schemes solved. They might solve situations similar to those solved by the preceding schemes in the context of the new situations, but the preceding schemes are still needed to solve their situations. Still, the new schemes can be regarded as reorganizations of the preceding schemes because operations of the preceding schemes emerge in a new organization and serve a different purpose. It is this second way of understanding the reorganization hypothesis that is relevant in our current work.

The Interference Hypothesis

There is an alternative to our basic hypothesis that has historical roots in educational practice in the elementary school. In contrast to the reorganization hypothesis, there is a widespread and accepted belief that whole number knowledge interferes with the learning of fractions. For example, Post et al. (1993) commented that:

> Children often have difficulty overcoming their whole number ideas while working with fractions or decimals.... To order fractions with the same numerator as 1/3 and 1/2, fourth graders in the RNP[1] teaching experiment often asked the clarifying question, "do you want me to order by the number of pieces or by the size of piece?"... RNP instructors thought their original lessons adequately treated the issue relating to using the size of the piece as the criterion for ordering fractions, but the children's whole number strategies appeared to persist and temporarily interfere with the development of this new concept (p. 339).

The belief portrayed by these researchers is similar to a comment made by Streefland (1991) in a detailed report of children's fractional knowledge: "But the only alarming ailment is the following one, namely, the temptation to deal with fractions in the same manner as with natural numbers" (p. 70). Streefland believed that we must focus on forming a powerful concept of fractions that is resistant to distractions. The distractions are of the sort that children add numerators and denominators when adding fractions. According to Streefland (1991), "we must not only focus on producing fractions, but also on grounded refutations of such misconceptions, or simply, on overcoming these misconceptions" (p. 70).

There is no question that children's fractional knowledge involves ways and means of operating that are not available in their whole number knowledge. I essentially agree with the following comment by Kieren (1993) concerning rational number knowing: "Although intertwined with, sharing language with, and using concepts from whole numbers, rational number knowing is not a simple extension of whole number knowing" (p. 56). Even though Kieren makes a distinction between rational number knowing and fraction knowing in that the former is at a higher level than the latter, his comment concerning rational number knowing pertains as well to fraction knowing. I agree with Kieren that fraction knowing in any case is certainly not a *simple extension* of whole number knowing. Rather, our hypothesis is that it *can emerge as a reorganization of whole number knowing* in the sense that I have explained above. In a developmental analysis of the operations that produce discrete quantity and continuous quantity presented in

---

[1] Rational Number Project.

Chap. 4, I explore whether the operations that produce each type of quantity can be regarded as unifying operations. The absence of such unifying operations would strengthen the interference hypothesis and the separation between the study of whole numbers and fractions that it seems to imply. In contrast, I find substantial similarity in the quantitative operations that produce continuous quantity and discrete quantity and provide reasons for why the quantitative operations that produce discrete quantity should be used to reconstitute (not replace) the operations that produce continuous quantity.

In a reaction to a preliminary draft of one of the chapters of this book, Kieren commented that, "while I, like you, decry the separation of whole number and fraction number knowing, I do not think that fractions as a reorganization of whole number based schemes is a necessary path or solution." Whether or not fraction knowing is *necessarily* a reorganization of whole number knowing depends on whether the operations involved in fraction knowing can emerge in the continuous context[2] with only minimal involvement of the operations that are involved in whole number knowing. The thrust of the research in this book is not to investigate this question. Children do construct whole number knowing, and the quantitative operations that unify discrete and continuous quantity support the idea that children can and do draw on that knowing in their construction of fraction knowing.

In our work, we focus on children's quantitative operations. Various researchers have found that these operations differ significantly from conventional ways of mathematical knowing. When presented with arithmetical problems, children use their current schemes in an attempt to solve them (Booth 1981; Erlwanger 1973; Ginsburg 1977; Hart 1983) in spite of emphasis on teaching practices that are not based on children's methods (Brownell 1935). Using phrases like "rational number knowing," "whole number knowing," "decimals," and "fractions" as if they refer to adults' more or less conventional ways of knowing suppresses children's quantitative operations in favor of the conventional ways. In fact, one of my primary goals is to formulate a language that can be used to refer to children's mathematical concepts and operations without conflating them with the more sophisticated concepts referred to by standard mathematical language.

To illustrate why I regard the development of a language to refer to children's mathematical concepts and operations as necessary, I use an example from a study by Nik Pa (1987). When interviewing nine 10- and 11-year-old children, Nik Pa found that they could not find 1/5 of ten items because "one-fifth" referred to one in five single items. The children separated a collection of ten items into two collections of five and then designated one item in a collection of five as "one-fifth." Nik Pa's

---

[2]By "continuous context" I refer to experiential episodes that contain items that are produced by moments of focused attention that are not interrupted by moments of unfocused attention, but which may be bounded by such moments of unfocused attention [cf. von Glasersfeld (1981) for the meaning of attention]. Scanning the sky from one horizon to the next on a perfectly clear day produces what I think of as an experiential continuous item as well as scanning a blank sheet of paper.

finding is quite similar to what "sixths" meant for a 9-year-old child named Alan who thought "sixths" meant "six in each pile" (Hunting 1983). Hunting found Alan's case to be representative, in its broad outlines, of those of the other 9-year-old children he studied. These findings indicate that children's fractional language differs substantially from the observer's meaning that "one-fifth," say, can refer to making five composite units[3] of indefinite but equal numerosity and then designating one of these composite units as "one-fifth."

The children that Nik Pa and Hunting interviewed obviously gave meaning to fractional words using their numerical concepts.[4] The constituting operations of these numerical concepts were left unspecified by Nik Pa, but whatever these operations might be, one interpretation of his findings could be that they interfered with or were detrimental to the children's learning of fractions. This interpretation, however, does not take assimilation into account. Instead the importance of Nik Pa's and Hunting's findings is that the children used their numerical concepts in assimilation. With this view, the orientation to the relationship between children's numerical concepts and fractional schemes changes. Even though the children's numerical schemes in the main did not qualify as fractional schemes, their schemes constituted the current *sense making constructs* of the children. In our view, the problem is not one of trying to avoid these assimilating schemes nor of considering these schemes as misconceptions. Rather, the problem is to understand the children's schemes and to learn how to help the children modify their current ways and means of operating.

For example, if the children in Post et al.'s (1993) experiment learned to reason in a way that one could attribute the inverse relation between the number and the size of the pieces of a whole to them, this would be a modification of their connected number[5] concepts. To form an inverse relation between the number and the size of the parts, the children would have needed to at least compare the number and size of the parts on two different occasions, which involves the use of their whole number knowing. Of course, they would need to do more, but this is enough to establish our position that children's whole number knowing is constitutively involved in their fractional knowing and should not be regarded as interfering with it. In this, I am in accord with McLellan and Dewey's (1895) belief that "fractions are not to be regarded as something different from number – or at least a different kind of process" (p. 127).

---

[3] By "composite unit" I mean a unit that is produced by uniting simple units into an encompassing unit. An example is uniting a regeneration of the chimes of a clock into a composite whole.

[4] The use of "concept" rather than "scheme" is intentional. The meanings of these terms will be commented on in Chap. 2.

[5] A connected number is constructed by the child by using the units of a numerical concept in partitioning a continuous item into parts and then uniting the parts together.

The Separation Hypothesis

The interference hypothesis would imply that the study of whole numbers and fractions should be separated in the mathematics education of children. In the main, the study of whole numbers occurs in the context of discrete quantity and the study of fractions occurs in the context of continuous quantity (cf. Curcio and Bezuk 1994; Reys 1991). Some may see this separation between whole numbers and fractions as compatible with Confrey's (1994) view of splitting and sequencing, so I will explore Confrey's (1994) splitting conjecture and some of its implications. She has focused on splitting as a basic and primitive action and set it in opposition to sequencing.

> In its most primitive form, *splitting* can be defined as an action of creating simultaneously multiple versions of an original, an action often represented by a tree diagram. As opposed to additive situations, where the change is determined through identifying a unit and then counting consecutively instances of that unit, the focus in splitting is on the one-to-many action. Closely related to this primitive concept are actions of sharing and dividing in half, both of which surface early in children's activity. Counting need not be relied on to verify the correct outcome. Equal shares of a discrete set can be justified by appealing to the use of a one-to-one correspondence and in the continuous case, appeals to congruence of parts or symmetries can be made. (p. 292)

In further elaboration of a split, Confrey (1994) commented that: "A split is an action of creating equal parts or copies of an original" (p. 300). She goes on to say that the number concept derived from splitting is independent from the number concept derived from addition in its successor action (Confrey 1994, p. 324). She clearly considers splitting and sequencing to be independent in their origins.

It is important to note that splitting as Confrey defines is not restricted to the continuous case. Nevertheless, splitting is seemingly rooted in the continuous case and sequencing is seemingly rooted in the discrete case.[6] As my own analysis will show, partitioning in continuous contexts is necessary for the construction of fraction schemes. Hence one could argue, based on Confrey's work, that fractions are based on splitting in continuous situations, while whole numbers are based on counting in discrete situations, i.e., the fractional and whole number learning could, and perhaps should, take place separately. In addition, Confrey's contention that splitting and counting have different experiential bases countermands the reorganization hypothesis.

In fact, Confrey (1994) herself sees the integration and coordination of the splitting and counting as essential. One reason for integration that she cites is to coordinate the number names of the quantities that arise through splitting with the number names the children construct in their (additive) number sequences. This consideration will emerge implicitly in my discussion as well. In addition, I will explain how the basic operations of splitting by n and sequencing by one are

---

[6] Confrey (1994) cited sharing, folding, dividing symmetrically, and magnifying as the basis for splitting.

not primitive operations, but are constructed from the same basic operations. In particular, I will argue against the primitive nature of simultaneity in Confrey's splitting conjecture, showing how the seeming sequentiality of number sequences and simultaneity of splitting are inextricably related. In my analysis of children's fractional schemes in future chapters, I will give an account of the construction of a basic splitting operation engaged in by the children in the study using both partitioning and iteration, which is a form of sequencing. Although this operation certainly emerges as a new and powerful operation, it is not isolated from the operations involved in number sequences.

A Sense of Simultaneity and Sequentiality

In my examination of the origins of splitting and sequencing, I consider the child's sense of simultaneity and sequentiality in the construction of two as a composite unit. I focus on simultaneity because Confrey said that the child creates simultaneously multiple versions of an original in a split, and this sense of simultaneity seems to be at odds with the sequential nature of operations when counting. My reason for choosing the construction of two as a composite unit is to demonstrate that both a sense of simultaneity and of sequentiality are involved in the construction of this most basic numerical concept. From this, I go on to argue that it follows that both a sense of simultaneity and of sequentiality are involved in number sequences as well as in splitting operations.

Confrey's restriction that the action of splitting creates *equal* parts or copies of an original implies a level of cognitive functioning that is constitutively not primitive in the sense that it would not be present in very young children (Piaget et al. 1960). The splitting action in Confrey's framework is at the level of mental operations rather than at the level of sensory-motor activity. Relaxing the restriction of creating equal parts does not diminish the value of her analysis for me because it provides a way of making a solid connection between partitioning operations and children's number sequences. I will use the term "fragmenting" to refer to simultaneity in breaking without the restriction of there being equal parts. Of course, I include Confrey's idea of splitting in fragmenting. Breaking off parts sequentially seemingly stands in contrast to fragmenting, and I use "segmenting" to refer to sequentiality in breaking without restriction on the size of the parts.

According to Confrey's conjecture, segmenting would lead to counting and fragmenting to splitting. I interpret Confrey as meaning that children's number sequences are based on segmenting rather than fragmenting and that fractional schemes are based on fragmenting rather than segmenting. In contrast, I advance the hypothesis that neither fragmenting nor segmenting is the more primitive and that both are involved in the construction of number sequences as well as in the construction of fraction schemes.

Establishing Two as Dual

If fragmenting were not involved in the construction of number sequences, sequentiality would need to be observed as the basal mechanism. Menninger (1969) claims in a historical study of the number sequence, however, that two and three did not develop in such a way that they were elements of a sequential order. In the development of two, he distinguished between two as dual and two as unity. His analysis of two is quite consistent with Brouwer's (1913) analysis of "two-oneness."

> This neo-intuitionism considers the falling apart of moments of life into qualitatively different parts, to be reunited only while remaining separated by time as the fundamental phenomenon of the human intellect, passing by abstracting from its emotional content into the fundamental phenomenon of mathematical thinking, the intuition of the bare two-oneness, the basal intuition of mathematics. (p. 85)

This "falling apart of moments of life into qualitatively different parts" is an expression of fragmenting. It is related to Menninger's idea of an awakening of consciousness – the isolation of self in an environment – "the I is opposed to and distinct from what is not I...." (Menninger, p. 13). The "falling apart" of moments of life has two necessary components. The first is an experiential awareness of one's experiential self on two distinct occasions.[7] But more is needed because the individual has to regenerate a prior experiential awareness of self in the current moment of awareness, which introduces the possibility of an awareness of precedence. Regenerating a preceding moment of awareness in the present makes possible the co-occurrence of the two moments of life. In this, there is also an awareness of one moment of life preceding the other; i.e., an awareness of sequentiality.

The second aspect is the awareness of recognizing an experiential item other than self, and then to be aware of recognizing another such experiential item while remaining aware of the first. Continued awareness of the recognition episode in the present act of recognition is again necessary for the two moments of life to be distinguished one from the other. Otherwise, the current fragment of experience would be the only item of awareness. The current awareness of the preceding fragment of experience makes possible copresent moments of life. Two moments of life are copresent if both are accessible to reflection or other possible operations.

An act of recognition is essentially an act of segmentation in that one separates what is being recognized from a background of possibilities. Recognizing an experiential item followed by recognizing another experiential item are *sequential* acts of segmentation. When the preceding recognition episode remains within awareness in a current recognition episode, this opens the possibility of becoming aware that the preceding moment of life comes before the current moment.

---

[7] See von Glasersfeld (1995a) for a discussion of the distinction between the self as center of subjective awareness.

So, the separation of two fragments of experience by one preceding the other in experience does not necessarily mean that the two fragments are experienced only sequentially. By bringing the preceding recognition episode into the present through a regeneration of the preceding fragment, the individual can experience a copresence of the two fragments. It is in this conceptual sense that the breaking apart of moments of life can be regarded as simultaneous as well as sequential. Both the sense of simultaneity and the sense of sequentiality are the results of conceptual acts and both involve bringing forth a preceding fragment of experience into the present by means of a regeneration of the preceding experience. I emphasize that becoming aware that one moment of life precedes another opens the possibility of experiencing both moments of life together, and vice versa. Neither is the more primitive. That is not to say that one does not experience simultaneous or sequential events. But these experiences are the results of an active intellect that organizes the experiences in a particular way.

Establishing Two as Unity

Regenerating a preceding fragment of experience in a current recognition episode is different than uniting the two fragments into a composite unit. Regardless of whether a child experiences two items as occurring sequentially or simultaneously in a way that I have been speaking, the items would remain separate and distinct rather than be reunited into a two-oneness if the uniting operation has not emerged. What this means is that Brouwer's basal intuition of mathematics depends on the construction of the uniting operation rather than the other way around. From a developmental perspective, Brouwer's basal intuition of mathematics is produced by the uniting operation – that is, it is a construction and not a given intuition.

I consider two as dual to be a dyadic pattern. Such patterns have their origins in moments of life that break apart as well as in the bilateralisms of the body and in patterns such as spatial or rhythmic patterns. A dyadic pattern is not any particular spatial or rhythmic pattern nor is it any particular bilateralism. Rather, regardless of its original source, a dyadic pattern is a recognition template that can be instantiated, and further modified and generalized in its use. In this, the recognition template that constitutes the dyadic pattern can be used to recognize any pair of experiential items as co-occurring if they are experientially contiguous. When the dyadic pattern is activated, the child has a sense of simultaneity – of the copresence of a pair of perceptual unit items.

A dyadic pattern is not the numerical structure implied by the phrase "bare two-oneness." But it is a composite structure. According to Menninger (1969), in two as a unity, "we experience the very essence of number more intensely than in other numbers, that essence being to bind many together into one, to equate plurality and unity" (p. 13). At the very core of the construction of number,[8] then, we see the essentiality of the operation that binds many together into one – which is an experience of copresence. I call this operation the uniting operation. It is simply the unitizing operation applied to two or more items. In that I regard these items as being products of the unitizing operation, *we see how two as unity is produced by the recursive use of the unitizing operation.* In this, it is essential to understand that for the uniting operation to produce two as unity, the items being united are produced by a regeneration of fragments of experience and these regenerated items occur in visualized imagination.[9] So, number is constructed through a coordination of the operations of unitizing and re-presentation (regeneration of a preceding item of experience).

Recursion and Splitting

The recursive use of unitizing in the construction of number is recapitulated in Confrey's (1994) concept of splitting. Confrey (1994) defines the concept of unit "in any world as the invariant relationship between a successor and its predecessor; it is the repeated action" (p. 311). She goes on to elaborate, "From 1, with our first split, we create the unit of n and the first number in the sequence as n" (p. 312). This particular kind of unit produces the geometric sequence with constant multiplier n as opposed to the arithmetic sequence, with constant addend n. In either case, according to Confrey, the unit is a repeated action.

For our current purposes, I focus on how recursion is involved in Confrey's concept of a split. In the most elementary case of a split, the breaking apart of two moments of life, I did not find it necessary to use recursion in our analysis of the copresence of the two moments. It was essential, however, for the experiencer to bring a preceding experience of self or thing into awareness in a current recognition episode. In the case of an intentional three-split, I do find it necessary to use recursion in explaining a sense of simultaneity of the results of the split. In the more general case where a unit of n is created, the splitting agent must begin with a unit and then fragment that unit item into n equal pieces. Creating n equal pieces itself involves recursive splitting actions, but Confrey goes beyond creating equal pieces and posits the creation of a unit of n.

She explicitly says that from 1 (which I take to indicate a unit item of some kind), "with our first split, we create the unit of n." What this means to us is that

---

[8] The operations that produce composite units often appear precociously in the case of two as unity.

[9] "Visualized imagination" is not restricted to visual imagery. It includes also regeneration of perceptual items in any sensory mode.

the splitting agent must begin with a unit and then fragment that unit item into n equal pieces. But that is not sufficient, because "we create the unit of n." Creating a unit of n is quite similar to the construction of the bare two-oneness as explained by Brouwer. The "falling apart of moments of life" can be thought of as analogous to starting with a unit and simultaneously breaking that unit into n parts if the necessity to make equal parts is relaxed.[10] The difference is that in the construction of two-oneness, as observer I conceived of the moments sequentially as well as simultaneously. In Confrey's analysis, there is no assumption of sequentiality. This is made possible, I believe, by the construction of splitting as an operation. If the splitting agent has already constructed what I call a partitioning structure, I indicate in a later chapter how such a structure could bring an intention of splitting forth without bringing a sense of sequentiality forth because sequentiality is symbolized. Being symbolized, however, does not eliminate sequentiality in splitting.

A similarity between the bare two-oneness and splitting is also present in the assumption that the parts produced by the split are united together into a composite unit. Further, to intentionally split the unit, 1, into n fragments, some composite unit structure has to be available to the acting agent. Early on, these composite unit structures may be dyadic, triadic, or quadriatic attentional patterns. Regardless of their nature, according to Confrey's definition of a split, the n fragments are reunited into a composite unit of numerosity n. Thus, using the results of prior operating as input in further operating is constitutively involved in a split just as it is involved in Brouwer's basal intuition of mathematics, albeit in a more advanced form. In our developmental analysis, the uniting operation would need to be present for Confrey's notion of a split to make sense, and therefore, it, the split, should not be thought of as a "primitive operation." Rather, it has to be categorized as a conceptual act.

Using the results of prior operating in further operating is even more dramatically involved in the second split. There has to be a new unit, not 1, but 1 reconstituted as a unit of n fragments which, from the observer's perspective, could be recombined to form the original unity. In other words, a composite unit with n elements has been created as a result of the first split, which is now used as input for further operating. How this second split might occur highlights the essentiality of both fragmenting and segmenting in making a split.

Distribution and Simultaneity

We have to always operate on something, and this "something" in Confrey's analysis of splitting is a unit of some kind. At the point of the second split, I find it necessary to introduce a new operation of distribution because it is quite unlikely that anyone can simultaneously split each of n things into n parts (Steffe 1994b, p. 21). Rather,

---

[10] We focus on breaking into n equal parts because of our interest in partitioning. Confrey's analysis of splitting is not restricted in this way.

the splitting agent would need to sequentially distribute the operation splitting by n, across the n elements produced by the first split rather than simultaneously split each of the n elements.[11] I argue that the distribution operation makes it possible to be aware of splitting each of n things into n fragments *before the splitting action is implemented*. This anticipation makes possible an awareness of simultaneously splitting all n elements. It also makes it possible to actually carry the operation out sequentially.

Splitting as a Recursive Operation

In Confrey's analysis, I understand splitting as an operation. That is, Confrey's definition of a unit as a repeated action involves input and output as well as mental action, which is to say that it fits well with our concept of an operation. In fact, Confrey's idea of splitting includes our idea of a recursive operation, an operation that, in our model, yields number sequences as well as multiplicative operations. This helps to place our reorganization hypothesis on a firm conceptual foundation.

In any event, I need to reconcile Confrey's notion of a repeated splitting action as being a unit, and our notion of a splitting action as an operation. Although the latter involves unitizing[12] in its construction, the constructive process must be distinguished from an awareness of an operation after it is constructed. von Glasersfeld (1995b) has made a similar point with respect to Confrey's idea of a repeated splitting action as being a unit: "Operational awareness of carrying out a repetition is indispensable in generating pluralities, but it is not a requisite for the conceptual construction of units" (p. 120). In other words, children do engage in the conceptual construction of units without being aware of the involved unitizing operation. Operational awareness comes later and involves the ability to step out of the stream of direct experience, to re-present a chunk of it, and to look at it as though it were direct experience, while remaining aware of the fact that it is not (von Glasersfeld 1991, p. 47). To look at the re-presented chunk of experience as though it were direct experience involves taking what is being looked at as an experiential unit. So, operational awareness indicates that the operation has been grasped as a unit, or, in other words, the operation has become the focus of attention. In her statement, Confrey seemed reflectively aware of the operation of splitting and, in this, constituted it as a unit. But this does not mean that children cannot engage in splitting a unit into subunits without being aware of the splitting operations in which they engage.

---

[11] In Chap. 6, we argue that distribution is a fundamental operation in constructing a multiplicative concept. Distribution is found in what is referred to as a coordination of two composite units.

[12] The unitizing operation is explained in Chap. 3.

Next Steps

The rationale for the reorganization hypothesis is still far from complete. Establishing that both simultaneity and sequentiality are involved in the construction of two as unity is a start, but it does not complete the argument for the number sequence. Following Menninger, who believed that two and three did not develop as part of a sequential order, I have argued that the initial experiences of three as trio does not include the experience of two as dual (Steffe 1988). The former experience excludes the latter in that three is not initially conceptualized as one more than two. Rather, three as trio is a triadic pattern in the same sense that two as dual is a dyadic pattern. I believe that the construction of the triadic[13] pattern involves both a sense of simultaneity and of sequentiality and that three as unity involves a recursive use of the unitizing operation. With the possible exception of quadratic patterns, I am still to make an argument that both a sense of simultaneity and of sequentiality are involved in the construction of children's number sequences. I am also yet to argue that children's number sequences are relevant in the case of "continuous quantity" as well as in the case of "discrete quantity." This is an important argument for us, because it is our goal that children, upon seeing, say, a blank stick, will regard the situation as a situation of their number sequence. That is, if fractional schemes are to be realized as reorganizations of children's number sequences, then the latter must be used in situations that later will be regarded as fractional situations. Finally, I am yet to argue that partitioning or splitting operations and iterable units can be integrated into the same psychological structure. This argument is critical because it countermands the assumption that splitting and sequencing are built from distinctly different experiential foundations.

Acknowledgment I would like to thank Dr. Thomas Kieren and Mr. Ernst von Glasersfeld for their comments on the first four chapters.

---

[13] Triadic patterns are explained in Chap. 4.

Chapter 2
Perspectives on Children's Fraction Knowledge

Leslie P. Steffe

The separation of the study of whole numbers and fractions is historical and contributes to the legendary difficulty children experience in the learning of fractions that inspired Davis et al. (1993) to comment that "the learning of fractions is not only very hard, it is, in the broader scheme of things, a dismal failure" (p. 63). I cite Davis et al. not because I believe that the teaching and learning of fractions is by necessity a dismal failure, but rather to accentuate the historical difficulties children experience in learning fractions in mathematics education. These difficulties are quite unsettling because they have been known for a long time. For example, in his famous study on the grade placement of arithmetical topics, Washburne (1930, p. 669) reported that a mental age level of 9 years should be attained by children if at least three out of four of them are to make the very modest mastery represented by a retention test score of 80% on the meaning of "nongrouping" fractions. But, in the case of "grouping" fractions, the analogous mental age was 11 years 7 months.[1]

An assessment of children's mathematical development conducted 50 years later in England and Wales (Foxman et al. 1980) also indicates the difficulty children experience when learning fractions. In their Primary Survey Report, it is reported that only 42% of the 11-year-olds of the study could say that one-fourth of one-half of a piece of string was one-eighth of the whole string. And only 61% could make a reasonable estimate of what fraction of the pegs in a bag were white, where 15 where white and 45 colored. Any estimate between and including one-tenth and one-half was accepted. Moreover, Kerslake (1986) found that 13 and 14-year old students in England had a good idea of fractions as part of a whole, which is compatible with Washburne's findings concerning "nongrouping fractions," but only a fragile

---

[1] A "nongrouping" fraction did not involve a composite part of a unit. For example, the children were asked "A pint is what part of a quart"? or instructed "Draw a line one-fourth as long as this line." A "grouping" fraction did involve a composite part of a unit. For example, when showing children a picture of three piles of five pennies each, they were asked what part of the pennies were in each pile. In another example, when showing a picture of five piles of three pennies each with a ring around two piles, the children were asked what part of the pennies had a ring drawn around them.

notion of fraction equivalence, which is compatible with Washburne's findings concerning "grouping" fractions and with the findings from the Primary Survey Report (Foxman et al. 1980).

In spite of Kerslake's findings that the concept of fractional equivalence is fragile even for 13 and 14-year olds, Smith's (1987) mathematically appropriate belief that the equivalence class is the central concept in the mathematics of rational number unproblematically drives not only what is taught about fractions in the elementary school, but also what is taught about fractions in mathematics courses designed for elementary school teachers (Long and DeTemple 1996, p. 374). Believing that children's mathematical knowledge corresponds to and can be explained by conventional mathematical concepts and operations, mathematics educators traditionally have regarded the content of children's mathematical knowledge as fixed and a priori. In our view, this belief constitutes what Stolzenberg (1984) called a trap. According to Stolzenberg, a trap is a:

> *Closed system of attitudes, beliefs, and habits of thought for which one can give an objective demonstration that certain of the beliefs are incorrect and that certain of the attitudes and habits of thought prevent this from being recognized* (p. 260).

Because children's mathematical learning in school occurs in the specific context of teaching, it might seem to be reasonable to regard the content of children's mathematical knowledge to be explained by conventional mathematical concepts such as fractional equivalence. However, several researchers working within a constructivist view of knowledge and reality have found it necessary to explain what students learn using constructs that differ significantly from standard mathematical concepts and operations (e.g., Confrey 1994; Kieren 1993; Thompson 1982, 1994; Steffe and Cobb 1988). I therefore seriously question the belief that school instruction should be based on concepts such as fractional equivalence.

On Opening the Trap

According to Stolzenberg (1984), it is indisputable that the contemporary mathematician operates within a belief system whose core belief is that mathematics is discovered rather than created or invented by human beings. This belief is equivalent to believing, as did Erdös, in a transfinite Book that contains the best proofs of all mathematical theorems (Hoffman 1987). Of course, this is a "mathematicians' book," and a belief in its existence apparently supports and sustains mathematical research for those who believe in it: "Mathematics is there. It's beautiful. It's the jewel we uncover" (p. 66).

Stolzenberg's (1984) contention that mathematics is not discovered but invented is according to Watzlawick (1984), "one of the most fascinating aspects of Stolzenberg's essay" (p. 254). It constitutes a shift in belief that is needed to open the trap because mathematicians' belief in the Book is reenacted by mathematics educators concerning the books of contemporary school mathematics.

The mathematics that is recorded in these books is usually regarded as a priori and as constituting what children are to learn. This assumption places the mathematics of schooling outside of the minds of the children who are to learn it, and it is manifest in the practice of separating the study of whole numbers and fractions between the discrete and the continuous as well as in the acceptance of concepts such as equivalent fractions as what children are to learn in school mathematics. School mathematics is regarded as a fixed nucleus, and one searches the school mathematics books in vain for mathematics of children.

Invention or Construction?

Stolzenberg's view is compatible with an assumption I make in our work with children, but it is not identical. The first difference resides in the meaning of the terms "invented" and "constructed." "To invent" implies the production of something unknown by the use of ingenuity or imagination. An invention certainly falls within the scope of what is meant by a construction, but the latter term implies conceptual productions within or as a result of interactions that I would not want to call inventions. Although any construction implies the production of a novelty, I would hesitate to call, for example, an association between two contiguous perceptual items (Guthrie 1942), an invention if for no other reason than many such associations are formed without forethought and sometimes even without the awareness of the associating individual. But I do regard associations as constructions regardless of the conditions of their formation (Steffe and Wiegel 1996). The boundary between the meanings of "to invent" and "to construct" is quite fuzzy, and it would certainly be counterproductive to insist that creative acts within this fuzzy boundary are exclusively one or the other. Nevertheless, making a distinction between the two provides a basis for a critique of the following rather restricted interpretation of constructivist learning.

> *In promulgating an active, constructive and creative view of learning, however, the constructivists painted the learner in close-up as a solo player, a lone scientist, a solitary observer, a meaning-maker in a vacuum.*
>
> (Renshaw 1992, p. 91)

Renshaw's interpretation of constructivist learning is based almost exclusively on the interaction of constructs within the individual. Social interaction seems excluded, so his characterization of constructivist learning is more or less compatible with the perhaps restricted view that mathematicians invent mathematics without the benefit of interacting with other mathematicians. In contrast, I emphasize the constructing individual as a socially interactive being[2] as well as a self-organizing and maturing being (Steffe 1996).

---

[2] Interaction here includes, but is not limited to, social interaction.

First-Order and Second-Order Mathematical Knowledge

The second difference between Stolzenberg's view and the constructivist view is apparent in Stolzenberg's comment that "when I stress the importance of standpoint, I am not preaching any brand of relativism. I do not say that there is your truth and my truth and never the twain shall meet" (p. 260). In taking this position, Stolzenberg seemed to be saying that the mathematics produced by one mathematician could be judged by other mathematicians concerning its fallibility or viability. That is, he was basically concerned with first-order mathematical knowledge – the models an individual construct to organize, comprehend, and control his or her experience, i.e., their own mathematical knowledge. In our work, we are mainly concerned with second-order mathematical knowledge – the models observers may construct of the observed person's knowledge (Steffe et al. 1983, p. xvi).

Distinguishing between first- and second-order mathematical knowledge (or models) is critical in avoiding a conflation between children's mathematical concepts and operations and what has been established as conventional school mathematics. Traditionally, there has been little distinction between these two kinds of knowledge, and school mathematics is considered as first-order mathematical knowledge. In our framework, we regard "school mathematics" as a second-order mathematical knowledge – a model of children's mathematics – rather than as the first-order model constituted by conventional school mathematics. Second-order models are constructed through social processes and I thereby refer to them as social knowledge. Regarding school mathematics as social knowledge is a fundamental shift in belief that is yet to be fully appreciated.

Mathematics of Children

We, as constructivist researchers, attribute mathematical knowledge to children that is independent of our own mathematical knowledge (Kieren 1993; Steffe and Cobb 1988; Steffe and Thompson 2000). Although the attribution of such knowledge to children is essential in their mathematical education, the first-order knowledge that constitutes children's mathematics is essentially inaccessible to us as observers. By saying this, we do not mean that we do not try to construct children's mathematical knowledge. Quite to the contrary, we spend a substantial part of our time, during and after teaching children, analyzing the mathematical knowledge that they bring to the learning situation as well as their evolving mathematical knowledge within the learning situation. What we do mean is that regardless of what the results of those analyses might be, we make no claim that the first-order models that constitute the children's mathematics correspond piece-by-piece to what we have established as second-order models.

We will use the phrase "children's mathematics" for whatever constitutes children's first-order mathematical knowledge and "mathematics of children" to mean our second-order models of children's mathematics. We regard the mathematics of children as legitimate mathematics to the extent that we find rational grounds for what children say and do mathematically. A shift in the belief of what should constitute school mathematics from conventional school mathematics to the mathematics that children do construct is foundational in opening the trap that has contributed to the historical difficulty in children's learning of fractions. In fact, a primary goal of our work is to construct second-order models of children's fractional knowledge that we are able to bring forth, sustain, and modify.

Mathematics for Children

I usually find it inappropriate to attribute even my most fundamental mathematical concepts and operations to children. For example, a set of elements arranged in order is a basic element in ordinal number theory. For a given number word, although children might establish a unit of units that they associate with that number word, I have not found sufficient warrant to infer that children constitute these composite units as ordered sets in the way I understand ordered sets (Steffe 1994a). The observer might regard the composite units that are attributed to the children as an early form of ordered sets, but to regard them as *ordered sets* would be a serious conflation of the conventional idea of an ordered set and our idea of a composite unit, which I have found useful in understanding children's mathematics. Conventional mathematics, such as ordinal number theory, can be orienting, but it is not explanatory; it alone cannot be used to account for children's numerical concepts and operations.

It might seem that the mathematics adults intend for children to learn remains unspecified. However, I regard *mathematics for children* as consisting of those concepts and operations that children might learn (Steffe 1988). But rather than regard these concepts as being a part of my own mathematical knowledge, I base mathematics *for* children on the mathematics that I have observed children actually learn. Essentially, mathematics *for* children cannot be specified a priori and must be experientially abstracted from the observed modifications children make in their mathematical activity. That is, mathematics for children can be known only through interpreting changes in children's mathematical activity. Specifically, the mathematics for a group of children is initially determined by the modifications that other children have been observed to make whose mathematical behavior is like the current children. I call these observed modifications zones of potential construction for the children whom I am currently teaching.

A teacher may not have constructed zones of potential construction suitable for the children he or she is currently teaching. Even in that case, a hypothetical zone of potential construction can be posited by the teacher to serve as a guide in the selection of learning situations. As a result of actually interacting with the particular children, the hypothetical zone of potential construction is reconstituted to form a zone of actual construction. The two zones usually diverge, because in the course of actually interacting with the children, they may make unanticipated contributions and new situations of learning may need to be formulated. Through establishing actual zones of construction, new possibilities may arise and a new zone of potential construction may be posited.[3] It is through such experimentation in teaching that children's mathematics may emerge in the experience of the teacher. In short, we recognize the necessity to modify our models of children's mathematics according to the children's work. Teaching mathematics is adaptive: It is the responsibility of the teacher to construct mathematics of and for children in the teaching context.

Our own first-order mathematical knowledge does play fundamental roles in formulating the second-order models that we call the mathematics of children. Perhaps the most fundamental of these roles is in orienting us as we formulate mathematics for children and decide how to interact with them. Rather than elaborate on these roles here, we discuss them throughout the remainder of the book because the discussion is concentrated and content specific. My focus in the next sections of this chapter is on developing a central conceptual construct – scheme – that I use in building models of children's mathematics.

Fractions as Schemes

Our use of the concept of scheme in building models of children's fractional knowledge is essential if Freudenthal's (1983) distinction between fractions and rational numbers is taken seriously.

> *Fractions are the phenomenological source of the rational number – a source that never dries up. "Fraction" – or what corresponds to it in other languages – is the word by which the rational number enters, and in all languages that I know it is related to breaking: fracture.* (p. 134)

Freudenthal's emphasis on fractions as the phenomenological source of the rational number is similar to Kieren's (1993) idea that ethnomathematical knowledge is at the center of mathematical knowledge building. In Kieren's (1993) words, ethnomathematical knowledge is that kind of knowledge that children possess "because they have lived in a particular environment. For example, children have shared continuous quantities and described such shares; they have seen measurements being made using fractional numbers" (pp. 67–68).

I believe that ethnomathematical knowledge includes Freudenthal's idea of fractions as the phenomenological source of rational number because, as Kieren emphasizes, it is a kind of *knowing*. In other words, to construct meaning for the term "fraction," we look to what children say and do as a source of our construction of such meaning. We bring Freudenthal's and Kieren's emphases together through the notion of the scheme, which is a conceptual tool that we use to analyze children's language

---

[3] The teacher may be yet to construct even a working model of the children's mathematical knowledge.

Fig. 2.1. A sharing situation.

and actions as they interact with us. That is, in our view, the evolving fractional knowledge of children consists of the construction of schemes of action and operation in their environments.

We describe schemes through observing children recurrently engage in goal-directed activity on several different occasions in what to us are related situations.[4] These descriptions are usually interesting and often contain insightful behavior on the part of the child. For example, Kieren (1993) described three 7-year-old girls as characterizing one of seven children's share of four pizzas in Fig. 2.1 as "a half and a bite."

As researchers, it is our intention to go beyond this description in an attempt to understand and formulate plausible conceptual operations used by the children as they established one child's share as "a half and a bite." In this, I infer that the children's assimilated situation, which involved a question of how much pizza one child would get as well as the picture of the seven children and the four pizzas, constitutes what I interpret as a sharing situation. This inference is based on the result of the children's activity – "a half and a bite." I infer that the children would need to establish a goal and engage in a sharing activity to reply as they did.

This intuitive understanding of the mental operations involved in sharing is enough to qualify the sharing activity as a scheme in the Piagetian sense if I could observe the three children engage in similar sharing activity in other situations. The necessity of inferring schemes based on repeatable and generalized action is based on Piaget's (1980) definition of scheme as action "that is repeatable or generalized through application to new objects" (p. 24). Focusing only on the activity of sharing, however, does not provide a full account of the concept of scheme. von Glasersfeld (1980), in a reformulation of Piaget's concept of scheme, described a scheme as *an instrument of interaction* and elaborated the concept in a way that opens the possibility of focusing on what may go on prior to observable action. It also opens the possibility that the action of a scheme is not sensory-motor action, but interiorized action that is executed with only the most minimal sensory-motor indication. Finally, it opens the possibility to focus on the results of the scheme's action and how those results might close the child's use of the scheme.

---

[4] It is essential to know the boundary situations of a scheme; that is, those situations in which the child's scheme proves to be inadequate from our point of view.

The Parts of a Scheme

According to von Glasersfeld (1980), a scheme consists of three parts. First, there is an experiential situation; an activating situation as perceived or conceived by the child, with which an activity has been associated. Second, there is the child's specific activity or procedure associated with the situation. Third, there is a result of the activity produced by the child.[5]

> "Schemes" are basic sequences of events that consist of three parts. An initial part that serves as trigger or occasion. In schemes of action, this roughly corresponds to what behaviorists would call "stimulus," i.e., a sensory motor pattern. The second part, that follows upon it, is an action ("response")... or an operation (conceptual or internalized activity). ... The third part is ... what I call the result or sequel of the activity (and here, again, there is a rough and only superficial correspondence to what behaviorists call "reinforcement"). (p. 81)

Unlike the stimulus in the stimulus-response theory, then, the situation of a scheme is an experiential situation as perceived or conceived by the child rather than by the observer. In Piaget's (1964) view, a stimulus:

> Is a stimulus only to the extent that it is significant, and it becomes significant only to the extent that there is a structure which permits its assimilation, a structure which can integrate this stimulus but which at the same time sets off the response. (p. 18)

For Piaget (1964), assimilation rather than association constituted the fundamental relation involved in learning, and he defined it as follows:

> I shall define assimilation as the integration of any sort of reality into a structure, and it is this assimilation which seems to me fundamental in learning, and which seems to me the fundamental relation from the point of view of pedagogical or didactic applications. ... Learning is possible only when there is active assimilation. (p. 18)

When I speak of assimilation, I do not assume that an experiential situation "exists" a priori somewhere in the mind in its totality as an object that a child retrieves. Rather, I assume that *records* of operations used in past activity are activated in assimilation. I further assume that the activated operations produce a "recognition template," which is used in creating an "experiential situation" that may have been experienced before.

So, in the first part of a scheme, records of operations from past activity, when activated, produce a "recognition template" that is used in establishing an experiential situation. When it is clear from context, I refer to the recognition template as an assimilating structure and to the operations that produce it as operations of assimilation. The experiential situation may be created by means of visualized imagination as well as perception. It may in turn activate the scheme's activity, which, in the case of a cognitive scheme, may consist of an implementation of the assimilating operations in the context of the experiential situation. The result of the cognitive scheme consists of whatever modification of the experiential situation is induced by the activity.

---

[5]The goal of a scheme is discussed in the section on the sucking scheme.

Learning as Accommodation

In Piaget's quotation concerning assimilation, he commented that "learning is possible only when there is active assimilation." Learning, however, is not to be equated with assimilation. Rather, when there is an irregularity or disturbance in the functioning of an established scheme, only then can accommodation take place, and not otherwise (von Glasersfeld 1980, p. 82). In our work, learning is construed as accommodation, that is, the modification of schemes.

> This feature of the Piagetian model, as I see it, constitutes its main basis as a constructivist theory of cognition in which "knowledge' is no longer a true or false representation of reality but simply the schemes of action and the schemes of operating that are functioning reliably and effectively
>
> (von Glasersfeld 1980, p. 83)

There is indeed interaction between schemes and experiential events, but as von Glasersfeld points out, the child does not get to know the observer's situations; in the sense that its schemes come to match or in any sense reflect structures as they might be to the observer in his or her situations. So, although an observer may have the observed child and the child's environment in his or her experiential field, and observe the child using schemes while interacting with events, perhaps including other people, the interaction from the point of view of the interacting child is between schemes and experiential events within the system that constitutes the child.

The Sucking Scheme

Glasersfeld uses the sucking scheme in illustrating his idea of scheme. He uses it not only because of its essentiality in the survival of *Homo sapiens*, but also because of its importance in the construction of object concepts (Piaget 1937) and, thus, eventually in the construction of numerical concepts and schemes (Steffe and Cobb 1988). I have chosen the sucking scheme of newborn infants to illustrate the possibility of nonlinearity among the parts of a scheme in that the parts do not always proceed one way from the scheme's situation to activity to result. One may regard the activity of the sucking scheme as being involved in assimilating objects in that case where the sucking action is driven by a sensation of hunger rather than by some sensory experience like touching the infant's cheek. In the case of the sensation of hunger, the activity of sucking is activated and the baby searches for something on which to suck, and often it is a part of the baby's hand. Here, the baby establishes a possible situation of the scheme by means of the activity that is driven by the gnawing sensation of hunger.

The possibility that a scheme's activity can be the primary operation of assimilation solidly differentiates a scheme from the classical S → R schema. In the latter, it is the observer's stimulus that sets off a response. In the former, the activity of the scheme may be triggered by disequilibria internal to the scheme and only then is a situation created by the actions of the child. I interpret Piaget's (1964) comment that "the response was there first" (p. 15) as meaning that the activity of a scheme can be involved in establishing a situation of the scheme as well as the other way around, which is an important consideration in self-generated mathematical activity.

In the case of the sharing scheme, I can imagine the sharing operations as being activated by the question of how much pizza one person gets and by being involved in establishing the situation as a sharing situation. In the case of the sucking scheme, the sucking activity (activated by a sensation of hunger) can be involved in establishing a situation of the scheme. In that case where the baby sucks its hand, the situation may be the only result of the scheme's activity. Unlike the sharing scheme, which is closed by implementing the sharing operations within the situation of sharing, the sensation of hunger would not be reduced in intensity by the activity of sucking. And yet, the infant may achieve some sense of satisfaction by implementing the activity of the scheme – the infant is temporarily "pacified."

In other cases, the recognition template may be used in assimilation without the scheme's activity being implemented. An example is where one observes the people in a large stadium. The question of how many people are in the stadium could be answered by counting the people as they exit the stadium or by counting them by counting the number of tickets sold. But as one sits in the stadium without recourse to either possibility, the activity of counting usually remains only minimally implemented, even though it may be evoked. The question of how many people are in the stadium is meaningful in that the activity of counting could be implemented given an appropriate situation. But the constraints in implementing the counting activity leave the individual without an activity, so the individual has a goal but no activity to reach the goal. In such cases, I would say that the individual has established a problem.

The Structure of a Scheme

Fig. 2.2 is a diagram of the idea of a scheme. This diagram is static and as such it can be grossly misleading in interpretation. But it does help to highlight the essential aspects of a scheme. The *Generated Goal* can be regarded as the apex of a tetrahedron. The vertices of the base of the tetrahedron constitute the three components of a scheme. The double arrows linking the three components are to be interpreted as meaning that it is possible for any one of them to be in some way compared or related to either of the two others. The dashed arrow is to be interpreted as an expectation of the scheme's result.

In the case of the sucking scheme, I have already indicated how the scheme's activity can lead to an establishment of a situation of the scheme. This situation along with the activity can in turn lead to a full stomach as a result. The result in turn can engender a feeling of satisfaction usually manifest as a sleeping baby and the scheme's activity is discontinued, which is indicated by the double arrow between the scheme's result and the generated goal.

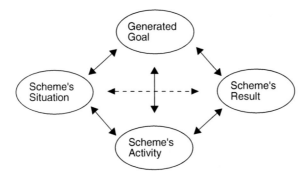

Fig. 2.2. A diagram for the structure of a scheme.

Given a generated goal and a result of a scheme, in some cases, it is possible for a child to establish a situation of the scheme if the scheme's activity is reversible. For example, a basic reason why 58% of the 11-year-old children in the study of mathematical development conducted by the National Foundation for Educational Research in England and Wales (Foxman et al. 1980) could not say that one-fourth of one-half is one-eighth is understandable when considering the possibility that their fractional schemes were not reversible schemes.[6] The children were first given a piece of string and then were asked to cut it in half. The children were then presented with one of the halves and were asked to "cut off one-fourth of this piece." The question "what fraction of the whole string that you started with is that little piece"? was then asked.

The children who were successful in cutting off one-fourth of one-half of the whole string had produced a result of their fractional scheme and their goal of making one-fourth had been satisfied. When the last question of the series of three was asked, this would serve to establish a new goal and a new situation using the results of the old scheme. To find one-eighth, the children might first reassemble the four pieces in thought and see them as one-half of the string partitioned into four equal pieces. The children could then produce another one-half of the string in thought also partitioned into four equal pieces, which would produce the whole string as two equal pieces each partitioned into four equal pieces. To do this, the children would need a fractional scheme that is reversible, in that they would be able to start from a result and reestablish the situation using inverse operations. So, in the case a scheme is reversible, its result can be used in establishing a situation of the scheme via the scheme's reversible activity. I stress, however, that these relations are only possible for some schemes. They are not a necessary aspect of all schemes. Some schemes are entirely "one-way" schemes that proceed from situation to activity to result.

---

[6] We intensively study the construction of such a scheme in later chapters where we explain how recursion is involved in such a reversibility.

The diagram also indicates that the goal of a scheme can be generated in the process of assimilation. In Fig. 2.1, a child may see the picture of the stick figures, but form no immediate goal for further action. The child may simply recognize the stick figures as indicating people. In this sense, there is an assimilation using concepts constructed at an earlier time. In the process of assimilation, the child may form a goal of finding how many stick figures because it may establish an awareness of more than one figure – an awareness of plurality – which in turn may activate counting activity. The arrows between the scheme's situation and goal and between the scheme's activity and goal indicate these possibilities.

It would be unlikely, in the process of assimilation, for the child to form the goal of finding how much of one pizza each stick figure would have if the pizzas were shared equally. But if another person were to ask an appropriate sharing question about an assimilated situation, the activated sharing operations would constitute a reinterpretation – a further assimilation – of the situation as originally established by the child. The resulting goal to find the share for one of the seven stick figures drives the sharing activity during the activity. One might say the goal frames the activity. Partial results (partial from the point of view of the goal) feed back into the goal and I assume that they are compared with the generated goal. The connecting line between the generated goal and the activity indicates this feedback system.

The connecting line between the scheme's activity and the scheme's results indicates that the results or partial results may modify the activity, which in turn may modify the results. Likewise, a modification of either of the scheme's activity or results may lead to a modification of the recognition template. In further uses of the scheme, the latter modification may in turn lead to a change in the scheme's activity. Of course, the generated goal may also change as the scheme is being used.

Seriation and Anticipatory Schemes

Operations of a scheme are basic in our construction of children's fraction schemes. For Piaget (1964), "An operation is ... the essence of knowledge: it is an interiorized action which modifies the object of knowledge" (p. 8). An operation, for Piaget, was always a part of a structure of operations. A key example of such a structure is seriation, the setting of elements in order. Piaget wrote that "an asymmetrical relation does not exist in isolation. Seriation is the natural, basic operational structure" (pp. 9–10). Seriation should be regarded as a basic mechanism of intelligence and as a product of spontaneous development, and it can be profitably considered as a scheme in von Glasersfeld's terms.

The seriation scheme can be used to portray what I mean by an anticipatory scheme. A child might form a goal of placing a collection of sticks in order from the shortest to the longest upon recognizing a collection of sticks. Prior to the activity of ordering the sticks the child might imagine the activity by imagining several sticks aligned in order. In this case, I say that the scheme is *anticipatory* as well as operative because the child can imagine the scheme's activity or result without carrying out the activity.

An ordering of the sticks is contributed to the collection of sticks by the seriating child. By activating the conceptual structure of seriation, a child can formulate an expectation that a collection of sticks be ordered. The child does not abstract seriation from the sticks; rather, the child contributes it to the collection of sticks.

Mathematics of Living Rather Than Being

Scheme is an observer's concept and, in the case of schemes that are mathematical, it refers to children's mathematical language and actions. As observers, we can make a distinction between our concept of scheme and the children's mathematical activity to which it refers, just as we can make a distinction between our concept of tree, and something "out there" to which we can point. One may object because the goal-directed activity of children is of a different nature than a tree. I agree they are of a different nature, but our concept, tree, can go beyond the concept that we initially constructed using the sensory material that was available to us. It can include our understanding of a tree as a dynamic living system and include such properties as photosynthesis.

Like our concept of tree, we also have a concept of children as physical objects. But that is only a beginning. We form the goal to understand children's mathematics as a constitutive part of a living conceptual system. This way of understanding their mathematics has great advantages for mathematics education and puts us in education, we think, in an appropriate frame of reference. No longer is the sole focus on the abstracted adult concepts and operations, and no longer is children's mathematical development conflated with those abstracted concepts and operations. Rather, the focus is on the living systems that children comprise and the problem is to understand how to bring the subsystems called mathematics of these living systems forth, and how to bring modifications in these subsystems forth. In this way, we may escape from Stolzenberg's trap.

Chapter 3
Operations That Produce Numerical Counting Schemes

Leslie P. Steffe

The primary goal of this chapter is to present a model of important steps in children's construction of their numerical counting schemes because the basic hypothesis that guides our work is that children's fraction schemes can emerge as accommodations in their numerical counting schemes. I consider a *number sequence* to be the recognition template of a numerical counting scheme; that is, its assimilating structure. This way of thinking of a number sequence was basic in the formulation of the reorganization hypothesis. A number sequence is a discrete numerical structure; it is a sequence of arithmetical unit items that contain records of counting acts. At all stages of construction, children use their number sequences to provide meaning for number words. A number word such as "twenty-one," say, can refer to a sequence of arithmetical unit items from "one" up to and including "twenty one." It is the operations that children can perform using their number sequences that distinguish among distinct stages of the number sequences. In what follows, I explain the operations that produce two prenumerical counting schemes as well as three distinctly different number sequences and, hence, three distinctly different numerical counting schemes. I also explain discrete structures that precede number sequences in development that I refer to as perceptual and figurative lots. These lot structures are produced by the operation of categorizing discrete items together, where categorizing is based on reprocessing sensory-motor items of experience using an operation called unitizing. In categorizing, when reprocessing is coordinated with re-presenting discrete items of experience, recursive unitizing emerges. Recursive unitizing is that operation which produces arithmetical units and numerical structures. I start by presenting an attentional model of unitizing and the different levels of units that this operation produces.

Complexes of Discrete Units

In formulating a model on the conceptual construction of unitizing, von Glasersfeld (1981) drew on his work with Silvio Ceccato whom he credits as the first to interpret the structure of certain abstract concepts as patterns of attention (Ceccato 1974). The basic problem was to develop a model of how human beings establish object concepts using the multitude of sensory material that is available through the various sensory channels.

According to von Glasersfeld (1981),

> Attention is not to be understood as a state that can be extended over longish periods. Instead, I intend a pulse like succession of moments of attention, each one of which may or may not be "focused" on some neural event in the organism. By "focused" I intend no more than that an attentional pulse is made to coincide with some other signal (from the multitude that more or less continuously pervades the organism's nervous system) and thus allows it to be registered. An "unfocused" pulse is one that registers no content (p. 85).

Von Glasersfeld's model of pulsating moments of attention provided an explanation of the mental operation that is involved in the construction of object concepts and the role these concepts play in the construction of numerical units. He called the operation as "unitizing." A group of cooccurring sensory-motor signals becomes a "whole" or "object" when an unbroken sequence of attentional pulses is focused on these signals and the sequence is framed or bounded by an unfocused pulse at both ends. The unfocused pulses provide closure and set the sequence of contiguous focused pulses apart from prior and subsequent attentional pulses. A focused moment of attention registers sensory material and an unfocused moment of attention can be regarded as a blank space. The records of making a sensory-motor item, or an item of experience, were graphically illustrated in terms of an *attentional pattern* as shown in Fig. 3.1 (p. 87).

The unfocused moments of attention are designated by "O" and bound the focused moments of attention designated by "I." The letters a, b,..., k designate sensory material selected by attention and this sensory material is registered as records of experience. I emphasize that the *attentional pattern* or *recognition template* is established as a result of individual–environment interaction, and the process it symbolizes constitutes a model of the operation that is involved in compounding sensory-motor signals together in the immediate here-and-now to form items of experience – the unitizing operation. Beginning with sensory-motor items, von Glasersfeld specified two more types of items that he called the unitary item and the abstract unit item, the latter of which is an interpretation of Piaget's arithmetical unit. These two items are the product of applying the unitizing operation to its own products. This recursive use of unitizing "strips" sensory content from sensory-motor items.

Sensory-motor items are isolated in experience, and there may be no element of recognition in their establishment. If a child does recognize an experiential item as having been experienced before, this involves the activation of records of operating that produced a previous experiential item. Because these records are activated, the child focuses on the sensory material that the activated records point to in immediate experience and compounds this sensory material together. It is in this way that the child is said to *reprocesses a current sensory-motor item* using the activated attentional pattern, which is an act of reflective abstraction. The activated attentional

Fig. 3.1. An attentional pattern: Sensory-motor item.

pattern used in recognition rerecords at least some prior records of operating and perhaps records novel sensory material. In this case, assimilation involves selection and variation and opens the possibility for modification of the recognition template (the activated attentional pattern). Realizing that selection and variation can be involved in assimilation highlights the importance of repeated experience in the construction of object concepts (Cooper 1990).

When an attentional pattern is activated by means other than relevant sensory material, Piaget's (1937) studies on object permanence indicate that children eventually develop the capability to use attentional patterns in the production of visualized images of those sensory-motor items they can recognize (von Glasersfeld 1995a). These images provide the child with an awareness of an experiential item. If the emergence of such awareness is coordinated with an awareness of its location in immediate experience, the child is said to have constructed an externalized, permanent object, or an object concept.[1]

Recognition Templates of Perceptual Counting Schemes

Following the development of object permanence, the process of categorizing can produce an attentional structure called a perceptual lot. Let us say, for example, that a child recognizes a perceptual situation using an attentional pattern (recognition template), an observer would associate with the word "cup." The child may continue to explore its visual field, assimilating another combination of sensory signals, then another, then another, and so on, all using the same attentional pattern. The process may lead the child to experience recurrence in assimilation if the immediate past experience of using the recognition template is available to the child in a current recognition episode. Those sensory features that are experienced as recurrent may be sufficient to activate reprocessing the established sensory-motor items by refocusing attention on the recurrent sensory features of each item using the same recognition template that was used in initial processing. Reprocessing forms the basis of categorizing the items together into a *collection of perceptual items*. This process of empirical abstraction produces a composite whole, but it is only experientially bounded.

Collections of Perceptual Items

An example of a collection of perceptual items is "the houses on the hill" if the restriction is accepted that an actual experience of the houses on the hill is required to establish the collection. The ability to visualize the houses on the hill [the collection] is yet to develop (Steffe et al. 1983). According to Inhelder and Piaget (1964),

---

[1]The ability to regenerate an experience of an item makes possible the falling apart of moments of life.

> When we see an orange and say, "This is an orange", the connexion between the object and the class is not directly perceived. What we are doing is to assimilate the orange we see to a perceptual schema of the sort described by Brunswick as an "empirical Gestalt". The orange is perceived as presenting the familiar configuration of an ovoid with a corrugated skin and an orange color. This configuration has acquired its stability as a result of previous perceptual experience. But it is closely linked to a number of sensori-motor schemata: peeling the fruit, cutting and chewing it, squeezing out the juice, etc. The class of oranges is based on schemata of this kind ... (p. 10).

At the level of perceptual collections, no class has been constructed to which a particular orange might belong. Still, there is a "perceptual schema" that has been constructed, and the concept of attentional pattern is used to account for it. Further, I can see no reason whatsoever from excluding such experiences as peeling the fruit, cutting and chewing it, or squeezing out the juice from being recorded in the attentional pattern implied by saying, "This is an orange." That is, the actions of the subject are included in the construction of object concepts. In fact, the actions of the subject are essential in producing the sensory material that is recorded in the attentional pattern, and this includes the sensory material produced by visual perception. That the subject's actions must be abstracted as properties of the object concept is made clear by Inhelder and Piaget (1964) in the following comments on the construction of the intension of a class.

> A class cannot be constructed by perception, for it presupposes a series of abstractions and generalizations, from which it derives its meaning or intension ... . A class can never be perceived as such, since it is generally of indefinite extension; and even when its extension is restricted, what the subject perceives is not the class itself but a certain spatial configuration of the elements which compose it (p. 10).

A perceptual collection is not yet a class and includes the graphic collections of Inhelder and Piaget (1964).

> The term "graphic collection" will therefore be used to describe a spatial arrangement of the elements to be classified where it seems clear that such a configuration plays an essential part in the eyes of the subject (p. 18).

Inhelder and Piaget pointed to the instability of graphic collections that children formed in the immediate here and now. In forming graphic collections, children do not intentionally choose to use a particular sensory feature on which to base their classifications. Rather, their use of recognition templates is in the moment at hand, and the recognition templates could change in the next moment depending on the features of the material at hand on which the children focus their attention. Inhelder and Piaget (1964) commented that at the level of graphic collections, "There is no question that children are perfectly well able to discover relations of similarity and difference..." (p. 45), but these relations are ephemeral in the case of graphic collections.

Perceptual Lots

Because of von Glasersfeld's (1981) distinction between a perceptual item and a perceptual unit item, it is possible to make distinctions that can be used in explaining the differences Inhelder and Piaget (1964) made between graphic collections and (non graphic) collections.

We use the term "collection" rather than "class" in the strict sense, because the former term carries no implication of a hierarchical structure of class-inclusions. However, these collections are no longer graphic, and objects are assigned to one collection or another on the basis of similarity alone (p. 47).

For us, assigning objects to Inhelder and Piaget's collection requires an abstraction beyond the empirical abstraction that is involved in refocusing attention on the recurrent sensory features of several items. It involves more, because when establishing a collection of perceptual items, there may be no intention on the part of the child to form such a collection. Rather, perceptual collections are formed in the moment. Nevertheless, reprocessing perceptual items as a means of taking them together opens the way to focus attention on the *unitary wholeness* of each sensory-motor item, which is *an operation of unitizing the sensory-motor items*. This is the process that produces perceptual unit items. The attentional pattern of a perceptual unit item is diagrammed in Fig. 3.2.

The notation in Fig. 3.2 is used to designate a single attentional moment focused on the unitariness of a sensory-motor item. In this, "n" is used to denote the necessity of having some, but no particular, sensory-motor material on which to focus. This development of the unitizing operation opens the possibility of the child categorizing nonhomogeneous items together on the basis of their unitariness – "things" that go together *because they are put together*. This fits well with how Inhelder and Piaget (1964) in part described nongraphic collections:

However, he [Eli 5; 6] then goes on to make collections based on similarity alone: the fish with the birds, etc., "because they're all animals", then the people, then pots, etc. "because they're all things for making supper" (p. 56).

Once established, if a perceptual unit item is used to reprocess a collection of perceptual items, this produces a collection of perceptual unit items. Records of the reprocessing action that categorizes perceptual unit items together are contained in an attentional pattern as diagrammed in Fig. 3.3. We call this pattern a perceptual lot – a collection of perceptual unit items – and use parentheses to denote that the action of categorizing occurred within experiential boundaries.

When a perceptual lot is activated in an experiential situation, the unit items of the pattern are activated, which provides the child with an awareness of more than one perceptual unit item. We refer to this awareness as an awareness of perceptual plurality.[2] In a perceptual plurality, the child senses the cooccurrence of more than

Fig. 3.2. The attentional pattern of a perceptual unit item.
$$\begin{array}{c} O\ I\ O \\ n \end{array}$$

Fig. 3.3. The attentional pattern of a perceptual lot.
$$(\underset{n}{OIO}\ \underset{n}{OIO}\ \underset{n}{OIO}\ \underset{n}{OIO}\ \ldots\ \underset{n}{OIO})$$

---

[2] We use "perceptual plurality" to refer to an awareness of the frequency of instantiation of a perceptual unit item within experiential boundaries. Our choice of "plurality" rather than "collection" is made to accentuate the child's awareness of more than one unit item rather than a unitary whole containing the unit items.

one perceptual unit item in immediate experience. Such an awareness of perceptual plurality can produce a sense of indefiniteness, or a lack of closure, which in turn can serve as a goal that activates the activity of counting. The child counts for the purpose of making definite the sense of indefiniteness.

I regard an awareness of perceptual plurality as a "quantitative"[3] property of a perceptual lot. It is what permits children who are restricted to establishing a lot of perceptual unit items as countable items to engage in purposeful counting activity in that it is their goal to make definite their sense of indefiniteness induced by their awareness of more than one perceptual unit item. However, as soon as the lot of perceptual unit items is hidden, the child is unable to count. Children for whom perceptual unit items must be available to count are called counters of perceptual unit items. They know how to count, but they need a lot of perceptual unit items in their visual field in order to carry out the activity.[4]

Recognition Templates of Figurative Counting Schemes

An important question concerns why children who are counters of perceptual unit items are aware of the items of a perceptual lot that are not in their visual field, but still cannot count the hidden items. For example, if such a child puts five tiles beneath a cloth and then three more tiles beneath the cloth, the child would not be able to count starting from "one" proceeding to "eight" to find how many tiles are beneath the cloth. The child would be aware that tiles were hidden because the child can produce a visualized image of a perceptual unit item.[5] The crucial factor in a child's being able to count hidden items is that the child can use its recognition template for a perceptual unit item to repeatedly produce visualized images of the perceptual unit item in re-presentation. This process is illustrated in Protocol I involving a child named Greg who was presented with a card that had the first three of a row of eight wooden discs covered (Steffe et al. 1983).

Protocol I.
I: I'll let you feel that one right here. (The first covered disc.) That is the first one. (Tells Greg which is the fifth disc in the row.)...How many are there in all?
G: (Attempts to touch the covered discs, but is stopped by the interviewer. Looks successively at the cover, focusing his gaze on specific places over the cloth where he thinks discs are hidden. As he looks at each place, he subvocally utters.)1-2-3-4-5-6. (Continues, gazing successively at the visible discs.) 7-8-9-10-11. Eleven.

---

[3] Quotation marks are used to indicate that we are speaking of quantity on a perceptual level.
[4] See Steffe et al. (1983), for examples of counters of perceptual unit items.
[5] An inability to create an image of a perceptual unit item would exclude experiencing perceptual unit items in their immediate absence, for then the child could only recognize the item in its immediate presence.

3 Operations That Produce Numerical Counting Schemes

Focusing his gaze on specific places over the cloth is an indication that Greg repeatedly imagined discs under the cover. His uttering the number words "1-2-3-4-5-6" indicates a second process because, when a child performs an act of counting, they coordinate *producing a countable item* and *uttering a number word*. To produce a countable item in this case means that Greg generated an experience of a disc in re-presentation and took that visualized image as his countable item in each act of counting over the cloth. This is indicated by Greg trying to touch the covered discs and by his focusing his gaze on those places where he believed a disc was hidden. Creating figurative items as countable was a novelty that Greg introduced in the context of his goal to count the discs. Of course, it "just happened" and it was not something that Greg intentionally introduced. Rather, it was a possibility created by Greg being able to use his object concept, disc, to sequentially generate disc images and to coordinate this activity with producing the words of sequence of number words. This is precisely what a child who is a counter of perceptual unit items cannot do.

Counting does provide a goal-directed activity that can be used to transform a perceptual lot structure into a higher-order structure. To experience a perceptual unit item in its absence, a child like Greg can generate a visualized image of the item, but the child is "in" the experience, not "outside" of it. That is, the child does not hold the visualized item at a distance and reflect on the image. An activity such as jumping rope might be clarifying. When the jumping action is re-presented, the child performs the jumping action in visualized imagination. When the jumping action is an object of reflection, the child holds the imagined action still by taking the visualized image of the action as a unit using its unitizing operation, which is what is meant by reflection in this case.

I assume that in producing a figurative lot structure, the child uses the perceptual lot structure in generating images of the items of the lot. Because the child can also use its concept (an item of a perceptual lot structure such as "disk") in re-presentation, this opens the possibility of the child using its concept of disc as an operation of unitizing each image of a disc in its visualized perceptual lot structure. When a child reprocesses the images of the discs by unitizing each image, this is the process that produces *figurative unit items*. By focusing attention on re-presented material, the child records the figurative experience in the recognition template, disc, and these records may override the previous sensory records. A result of this process is illustrated in Fig. 3.4. The broken line indicates records of figurative material that was reprocessed.

The essential difference between the figurative material produced in a re-presentation of a perceptual lot structure and a figurative unit item resides in what the child can take as an item of reflection. In the former case, the child can hold a *perceptual* unit item at a distance "out there" as an item of reflection. Similarly, when figurative material has its source in a figurative unit item, the image can be held "at a distance" as an object of reflection.

Fig. 3.4. The attentional pattern of a figurative unit item.   O̲I̲O̲

Fig. 3.5. The attentional structure of a figurative lot.   (OIO OIO OIO ... OIO)

A figurative unit item is not as abstract as Piaget's arithmetic unit because the child is still dealing with items of its ordinary experience and not with the more abstracted units of number. Still, the figurative unit item is the first *interiorized concept*. Records of reprocessing the images of the items of a perceptual lot produce a figurative lot structure, which is diagrammed in Fig. 3.5. The dotted lines indicate records of figurative material. A figurative lot is experientially bounded by the beginning and end of the process that produced it.

When the attentional structure of a figurative lot is activated in an experiential situation, such as covered discs, the unit items of the structure are activated and provide the child with an awareness of more than one figurative unit item, which I refer to as an awareness of figurative plurality. In this, the child has a sense of the cooccurrence of more than one figurative unit item in immediate experience. This awareness of figurative plurality can also produce a sense of indefiniteness, or a lack of closure, which, in turn, can serve as a goal that activates the activity of a counting scheme. I regard an awareness of figurative plurality as a quantitative property of a figurative lot introduced by the activity of the child.

The construction of figurative lots provides an account of the construction of *perceptual collections as permanent objects* analogous to the permanence of singular objects. This construction is crucial in the construction of number. An indication that a child has constructed the operations that produce figurative lots is where a child is observed counting by putting up fingers in synchrony with uttering "1, 2, 3, 4, 5, 6"–"7, 8, 9" to find how many items are hidden in two locations, six in one and three in another if the child by necessity counts from "one." The motor acts of putting up fingers stand in for the elements of the lot structure, and the child's counting scheme is judged to be figurative rather than numerical.

A child who can produce a collection of figurative unit items and substitute them for, say, covered squares, can give meaning to a phrase like "six squares" prior to counting. In this, it would be unlikely that the child would visualize six squares unless a pattern for "six" had been established. Rather, the child would produce visualized images of squares without specifying how many. This production of a figurative collection provides the child with a sense of the cooccurrence of more than one item in visualized imagination, and the goal to make this sense of indefiniteness definite is what drives counting. However, an awareness of figurative plurality should not be regarded as the quantitative property of a numerical structure that we call numerosity, because the child has to actually count to establish intensive meaning for number words.[6] In the case of a figurative counting scheme,

---

[6] Because the child is aware of figurative plurality, a child at this level can give meaning to a number word by producing visualizing images of hidden items. This extensive meaning of the number word remains indefinite, and to make it definite, the child has to actually count. Following Thompson (1982), counting is called the intensive meaning of number words.

the child is yet to run through the activity of counting the elements of the figurative plurality and produce its results in thought, which is essential for an awareness of numerosity. The figurative counting scheme is still a prenumerical scheme.

Numerical Patterns and the Initial Number Sequence

There is a type of reflective abstraction that comes in between the two levels identified above,[7] and that is the ability to reprocess the results of a scheme's activity to achieve a goal in the midst of those results being produced. An example is where Jason, a 6-year-old child, was observed counting as described in Protocol II (Steffe 1992). Jason interpreted the situation of the protocol as there being eight cookies beneath one cloth and ten beneath another, even though the interviewer had explicitly said ten cookies were beneath both cloths.

Protocol II.

I: (Places a cloth in front of Jason.) See those chocolate cookies under there? (The teacher and Jason were only pretending.) Put the number on the cloth that shows how many you would like to put under the cloth. (Jason puts "8" on top of the cloth; the teacher lifts an adjacent cloth.) See those chocolate cookies under there?

J: Uh huh. (no.)

I: Well, let's put some under there. (The teacher is still pretending.) Now, there are ten cookies under both cloths. (Places the numeral "10" immediately above both cloths.) How many are under here? (The adjacent cloth.)

T: (Touches the cloth with the numeral "8" on it eight times.) 1-2-3-4-5-6-7-8. (Continues touching the other cloth as if it hid ten cookies.) 9-10-11-12 (Completes a row of four points of contact and then continues touching the cloth immediately beneath the completed row.) 13-14-15 – 16. (Looks up at the teacher while saying "sixteen" and then continues touching the cloth immediately beneath the two completed rows, continuing to look at the teacher.) 17 – 18 (Touches the cloth emphatically when saying "eighteen" indicating that he was done.)

There were two significant aspects of Jason's counting activity. First, Jason seemed still obliged to establish meaning for number words by counting. This is indicated by counting to "eight" and by counting ten more times beyond counting to "eight." Second, when counting ten more times, Jason looked intently at the cloth as he completed two rows of four and then changed from looking intently at the cloth to looking at his teacher. This, along with stopping at "eighteen," indicates that Jason reprocessed his counting acts while he was in the activity of counting.

---

[7]The ability to establish figurative unit items as substitutes for hidden perceptual unit items is one level, and the ability to run through a counting activity and produce its results in thought without motor action and without given sensory material to act on is the other level.

To keep track of counting while in the activity involves taking a counted item (touching the cloth and saying a number word) as material for further unitizing. We could not say that Jason actually counted the reprocessed counted items, because he organized his counting activity into two patterns of four and a pattern of two. Nevertheless, to "step out" of counting while in the activity involves taking the counted items as material for further unitizing using the same object concept[8] that was used in establishing the situation of counting. Using his interiorized object concept, cookie, to compound the sensory material produced by touching the cover and uttering a number word into a unit item transforms the object concept into an interiorized attentional pattern that contains records of counting acts. We call these interiorized attentional patterns *arithmetical units*. They can be considered as the records of the unitizing operation, or "slots," and are diagrammed as follows: 0 I 0.

Jason's organization of his continuation of counting into patterns produced what I regard as *numerical patterns*. The result of this process is diagrammed in Fig. 3.6. A numerical pattern consists of a composite of cooccurring arithmetical unit items.[9] But rather than focus on the pattern as one thing, the child is aware of the cooccurring elements of the pattern. In Fig. 3.6, the solid segments under the moments of focused attention indicate records of counting acts. Even though experientially bounded, a numerical pattern is not yet a number sequence but it is a sequence of arithmetical units.

The numerical patterns of four that Jason constructed contained records of counting signified by "9, 10, 11, 12" and "13, 14, 15, 16." In this case, I think of the attentional structure of a numerical pattern as an abstracted segment of the child's sequence of counting acts. In this, I regard reprocessing counting acts as did Jason as a *disembedding operation* that interiorizes segments of counting acts ["9, 10, 11, 12"] and transforms sensory-motor counting acts into counting operations.

If the numerical patterns established were permanent, i.e., if the arithmetical units did not decay, then I would say that Jason had constructed *numerical* permanent object concepts for "four" and for "two." The problem of how a child constructs permanent object concepts for other number words of his or her number *word* sequence has been explained as autoregulation of the disembedding operation (cf. Steffe and Cobb 1988, pp. 310–311). The abstracted numerical patterns induce a systemic disequilibrium in the child's counting scheme that drives the process of disembedding until equilibrium is restored. This autoregulated process produces what I call an *initial number sequence* (INS), which is a sequence of

Fig. 3.6. The attentional structure of a numerical pattern.   (0 I 0 0 I 0 0 I 0 0 I 0)

---

[8] In this case, an object concept is a figurative unit item; that is, a particular type of attentional pattern. An attentional pattern, when implemented, is an operation of unitizing.

[9] A numerical pattern can also consist of a composite of abstract unit items; that is, interiorized figurative unit items that do not contain records of counting.

arithmetical units containing records of counting acts. It is at the same level of abstraction as von Glasersfeld's (1981) arithmetical lot. The arithmetical lot is more general than a number sequence simply because its elements do not necessarily contain records of counting acts. Both structures are essential because they mark an early divergence in the construction of mathematics; the arithmetical lot leads to set theory and the INS leads to number theory.

Using an arithmetical unit to produce a visualized image of a counting act can be thought of as an experience of discrete unity. The figurative material produced need not carry any resemblance to the figurative unit items from which the arithmetical unit was abstracted or to the counting acts that served in making the interiorized records of counting. It may be a visualized image of a fragment, a motion, or a sound depending on the context of activation. Upon activation of the INS, the crucial realization is that there is some figurative material produced that refers to counting a collection of perceptual unit items. This permits a number word to stand in for a collection of perceptual unit items that, if counted, could be coordinated with the utterances of the number words from "one" up to and including the particular number word. The child knows this and therefore does not have to run through the activity that would actually implement it at the level of sensory-motor experience.

A number word of the INS, like "seven," can activate a sequence of *arithmetical unit items* of specific numerosity. I consider such a sequence as an elaborated numerical pattern in the sense that the child is aware of the elements of the pattern but not the pattern as one thing – as an entity. Moreover, the child may be aware of elements at the beginning and the end of the pattern, but not necessarily of all of the elements in between. When activated, such a numerical pattern [which I call a *numerical composite*] in turn activates the production of figurative material that refers to counting. An activated numerical pattern is somewhat like a resonating tuning fork with the stipulation that its resonating creates an image of counting. The image might be simply some minimal re-presentation of the involved number words that symbolize counting, or it might be a plurality of flecks that symbolize countable perceptual items. The hypothetical *slots* that contain records of experience that are used in producing operative images are realized experientially only upon activation of the program of operations that produce them, and they are realized only as images produced by means of the records contained in them.

The results of constructing the INS are dramatic but easy to observe. By May 21 of his first grade in school, Jason could use his counting scheme in ways that were previously not possible.

Protocol III.

T: (After Jason places 16 chips in a cup.) Take out four.
J: (Takes out four.)
T: How many are left in there?
J: (Places his hands on his lap and concentrates on manipulating the poker chips. The teacher cannot see what Jason is doing.) There's twelve in there.
T: Tell me how you did that!
J: (Places chips on the table one at a time.) 16-15-14-13 – there's twelve in there.

Fig. 3.7. The attentional structure of the initial number sequence.

$$(0\underline{I}0 \quad 0\underline{I}0 \quad 0\underline{I}0 \quad 0\underline{I}0 \ ...)$$
$$1 2 3 4$$

Fig. 3.8. The attentional structure of the numerical composite "six."

$$(0\underline{I}0 \quad 0\underline{I}0 \quad 0\underline{I}0 \quad 0\underline{I}0 \quad 0\underline{I}0 \quad 0\underline{I}0)$$
$$1 2 3 4 5 6$$

The change in Jason's use of his counting scheme implies interiorization of the activity of counting. Because Jason's counting scheme was a figurative scheme prior to this teaching episode, it is not too much to say that his counting scheme had undergone a metamorphosis. It was now a numerical scheme whose attentional structure is illustrated in Fig. 3.7. The numerals under the solid bars are meant to designate auditory records of number words.

It is important to note that the attentional structure that is used in assimilation *contains records of the activity of the scheme*. In contrast, the activity of a perceptual or a figurative counting scheme is not contained in the recognition templates of these schemes. It is also important to note that number words of the INS refer to numerical composites. For example, "six" refers to the attentional structure in Fig. 3.8. We refer to this attentional structure as an *initial segment* of the child's number sequence as well as a numerical composite.[10]

The Tacitly Nested Number Sequence

The construction of the INS involves reprocessing actual counting acts and, thereby, it can be appropriately thought of as a sequence of interiorized counting acts. A number word of the INS symbolizes the initial segment of the number sequence from "one" up to and including the given number word. But it does not symbolize a composite unit containing that initial segment, which is a major advancement upon the construction of the tacitly nested number sequence (TNS).

We have illustrated in Protocol III how Jason could operate in May of his first grade in school when he had constructed the INS. In that Protocol it is understandable why Jason said "16-15-14-13" to find how many poker chips remained under the cloth because he had just counted to "sixteen" from "one" to place 16 poker chips in the cup. Experientially, he was at the end of his counting activity and could regenerate the four immediate past counting acts and use these regenerated counting acts in a backward sequence as he manipulated the poker chips. Jason monitored his counting acts as he counted backward four times because he knew that "twelve" referred to the numerosity of the hidden poker chips. Still, there was no indication in Protocol III that Jason took his numerical pattern, four, as input for making countable items. Rather, he used his numerical pattern, four, in monitoring reenacting

---

[10] A numerical composite is a sequence of arithmetical units. But there is no unit containing the sequence except for the beginning and end of actually counting.

3   Operations That Produce Numerical Counting Schemes   39

his forward counting acts in the backward direction. But by December of his Second Grade in school,[11] he could take numerical finger patterns as input for making countable items.[12]

Protocol IV.
T: (Places "17 – 8" in front of Jason.) What would that be?
J: (In synchrony with putting up fingers, utters.) 17-16-15-14-13-12-11-10. That would be (pauses) – nine!!

Jason's utterances of number words from "seventeen" down to "ten" in synchrony with putting up fingers solidly indicates that "8" referred to a numerical finger pattern. Jason had not counted poker chips from "one" to "seventeen" as he did in Protocol III, so "seventeen" referred to counting backward seventeen times. Further, the expression "17 – 8" referred to counting backward eight times starting with "seventeen." His finger pattern for "eight" provided him with a means of separating the first eight of these counting acts from those that would remain. That he took his finger pattern as a unit that contained his countable items is indicated when he said "nine" to indicate how many would be left. This act serves as an indicator that he took his numerical finger pattern, eight, as input for making countable items because "nine" referred to the result of the counting acts he would carry out had he continued to count backwards to "one." That is, the counted items he would produce had he counted backward nine times to "one" were identical to the counted items from "one" forward to "nine" – the elements of his numerical finger pattern for "nine." My inference that counting was reversible for Jason is based on his *not* counting backward from "nine" to "one" to produce a finger pattern that he could recognize to establish the numerosity of the counting acts.

Reversibility of counting implies that Jason was explicitly aware that counting backward nine times from "nine" and forward nine times from "one" would produce the same numerosity. What this means is that he set his numerical finger pattern for "nine" at a distance and took it as an object of reflection. Reflecting on his finger pattern in turn implies that he united the elements of the finger pattern into a composite unit – as an entity on which he could reflect. Uniting the elements of his finger pattern for nine into a composite unit certainly was not restricted to that particular finger pattern, so I infer that he could also unite his finger pattern for eight into a composite unit prior to counting in Protocol IV. This is what it means for Jason to take his finger pattern for "eight" as input for making countable items prior to counting because he was aware not only of his finger pattern for eight,

---

[11] The initial number sequence is an unstable number sequence that is only transitory to a more adequate number sequence.

[12] When the elements of a finger pattern become recorded in arithmetical units, the finger pattern is constituted as a numerical finger pattern and can be used as numerical meaning of number words.

but also that he could count backwards eight times starting from "seventeen" to implement eight counting acts.

It is the case that Jason did not explicitly count his backward counting acts by putting up a finger in synchrony with saying "seventeen" and then saying, "that is one". That is, he did not explicitly count his counting acts. However, putting up a finger served two functions. On the one hand, it served as a record of a counting act. And, on the other hand, it served as a countable item in a way similar to how putting up a finger is a countable item in the figurative counting scheme. In the latter case, putting up a finger is a substitute for a figurative unit item and, in the former case, putting up a finger symbolizes an arithmetical unit item.

The inference that Jason took his numerical finger pattern for eight as input for his uniting operation finds corroboration in his clear separation of backward counting acts into two parts prior to counting; into the first part containing the eight counting acts he actually carried out and the second containing the counting acts preceding "ten" that he did not carry out. This clear separation would be possible only if he understood the number sequence from seventeen down to one as breakable into two unitary parts prior to counting.[13] The first unitary part would be of numerosity eight starting with "seventeen" and proceeding backward, and the second would be of unknown numerosity and would begin with the number word preceding the last number word said when counting. What makes the inference concerning the unitariness of the parts especially compelling is that "nine" referred to the numerosity of the counting acts he never carried out as well as to the number word preceding "ten." We model his meaning for "nine" in Fig. 3.9 as attentionally bounded rather than as experientially bounded as was the case for a numerical composite. It is attentionally bounded (zero's at the ends) because it arises as the result of reprocessing a sequence of counting acts using the arithmetical unit.

Activation of the template of Fig. 3.9 permits the child to be aware of the individual elements of a composite unit formed at the experiential level as well as of the unitariness of the composite of elements. But it is not possible for the child to separate the experience of the unitariness of the elements from the experiential cooccurrence of all the elements. That is, the elements do not "drop out," leaving an empty shell that can be filled with elements upon iteration of any one of the elements.[14]

$$0(0\underline{1}0\ 0\underline{1}0\ 0\underline{1}0\ 0\underline{1}0\ 0\underline{1}0\ 0\underline{1}0\ 0\underline{1}0\ 0\underline{1}0\ 0\underline{1}0)0$$
$$\quad\ 9\quad\ \ 8\quad\ \ 7\quad\ \ 6\quad\ \ 5\quad\ \ 4\quad\ \ 3\quad\ \ 2\quad\ \ 1$$

Fig. 3.9. Attentional structure of a composite unit of numerosity nine.

---

[13] Note that Jason did not first count from "one" to "seventeen" in a way similar to how he counted in Protocol III.

[14] For a more complete analysis of the tacitly nested number sequence, see Steffe and Cobb (1988).

The Explicitly Nested Number Sequence

Like the INS, the TNS is transitional to a higher stage of the number sequence. This is indicated by the way Jason could use his number sequence during the latter part of his Second Grade in school. Jason could now estimate a missing addend, add the estimate to the first addend, and check the result to see if it was the sum.

Protocol V.
T: (Places the expression "27 + ___ = 36" in front of Jason.) We have twenty-seven and some more and that is thirty-six.
J: Twenty-seven – (Pause of about 20 seconds.) Let me see – (Another pause.) –twenty-seven plus seven – its nine more!
T: That's really good! Is there another way to solve that one?
J: Uh-uh. (no.)
T: Could you do it by counting backwards?
J: (Sequentially puts up fingers.) 36-35-...- 27. Nine.

To operate as he did in Protocol V requires that Jason disembed the remainder of 27 in 36 from 36. This required the use of using his uniting operation in the following way. He knew that he could count the elements of the remainder starting with "28" by one. But rather than count them, he made an estimate of *one seven*. That is, he gauged the size of the remainder, checked his estimate, and then added what was left of the remainder to the estimate. To do this, he would need to be aware of the unitary structure of the remainder without consideration of its elements. He would also need to be able to generate the elements at will and move back and forth from the level of the unit structure to the element level in a rather rapid succession, all of which indicates an awareness of the remainder as being included in 36 as well as a unit structure apart from 36. Correspondingly, we call his number sequence *explicitly nested*.

These operations indicate that his number words and numerals referred to composite units (e.g., seven) that not only *could be* disembedded from segments of his number sequence, but also embedded in segments as well (e.g., the segment from 27 to 36). But, the operations indicate more because of the way he made and used the estimate, seven. After making the estimate, he extended 27 by seven more and then used that unspoken result, 34, as an initial segment of his number sequence to 36. He maintained 34 as composed of 27 and seven more, and extended seven by how many more it took to reach 36. This demonstrated ability to interchange the sense of seven from being a product of operating to being material of further operating, and then to take the results of that operating, 34, as material in still further operating, is a strong indication of what we call a *recursive numerical scheme*.

We also regard Jason's way of proceeding as being made possible by an iterable unit of one. When the unit of one is iterable, a number word refers to a composite unit containing a unit which can be iterated the number of times indicated by the number word. This provides great economy in the child's reasoning. For example, understanding that 36 was composed of the two parts, 27 and some more, Jason knew that removing one of the two parts would leave the other because he treated the two parts of 36 *as if* they were singleton units. Of course, Jason could distinguish between the two parts both conceptually and experientially. Nevertheless, subtracting as the inversion of adding involves the child partitioning a numerical whole into two numerical parts and recombining the parts to produce the whole all without consideration of the elements of the parts. In this, the child reasons with composite units rather than with the elements composing them.

In the case of the iterable unit of one, it is easy to emphasize sequentially producing units of one and combining each unit produced with those preceding. Jason was quite capable of these operations. However, of primary importance is that number words like "five hundred thirty-two" can refer to a unit that *could be* iterated 532 times to "fill out" the composite unit to which it belongs.

In the construction of an iterable unit of one, the child reprocesses the symbolized counting acts using an arithmetic unit. This produces an abstracted unit that I call iterable because it stands in for counting acts that could be carried out. In fact, the iterable unit can be "held still in suspended animation" or "let go in iteration." In the first case, it symbolizes the structure from which it was abstracted, and in the second case, it is used in producing the structure of a unit of units. The iterable unit of one opens the possibility for a child to "collapse" a composite unit into a unit structure containing a singleton unit, which can be iterated many times. So, when using the iterative unit to give meaning to a number word like "seven," the iterable unit symbolizes a composite unit structure that can be filled out by iterating the iterable unit seven times. The diagram in Fig. 3.10 is meant to indicate an iterable unit within a composite unit structure. In the diagram, the red underlining is meant to indicate records of sensory material involved in iterative acts of counting.

An Awareness of Numerosity: A Quantitative Property

An awareness of numerosity, which is an awareness of the potential results of counting, changes dramatically across the number sequence types. In case of the INS, the child can form an image of a plurality of figurative unit items that are countable, and the child is aware of more than one of them. This differs from an awareness of figurative plurality in that the images are produced by means of activated arithmetical units and thus can consist of the most minimal of figurative material. Any figurative material will suffice for the establishment of countable items. We call an awareness of the results of counting these countable items an *awareness of figurative numerosity* to emphasize an awareness of the figurative unit items as countable items.

Fig. 3.10. The attentional structure of a composite unit containing an iterable unit.   0(0 1 0)0

Because the TNS consists of interiorized counting acts, the child can form an image of a sequence of counting acts prior to counting and constitute them as countable items. That is, the counting scheme can be used as input for its own operating and counting acts are objects that can be counted. We call the child's awareness of the results of counting such countable items an *awareness of arithmetical numerosity* to emphasize an awareness of arithmetical units as countable items.

At the stage of the explicitly nested number sequence (ENS), an iterable unit of one symbolizes unit items produced by iterating. This symbolic function of an iterable unit of one provides the child with an awareness of the unit items in a composite unit without the necessity of making images of them. Rather than introduce a new term, we call this symbolic awareness an *awareness of arithmetical numerosity* as well and rely on the interpretation of the child's particular number sequence to make a differentiation.[15]

The Generalized Number Sequence

Children who are in the stage of the ENS have constructed one as an iterable arithmetical unit, and they can operate with and on composite units. However, they are yet to construct composite units as iterable units in a way similar to their iterable units of one. When children have constructed composite units as iterable, this indicates that they are at least in the process of reorganizing their ENS into what we call the generalized number sequence (Steffe 1992). A generalized number sequence is a sequence of composite units of equal numerosity (including the unit of one) that can serve as countable items. In a generalized number sequence, all of the operations of the ENS can be carried out using composite units as well as other operations that are specific to the generalized number sequence. In this sense, the generalized number sequence supersedes the ENS in that it can be used in all of the ways that the ENS can be used as well as ways in which the ENS cannot be used. Speaking metaphorically, children are in a "composite units" world rather than a "units of one" world.

I do not present an attentional analysis of the construction of a unit containing sequence of units of units because the generalized number sequence was presented as a conjecture when analyzing the schemes of action and operation involving composite units of a third grade child named Johanna who had constructed the ENS (Steffe 1992). This conjecture was based on Johanna's ways and means of

---

[15] A child can be aware of the types of discrete quantity at the preceding levels of the number sequence. For example, a child who has constructed only the initial number sequence can be aware of perceptual or figurative pluralities because they can engage in the operations that produce them. But it is always understood that such a child could constitute these pluralities as perceptual or figurative numerosities if they produced them using their number sequence.

operating with composite units. Johanna was asked to find how many blocks were in five rows of blocks she had made with four blocks per row after the blocks were hidden from view (Steffe 1992, pp. 292 ff.). After the blocks were hidden, Johanna was first asked how many blocks were in the *first three* rows. After sitting silently for about 25 s with her hands resting in her lap in deep concentration, she said "twelve." So, she was asked how many more rows she had, and she said, "two" and that there were eight blocks in the two rows. In reply to the question, "How many blocks in all five rows?" Johanna again sat quietly for about 15 s and replied, "twenty!" In explanation, she said, "Because I added up. Twelve plus four is 16, and 16 plus 4 is 20!"

Whatever figurative material Johanna generated as meaning for a row of four, she produced five rows of four in visualized imagination. Producing such an image is within possibility for children who have constructed the ENS because a row of blocks can be re-presented as an entity – a row – without consideration of the elements of the row. This re-presentational capacity is made possible by the iterable unit of one. Once Johanna produced an image of five rows of four, she separated them into three rows and two rows, which indicates that she took the five rows as a unit as well as the three rows and the two rows. Her re-presentational capability is indicated by her reply "twelve" to the question of how many blocks were in the first three rows and the 25 s she sat quietly. This is a rather longish period of time to count by fours to twelve. Indeed, there was sufficient time for her to separate the five rows of four into three rows and two rows, a separation that is clearly indicated by her knowing almost immediately that there were two remaining rows of four after she said "twelve."

Johanna finding how many blocks in three of the five units of four and of finding how many blocks in all five units of four demonstrates children's potential who have constructed the ENS. Not only did Johanna assemble the operations to produce a unit containing a sequence of five rows of four, she disembedded three rows of four from the five rows and used the three rows of four as countable items in keeping track of how many units of one she counted. She also engaged in progressive integration operations[16] ("twelve plus four is sixteen, and sixteen plus four is twenty"). Johanna operated as if she produced a composite unit containing five units of four and then used these units in progressively integrating these units of four. But I could not infer that Johanna used the operations that produce a unit of units in assimilating the situation that I presented to her. Instead, all I could infer was that she could assemble operations that produced a unit of units in assimilation. I could also infer that once

---

[16] By an integration of two composite units we mean that the child, whose goal it is to find how many elements in the two composite units together, first unites the two composite units into a unit containing them, disunites each of the two composite units into their elements while maintaining an awareness of the two composite units as well as of their elements, and then counts the elements to establish their numerosity. The containing composite unit serves as background while the child is operating.

she assembled five rows in re-presentation, she could drop down to the rows and unpack their elements. This way of operating was definitely an accommodation in her operations with units of one, but it remains an open question whether these operations were sufficient to claim that she had constructed a generalized number sequence. That remains the subject of further investigation in the following chapters.

An Overview of the Principal Operations of the Numerical Counting Schemes

The Initial Number Sequence

In the case of the INS, number words refer to sequences of arithmetical unit items that called *numerical composites*. In saying, "seven," for example, a child in the stage of the INS knows that it could run through the individual items of a collection of *hidden* perceptual unit items and coordinate a number word from "one" to "seven" with each item. The child knows it can do this, and "seven" symbolizes the activity. If the child mentally runs through the activity, the result would be a sequence of counted unit items. I call such a sequence produced by counting a numerical composite as well as the sequence of arithmetical unit items that was symbolized prior to counting that the child used in producing an image of the activity.

INS children can count-on starting at seven to find how many items are hidden at two locations, one hiding seven and one hiding six (S-e-v-e-n; 8, 9, 10, 11, 12, 13 in synchrony with putting up fingers.). The child stops counting when recognizing a finger pattern for "six." Finally, INS children can re-present counting acts in their immediate past and use the re-presented counting acts as countable items in further counting, which is a crucial operation in the construction of the TNS. However, INS children are yet to disembed a numerical part from a numerical whole. For example, in the case of 12 items hidden at two locations, seven at one location and some more at the other, INS children might assimilate the situation using their INS and interpret it as 12 items hidden at the other location and try to count-on twelve times beyond seven to find how many items hidden in both locations. Others might not be able to engage in any counting activity at all.

The Tacitly Nested and the Explicitly Nested Number Sequences

In a situation, say, where a child has put a handful of pennies with 19 pennies, and then counts all of the pennies and finds that there are 27, a child who has constructed either the TNS or the ENS might become aware of the unknown numerosity of the added pennies and thereby form a goal to find how many pennies were added. For a TNS child, "twenty-seven" refers to a number sequence from "one" up to and including "twenty-seven" that the child can take as elements of a composite unit. Any number word to "twenty-seven" refers to an initial segment of this number sequence, so the child can separate the number sequence up to and including "twenty-seven" into two *embedded* parts. In particular, "nineteen" symbolizes an initial part of the number sequence to "twenty-seven," and this permits the child to form the goal of finding the numerosity of the remainder of nineteen in twenty-seven. The child is at nineteen in the number sequence to twenty-seven, and because the initial part of the number sequence to twenty-seven is symbolized by the number word "nineteen," the child can focus on continuing to count to "twenty-seven" starting with "nineteen" while keeping track of how many times she counts. In all of this, the child neither disembeds nineteen nor the remainder of nineteen in twenty-seven from twenty-seven. Rather, the child is operating with the elements inside of the composite unit comprised by twenty-seven.

In the case of an ENS child, the units of twenty-seven have been constituted as iterable units. So, the image of twenty-seven may be of a singular unit item that can be iterated twenty-seven times. This provides the child with great economy in thinking, because to regenerate an experience of counting twenty-seven items, the child does not need to visualize a plurality of counting acts. "Twenty-seven" may stand in for a composite unit item (or slot) that could be filled with twenty-seven unit items by counting to "twenty-seven." This iteration of a unit item by counting is indicated when the child counts its acts of counting from "nineteen" up to and including "twenty-seven" starting with "one." But this is only an indication because some children who are yet to disembed nineteen from twenty-seven can do as much. However, when a child reasons strategically as follows, this is a solid indication of the explicit nesting of one number in another. A child named Johanna was asked to take twelve blocks and the interviewer took some more and told Johanna that together they had nineteen. After sitting silently for about 20 s, Johanna said "seven" and explained, "Well, ten plus nine is nineteen; and I take away two – I mean, ten plus two is twelve, and nine take away two is seven" (Steffe 1992, p. 291). Johanna disembedded ten and nine from nineteen and then operated on the two numbers until she transformed them into twelve and another number that, when added to twelve, would make nineteen.

The strategic reasoning exemplified by Johanna is made possible by the ability of an ENS child to form an image of nineteen as a singular unit item that can be iterated by counting nineteen times to fill out a composite unit containing nineteen items. The child can "go" to the implication of its operative image and generate an image of component parts of nineteen and *operate on these images*. So, I do not regard a composite unit structure as an object the child has stored somewhere in memory. Rather, I regard it as a product of an ensemble of possible operations that is symbolized by number words. In regarding composite unit structures in this way, I am in agreement with Dörfler (1996) about not pursuing a theory of mind where the mind has or contains mental objects corresponding to numbers, natural or rational. I highlight the ability of the child to disembed an image of the continuation of ten in nineteen and to unite the items of this image into a composite unit without destroying the composite unit structure that contains it. Iterating a unit item and disembedding a numerical part from a numerical whole are two principal operations of the ENS. A third is taking parts or wholes as material of further operating as did Johanna in strategic reasoning.

Final Comments

I have already explained that there are two ways of understanding the hypothesis that *children's fraction schemes can emerge as accommodations in their numerical counting* schemes. The first way of understanding the hypothesis concerns the construction of the number sequences. The second is that children use their numerical counting schemes to construct fractional schemes by operating on material in continuous quantitative situations that are not a part of the situations of the numerical counting schemes. When operating on novel material using the operations of their numerical counting schemes, the hypothesis is that children will use their numerical operations in ways that are also novel. They may also use operations that are not a part of the numerical operations that originate from their constructed situations. The fractional schemes that are produced would solve situations that the numerical counting schemes did not solve, and they also would serve purposes that the numerical counting schemes did not serve.[17] But the fractional schemes would not supersede the numerical counting schemes in the way the ENS supercedes the TNS because they would not solve all of the situations the numerical counting schemes solved. The numerical counting schemes would be still needed to solve their situations. Still, the fractional schemes could be regarded as a reorganization of the numerical counting schemes because operations of the numerical counting schemes emerge in a new organization.

It is this second way of understanding the reorganization hypothesis that is relevant in our current work. Fractional schemes and numerical counting schemes are not of the same type if for no other reason than they are used by children for different purposes. If children do use their numerical counting schemes in situations that are novel to the schemes, this would entail more than using the schemes in assimilation. The children would also need to engage in generalizing assimilation.[18] So, if a counting scheme is used in generalizing assimilation, this means that the activity of counting is not implemented in the assimilating process. Rather, the operations that produce a situation of counting are evoked and generalized in the process of producing a situation of counting that is novel to the scheme.

---

[17] Tomas Kieren has always been fond of saying that fractional schemes are used to find how much and counting schemes are used to find how many.

[18] An assimilation is generalizing if the scheme involved in assimilation is used in a situation that contains sensory material or conceptual items that are novel for the scheme but the scheme does not recognize it, and if there is an adjustment in the scheme without the activity of the scheme being implemented.

Chapter 4
Articulation of the Reorganization Hypothesis

Leslie P. Steffe

When fractions are introduced in school mathematics, they are usually introduced in the context of continuous quantity. Number sequences are essentially excluded because, as quantitative schemes, they are thought to be relevant only in discrete quantitative situations. Even though I developed number sequences in Chap. 3 in the context of discrete quantity, I can see no principled reason to keep them separate from continuous quantity. Reserving number sequences for discrete quantity stands in opposition to the concept of the real number line in higher mathematics, and, in this chapter, I argue that it also stands in opposition to the development of quantitative schemes. In articulating the reorganization hypothesis, I establish that a composite unit of specific numerosity can be used to make a split in the way that Confrey (1994) explained. This involves more than simply indicating the possibility of transferring the operations involved in compounding discrete units together to splitting continuous units. I do a deeper developmental analysis of children's quantitative schemes in which I explore whether the operations that produce discrete quantity and the operations that produce continuous quantity can be regarded as unifying quantitative operations. If so, these quantitative operations would justify the reorganization hypothesis.

I start the chapter with analysis of the construction of continuous items of experience as well as connected but segmented items of experience and develop the notion of quantity as a property of an object concept that can be subjected to comparison. The question "What is quantity?" can be interpreted as a question about mathematical concepts that exist independently of the children who are to learn them rather than about children's quantitative concepts. As an adult, I can say that intensive quantity is nonadditive, that an extensive quantity is additive, that the quotient of two extensive quantities yields an intensive quantity, and that an extensive quantity arises as a result of counting or measuring (Schwartz 1988). These ideas of quantity are essential, but they do not specify the operations children use to generate quantity.

Davydov (1975), following Kagan (1963), formulated a definition of quantity that supports the idea of quantity as a property of a concept. According to Davydov (1975), a quantity is any set for which criteria of comparison have been established for the elements. The necessity to specify criteria of comparison assumes that some common property of the elements has been established. Davydov's (1975) idea of quantity, then, orients us to viewing the origins of quantity in properties of concepts in the way I regard numerosity as a property of a composite unit structure rather than as the composite structure. Quantitative properties of concepts such as the iteration of a unit structure are introduced by the knowing subject's actions in the construction of the concepts.

Perceptual and Figurative Length[1]

Children do construct "continuous" items of experience as well as discrete items of experience. Although I do not and cannot make a definitive distinction between these two kinds of items of experience, continuous items of experience involve motion of some kind.[2] Such motion might be moving the eyes, crawling along the floor, walking along a path, sweeping one's hand through a space, or scratching a path in the frost on a window with a fingernail. In so far that each of these motions have a beginning or an end, they can be isolated from the rest of one's experiential field and, along with sensory material from the visual or tactual mode, form what I call experiential continuous items.

I usually think of the path of the motion as being the experiential item when there are visual records of the path. However, I may overemphasize the visual perceptual records. For example, a unitary item that corresponds to something like what an adult would call "rod" [a rod template] contains records of the motion involved in moving the eyes in the construction of the unitary item. If the unitary item can be used in re-presentation, i.e., in "visualizing," the scanning motion that is recorded in the rod template could be reenacted and produce a regeneration of the recorded visual material that constitutes the path of the motion. If the child becomes aware of the visualized path including its endpoints, of the motion that produces the path, and of the duration of the motion, the child would be aware of figurative length.[3]

If an observer's rod is moved from one place to another, the child might know that it is the "same rod" because the rod template can be used in re-presentation as indicated above. This is nothing but object permanence – the rod "exists" for the child independently of its particular location in experience because the child is aware of the rod without it being in the immediate visual field. But, a child still might believe that a rod whose right most end-point, say, goes beyond that of

---

[1] The concepts of discrete and continuous quantity presented in this chapter have their origin in Steffe (1991).

[2] If a continuous item of experience is bounded, from that perspective it is also discrete. Similarly, if a discrete item of experience has an interior that would qualify it as a continuous item of experience. Of course, by an "item of experience" I refer to an implemented attentional pattern with the understanding that the experiential item is a permanent object.

[3] Excepting an awareness of duration would eliminate an awareness of the continuity of the scanning motion over the regenerated visual material.

another rod is "longer" regardless of the relative positions of the left most end-points. This judgment would be an indication that the child is aware at least of perceptual length. An awareness of perceptual length means that a child becomes aware of the duration of scanning as well as of the visual path of scanning.

Children do construct experientially connected but segmented items as well as experientially continuous items like a rod. A sidewalk is one example and a row of telephone poles is a complementary example.[4,5] In the case of a sidewalk, when establishing a row of sections a child might experience a section as an object concept in a way that is similar to any other object concept. If so, the child could use this concept in assimilation and become aware of scanning more than one section in the assimilation. This introduces repetition into the object concept and an awareness of a plurality of sections. When coupled with an awareness of the perceptual length units of the sections, this produces what I call an awareness of perceptual length of an experientially connected but segmented unitary item. Such an awareness of perceptual length is a gross quantitative property of a row of perceptual unit items called segments.[6]

When a child uses its concept of a continuous but segmented unitary item in re-presentation, this opens the possibility of the child using its concept, segment, in reprocessing a re-presentation of a continuous but segmented unitary item. This process produces a segment as a figurative unit item and a row of segments as a figurative lot structure that is analogous to the figurative lot structure of Fig. 3.5 in Chap. 3.

Generically, the property of a segment that I call length is an awareness of the scanning action over an image of the segment along with an awareness of the duration of the scanning action. If the child is aware of a figurative row of segments (a figurative plurality of segments), then that awareness, when coupled with an awareness of the scanning action over the segments of the row, is what I mean by an awareness of figurative length of a row of segments. The construction of a figurative row of segments illustrates the construction of operations that children could use in future occasions to project units into experientially continuous units and provides a basis for a synthesis of discrete and continuous units. It is meant to illustrate how children's construction of continuous quantity at the most elementary level involves operations that are also involved in their construction of discrete quantity.

Piaget's Gross, Intensive, and Extensive Quantity

Quantitative properties of concepts like the iteration of a unit structure are introduced by the knowing subject's actions in the construction of the concepts. Hence a quantitative property of a concept can be viewed as an abstraction of the records of

---

[4] They are complementary because the former leads to length and the latter to distance.
[5] See Steffe (1991) for a conceptual analysis of this construction.
[6] Here, "segment" is not to be interpreted mathematically.

actions that were involved in the construction of the concept. This fits well with Piaget and Szeminska's (1952) idea of a gross quantity. An awareness of perceptual length is a gross quantitative property of a segment or of a row of segments. It is a gross quantitative property because the child abstracts it from the activity of scanning a segment or a row of segments by means of pseudo-empirical abstraction. The types of quantitative comparisons explained by Piaget and Szeminska (1952) help to understand the difference between an awareness of perceptual length and an awareness of figurative length.

Gross Quantitative Comparisons

In Piaget's work, if a child judges that six blocks arranged in a row are "more than" seven blocks aligned in a shorter row as shown in Fig. 4.1, this would indicate an awareness of the intervals between the blocks. In this case, the child is aware of a row of blocks as a segmented but connected unitary item. If there is no further indication that the child is aware of producing a figurative row of intervals, the child's comparison would be judged as a gross quantitative comparison, where the gross quantity is an awareness of perceptual length[7] of a row of intervals.

Intensive Quantitative Comparisons

If two rows of blocks were of equal length as in Fig. 4.2, and if a child made a judgment that the bottom row has more blocks than the top row because they are closer together and the rows are of the same length, Piaget and Szeminska (1952) called this an intensive quantitative comparison. In my terminology, I would say that the

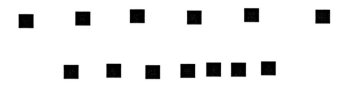

Fig. 4.1. Two rows of blocks: Endpoints not coincident.

Fig. 4.2. Two rows of blocks: Endpoints coincident.

---

[7] In this case, the child is aware of a row of blocks as a segmented but connected unitary item. Nevertheless, the child may be aware also of perceptual plurality.

child coordinates an awareness of the perceptual length of the rows and an awareness of the perceptual plurality of the counters in the rows. To make this intensive quantitative comparison, the child does not make a comparison based solely on the perceptual length of the rows as in Fig. 4.1. Instead, the child would need to extract herself from the immediate here and now and not make a quantitative comparison of "the same." To do this, the child would need to be aware of the intervals between the blocks of each row of blocks and of the perceptual density[8] of the blocks in each row. Furthermore, the child would need to be aware that the increase in perceptual density of the blocks in the bottom row is compensated for by the decrease in the length of the intervals. Hence, the child would need to operate at least one level above the perceptual level as well as at the perceptual level. Otherwise, the child would not reflect on the intervals between the blocks as well as on the density of the blocks in a row. What this means is that the coordination of two gross quantities does not happen at the level of the gross quantities. Rather, it occurs in re-presentation.

A child who can make an intensive quantitative comparison, then, must be able to visualize a row of blocks to coordinate two gross quantities. I do not assume that the child is aware of a visualized row of blocks when using its recognition template in making the comparison because the visualizing experience may occur outside of the awareness of the child. But the template for producing a figurative row of blocks must be active in re-presentation for the child to "step back" from the rows of blocks in the immediate here-and-now and coordinate the perceptual length and density at the level of re-presentation when making the comparison.

The dual use of the template at the re-presentational level and the perceptual level allows the child to focus on either the intervals between the blocks or on the blocks. This means that, in a way, the possibility of a separation between the intervals of a row of blocks and the blocks was already "contained in" the recognition template prior to operating in the immediate here-and-now. This differentiation between the intervals of a row and the blocks of the row allows for the intuitive understanding that the blocks are more frequent in the bottom row and the intervals between them are shorter, which in turn leads to the judgment of "more blocks in the bottom row." The child can now make judgments in comparing the rows that were not previously possible. So, to reiterate the central point, by using a figurative row of blocks, a child can focus on either the intervals or the blocks and coordinate the two.

An Awareness of Figurative Plurality in Comparisons

In the construction of the figurative lot structure (Chap. 3, Fig. 3.5), I did not account for the intervals between the figurative unit items in those cases where the elements are arranged in an identifiable path. These intervals are not empty

---

[8] By "perceptual density," I mean an awareness of the frequency of instantiation of the block concept within definite boundaries. By "perceptual plurality" I also mean an awareness of more than one instantiation of the involved template.

in that they contain no perceptual material. Likewise, in a figurative row of segments, what is in between the segments is not vacuous. So I regard the structure of a figurative row of blocks as a more complete account of both a figurative lot structure and a figurative row of segments in special cases. As a coordination of two gross quantities, intensive quantity, then, is nothing more than an awareness of both figurative length and figurative density. The way in which I defined an awareness of figurative length in the case of a figurative row of segments necessarily included an awareness of figurative plurality in that an awareness of figurative length includes an awareness of more than one figurative unit of length. This generalizes Piaget and Szeminska's (1952) idea of intensive quantity and constitutes a step in unifying discrete and continuous quantitative schemes because it opens the possibility that a connected but segmented unit can be a situation of the child's counting scheme.

This possibility is clearly indicated in some of the experiments conducted by Piaget et al. (1960). They aligned two rows of five matches parallel to one another and then made a zigzag path using the matches in one of the rows. One of the youngest children they interviewed fits my idea of an awareness of perceptual length as a gross quantity.

Protocol I.

CHA (4; 0) Five matches in parallel alignment with five more: "Is it the same length from there to there as from there to there or are they different? – The same length. And like this (The right-hand row is re-arranged in a series of zigzags)? – No.–Which is longer?–I don't know which but they aren't the same. –Is there the same way to go? – No. – Is one a longer road to go? –Yes, there (indicating turnings) no, that one (straight row), because it's long. Here, it's nearer the end (zigzag row, showing relation between extremities). (Piaget et al. 1960, p. 106)

CHA's indecisiveness concerning which row was the longer after the zigzag path was made indicates an awareness of moving over the segments (matches). First, CHA judged the zigzag path to be longer apparently because of the up and down oscillating motion as opposed to the linear motion. Here, CHA focused on traversing the paths. When CHA judged the straight row to be longer, the endpoints of the rows became relevant and CHA made a comparison of the perceptual unit lengths between the endpoints. That there were five matches was irrelevant in his comparisons.

The authors report that some older children begin to become aware of the plurality of the matches in each row, but this is lost if the change in shape is excessive or if one of the matches is broken. This is exactly what I would expect in the case of an awareness of figurative length. JAN, in the Protocol II, was judged to be a level IIA child whereas CHA was judged to be a level I child.

Protocol II.

JAN (5; 10) Two straight rows, each of five matches, in parallel alignment: "It's the same length. And like this (both outlines forming right angles with two matches in one limb and three in the other)? – Also. – And like this (one outline made up of right angles and the other in zigzag formation)? – (He counts.) Also the same because it's four and four. (The experimenter breaks two matches in half. Four whole matches are arranged in the shape of a right angle and the remainder forms the outline of a staircase.) And like this? – Here it's longer. There are eight of them: that's more. – Yes but to walk along, is the road longer, or are they the same? – It's longer. (Piaget et al. 1960, p. 108).

That Jan counted when one of the two rows was arranged in an L shape and the other in a zigzag path indicates that his concept of length included an awareness of how many matches were in the rows. His "correct" comparison of length after counting indicates that he coordinated the figurative length of a row of matches and the figurative plurality of the matches because children who are gross quantitative comparers would have judged that one of the rows of matches was longer. After two matches were broken into half, he seemed to rely on the relative density of the matches in the two configurations and based his judgment of "longer" on his judgment of "more." Breaking two of the matches into two parts apparently destroyed the length units he established for the matches. He apparently did not conceptually reunite the two parts of each match back together to form a whole match, which is necessary to conserve the length of a match. This nonconservation of the length of a match solidly indicates that he established at most figurative length. Still, I see the beginnings of a unification of discrete and continuous quantity in the way in which Jan compared the two configurations of matchsticks.

Extensive Quantitative Comparisons

Extensive quantity enters into Piaget and Szeminska's (1952) system when the child introduces arithmetic units into intensive quantities. In this case, the child might regard the comparison in Fig. 4.1 as indeterminate because the blocks in the upper row are more spread out but the row is longer. I believe that the units of which Piaget and Szeminska wrote in the case of his extensive quantity are the result of a specific kind of unitizing activity that gives rise to abstract units. I have already shown that there are unitizing activities that precede Piaget's specific kind and, as I have shown in Chap. 3, that follow. Children who are gross quantitative comparers make perceptual unit items, and children who are intensive quantitative comparers make figurative unit items. The extensive quantitative comparers produce Piaget's (1970) arithmetic units, where: "Elements are stripped of their qualities…" (p. 37). Although these units are at the same level of abstraction as the arithmetical units I have identified, there is no requirement that they contain records of counting acts. They are parallel to the abstract unit items of von Glasersfeld's arithmetical lots.

As before, I generalize Piaget and Szeminska's (1952) idea of extensive quantity in the case of a row of blocks to include the row of segments between the blocks. If a child reprocesses a figurative row of segments using the template that was involved in producing the row, this opens the possibility of "stripping" the figurative segment of its sensory-motor qualities and constructing an abstract unit segment on a par with our abstract unit. Contrary to what might be expected, the property of the segment that I called length is not eliminated. Rather, the records of motion are still contained in the abstract unit segment. These records of motion are now turned into an operation rather than a figurative action. That is, the child no longer has to run over the items, even in visualized imagination, to recreate the sense of motion and the resulting sense of length. Instead, the records of motion are implicit in the recognition template. This is reflected in the Protocol III of a child JEA, 6 years 3 months of age in the matches' experiment of Piaget et al. 1960. JEA was judged to be at stage IIB.

Protocol III.
JAE (6;3) They're the same because there's five and five. – And like this (straight row and staircase)? – It's still the same because it's still five and five. It's still just as far. – One match broken and the bits laid end to end.) – It isn't the same anymore because you've broken one. – But is one of the roads shorter than the other? – Yes, that one, because it has seven bits and this one has five. – And if I break more matches? – It'll be longer still. – And if I lay them out straight like this (two straight lines.)? – That path (with broken matches.) will be longer. (Piaget et al. 1960, p. 112).

In contrast to JAN, JEA maintained that there were still five matches in each row without counting after one row of matches was changed into a zigzag path[9] [JEA said that ,"It's still the same because it's still five and five. It's still just as far."]. This indicates that the conceptual structure that constituted a row of segments for JEA was at least at the same level as a numerical composite in the discrete case. Further, JEA's meaning of "five" included five units of length as indicated by the comment, "It's still just as far." Correspondingly, I call JEA's concept of "five" a connected numerical composite or, more simply, a connected number.

There was still a lacuna in JEA's reasoning as indicated by the experiment of breaking one match into pieces.[10] I interpret this lacuna in reasoning as indicating a lack of the uniting operation, which is that operation that JEA would have used to conceptually reunite the broken matches into the original matches. This indication of the lack of the uniting operation warrants interpreting JEA's connected number five at the level of a numerical composite in the discrete case. Her connected number consisted of a sequence of connected segments. Constructing connected numerical composites opens the way for the construction of a connected number sequence, which is a number sequence whose countable items are the elements of a connected but segmented continuous unit. The construction of a connected number sequence is an initial step in the construction of measurement as well as an important step that integrates discrete and continuous quantity. A child at this level has constructed an awareness of indefinite length as well as of indefinite numerosity as quantitative properties of a connected number. Hence, I regard both an awareness of length and an awareness of numerosity as extensive quantities, which generalizes the concept not only across the discrete and the continuous, but also across the schemes that constitute measurement and number.

---

[9] I assume that "staircase" is used for JEA rather than "zigzag path."
[10] I assume that one match in Protocol III was broken into three pieces because JEA said there were seven bits.

Composite Structures as Templates for Fragmenting

In my analysis, I found that of the operations that produce an awareness of indefinite length and indefinite numerosity are not of a different kind or a different genre. This fits well with the finding of Piaget et al. (1960, p. 149) that children's construction of the operations of measurement parallels their construction of number. To complete the analysis, I turn now to the construction of the operations involved in fragmenting a continuous unit.

Piaget et al. (1960) observed what they thought was a developmental lag in the construction of a unit of length. It is important to understand that a unit of length for Piaget et al. was an iterable unit.

> Unlike the unit of number, that of length is not the beginning stage but the final stage in the achievement of operational thinking. This is because the notion of a metric unit involves an arbitrary disintegration of a continuous whole. Hence, although the operations of measurement exactly parallel those involved in the child's construction of number, the elaboration of the former is far slower and unit iteration is, as it were, the coping (capping) stone to its construction (Piaget et al. 1960, p. 149).

They found that only one child in ten of those from six to seven years of age, half of those from seven to seven years six months, and three-quarters of those from seven years six months to eight years six months attained operational conservation of length. For them, operational conservation of length "entails the complete coordination of operations of subdivision and order or change of position" (Piaget et al. 1960, p. 114).

Although I have not presented such normative data for the construction of the explicitly nested number sequence, based on my observations in interviews, these ratios seem compatible with the ratios of children who have constructed the ENS at corresponding ages, with the possible exception of the children in the age range from 6 to 7 years (Steffe and Cobb 1988). For this youngest group, I would expect that a greater ratio than one in ten would have constructed the explicitly nested number sequence. Nevertheless, I find the compatibility striking and indicative of the common operations required for the construction of the ENS and subdivision (fragmenting) schemes.

In the context of the matchstick experiment, I have shown that children can construct an abstract unit of length and a connected number sequence, both integral to the construction of extensive quantity, which parallels the construction of the arithmetical unit and the initial number sequence. I will now show that operational fragmentation into equal fragments emerges for planar spatial regions at the level of the initial number sequence by interpreting the experiments of Piaget et al. (1960) on subdividing a continuous unit. That is, operational fragmentation of planar regions into equal units emerges upon the emergence of extensive quantity. For this reason, I would not say that there is a developmental lag in the establishment of a unit of length in the context of connected but segmented units. However, subdivision of nonlinear continuous units is apparently more difficult than the establishment of connected but segmented units. For this reason, I would expect to find that the operational segmenting of nonlinear continuous units lags behind establishing connected numbers. The main issue that is addressed in this section concerning arithmetical and length units in spontaneous development is the relation between the construction of operational fragmenting (up to and including partitioning) and the construction of number sequences.

Experiential Basis for Fragmenting

The construction of operational fragmenting is a product of spontaneous development in the same way that the number sequence is a product of "spontaneous development"[11] in that fragmenting has its own experiential foundations separate from the establishment of connected but segmented units. Consider the example of a plate dropped on a hard surface. I distinguish among an experience of the plate, of the plate shattering, and of the shattered plate. These experiences are separate and distinct one from the other, even though they are contiguous. The auditory and visual experience of the plate shattering would be an item of experience attentionally no different than any other experiential item. Nevertheless, experiences like the shattering of a plate would contribute to the spontaneous construction of fragmenting operations.

If a child establishes the fragments of the shattered plate as perceptual unit items, the attentional structure of a shattered plate would be no different than the attentional structure of a perceptual lot. In this, I assume that the child forms an attentional pattern that I would call "fragment," and uses this attentional pattern in reprocessing the fragments of the broken plate, creating a perceptual lot structure. This process of categorizing in fragmenting is the same process that makes possible the uniting operation and a unit of units.

The experiential contiguity of the plate, the shattering plate, and the shattered plate qualifies the records of the three experiences as possible elements of a scheme. A goal structure along with the possibility of enacting the breaking action upon the recognition of a plate might be missing, however. If a child intentionally drops a plate in an attempt to reenact the breaking action, I would say the child has established a fragmenting scheme! Forming a goal to break a plate and carrying out the actions necessary to break it are essential elements of a fragmenting scheme and thus of fragmenting operations.

In the establishment of a fragmenting scheme, there has to be a change in the breaking action from being simply observed to being intentionally executed. For the

---

[11] Number sequences develop during early childhood mathematics education, and so they do not develop independently of the children's mathematics education. For this reason, I have placed the phrase, "spontaneous development" in quotation marks to acknowledge the contribution of children's mathematics education in their construction of number sequences.

fragmenting scheme to be transformed into a fraction scheme, the breaking action has to change further to include fragmenting a continuous unit into so many equal fragments. I am interested in the development of the intentional fragmenting of a continuous unit into a specific plurality of equal fragments.

Using Specific Attentional Patterns in Fragmenting

My assumption is that fragmenting a continuous unit into a specific plurality of equal fragments originates in the context of using interiorized dyadic and triadic attentional patterns to project units into a continuous unit. Several studies do indicate the primacy of dyadic and triadic attentional patterns in children's development of number. For example, using 48 three-year-olds, 48 four-year-olds, and 48 five-year-olds, Gelman and Gallistel (1978) found that only after a 1 s exposure to two items, 33 three-year-olds, 44 four-year-olds, and all 48 five-year-olds recognized two items. For three items the corresponding numbers were 28, 37, and 43. In the case of four items, the numbers at each age level were 9, 23, and 33.[12] These data clearly indicate the primacy of dyadic and triadic attentional patterns, especially dyadic attentional patterns, in children's construction of discrete structures.

Fragmenting a circular cake: Two dolls. The emergence of dyadic attentional patterns in the context of fragmenting is not independent of the kind of continuous unit involved. Piaget et al. (1960) began their study of fragmenting continuous units using a circular slab of modeling clay together with two little dolls. The child was told that the clay is cake and the dolls are going "to eat it all up, but they've each got to have exactly the same amount as the other: how shall we do it?" (p. 302). The child was supplied with a wooden knife.

For 3-year-olds, the earliest response was cutting out little pieces of the cake without knowing when to stop; cutting the cake was an end in itself. "This fact is interesting because it shows that there cannot have been any anticipation of the aim, or if there was, it was not the sort of anticipatory schema which enables a

---

[12] When increasing the exposure time to 5 s for four items, the numbers at each age level were 21, 29, and 37, indicating that some children counted. Almost 50% of the 4-year-olds could recognize four items after an exposure of only 1 s. Recognizing four items after only a 1-s exposure is important because it is quite likely that the recognizing child would regenerate an image of the items after the exposure, which indicates that their quadriadic attentional patterns were constituted at least as figurative lots. However, the phenomenon of subitizing, instant recognition of numerosity, such as in the case of dyadic and triadic patterns, may have enabled many of these 4-year-olds to recognize the four items without their quadriatic attentional pattern being constituted as a figurative lot (von Glasersfeld 1981).

child to know in advance that he must cut two pieces, using up the whole cake" (Piaget et al. 1960, p. 304). Children's sensory-motor activity is essential for the formation of such an anticipatory fragmenting scheme, as shown in Protocol IV.

Protocol IV.
FRAN (4; 2) His final replies are at level IIA. ...he begins by cutting two small pieces out of a clay cake and handing them to the two dolls – this in spite of the fact that the experimenter had warned him: "The mummy says that they can have all the cake." As he stops short after cutting those two small bits, the experimenter remarks: "Now don't cut such tiny bits. You divide up the whole cake. He is given another cake and cuts off first two pieces, then two more, and finally divides the rest in two. With a third cake he again cuts out little pieces and goes on to give as many as fifteen such bits to each doll. But with a rectangle, he immediately divides the 'cake' into two pieces (unequal but leaving no remainder). A square cake is cut into four and each doll is given two quarters. Finally, when given another round cake, he cuts it into two." (Piaget et al. 1960, p. 305)

FRAN initially used a dyadic attentional pattern (two dolls) to establish a goal, as indicated by cutting two small pieces out of the cake. FRAN's dyadic attentional pattern served as a guide throughout the fragmenting activity, which indicates that it was constituted at least as a figurative lot. Because FRAN eventually cut the whole of a round cake into two, I infer that he reprocessed the implemented dyadic attentional pattern, i.e., abstracted the dyadic pattern from its implementation. At the end of the protocol, FRAN knew that he had to share the whole of the cake between two people and he knew how to do it. But FRAN, according to Piaget et al. (1960), did not think of a part as being included in the whole from which it was made and to which it still relates in thought. It was simply a piece removed from the whole. In fractional schemes, the child can preserve the part in the whole.

Fragmenting a rope: Two dolls. According to Piaget et al. (1960), fragmenting a circular cake into two parts does not emerge until approximately 4 years of age. However, most of the 3-year-olds in a study by Hunting and Sharpley (1991) were able to cut a rope into two pieces. Two hundred and six children ranging in age from 3 years 4 months to 5 years 2 months were individually asked to share a piece of string 300-mm long between two dolls. Nearly all of these children cut the string just once (89%). Of these approximately 183 children, the difference in the two parts was less than 15 mm for 79 of them, which solidly indicates sensitivity to the equality of the parts. The difference in the two parts was between 15 mm and 30 mm for 28 other children, which can be also interpreted as indicating sensitivity to the equality of the parts. In sum, then, the difference in the two parts was less than 30 mm (1.2 in.) for 117 children, which is approximately 57% of the children interviewed. This finding establishes that a majority of children in the age range studied can share a linear object approximately equally between two people.

Of the remaining children who cut the string into two parts, 49 of them cut it so that the difference in the two parts was between 30 and 60 mm and 27 of them cut it so the difference was greater than 60 mm. For the 27 children for whom the difference was greater than 60 mm, equality of the parts did not seem to be an issue.

Whether the equality of the parts mattered for the 49 children for whom the difference was between 30 and 60 mm is more ambiguous.[13]

On purely conceptual grounds, in order to intentionally cut the string into two parts, I infer that the child's dyadic attentional pattern would be at least a figurative lot. That would allow children to generate and maintain an image of two dolls in the process of cutting the string, allowing the children to intentionally cut the string into parts, one for each doll. But sensitivity to the equality of the parts would be lacking despite a sense of twoness. Therefore, I infer that the 66 children who cut the rope into two parts, but whose parts were more than 30 mm different in length had a figurative attentional pattern.

When the dyadic attentional pattern is figurative, activity is guided by the twoness of the dolls. The intention to make equal parts specifies a property of the ropes, not the dolls. Hence, in order for the children to show greater sensitivity to the equality of the parts, the activity must be guided by properties of the rope parts. Since the dyadic attentional pattern, the twoness, initially referred to dolls, to use it to make fair shares of the rope involves a substitution of the rope parts for the dolls. This kind of flexibility in the referent of the units of the attentional pattern requires an arithmetical unit, where, in Piaget's terms, "elements are stripped of their qualities." Being stripped of their qualities, any sensory material can be used to "fill" the slots that the arithmetical units comprise and can stand-in for other sensory material. Therefore, children would not be able to intentionally break a continuous item of experience into two *equal* parts until they had constructed two as a numerical pattern – a pair of arithmetical units. Hence I infer that the dyadic attentional pattern of the 117 children for whom the difference in the two parts was less than 1.2 in. was numerical.

Once the dyadic attentional pattern is an arithmetical lot, further sensitivity to the equality of the parts would seem possible because the children could unitize the rope parts after cutting them and make an extensive quantitative comparison between them. Uniting the pair of items together permits setting the items "at a distance," which in turn permits awareness of the items, not only as distinct and separate but also as fair shares. This would lead to an inclination to review the two parts in an attempt to verify their equality.

After the children cut the rope, Hunting and Sharpley (1991) did ask the children if the dolls would be happy with their ropes. Out of 216 children, checking behavior was observed in the case of 94 children. Forty-seven simply looked at the two parts of the ropes in a visual check, 41 placed the two parts side-by-side, and 5 placed the ropes end-on. The behavior of the 41 children who did place the two parts of the ropes side-by-side indicates a review of the units comprising the rope parts. Whether it indicates two as unity is an open question. Had the children spontaneously checked, that would be a stronger indication of the construction of two as unity.

Subdividing a circular cake: *Three dolls.* Piaget et al. (1960) and Hunting and Sharpley (1991) agree that subdividing a circular cake does not require a more

---

[13] There is a discrepancy of 10 children between the total number of children who cut the string into two pieces and the number of children reported in the subcategories.

sophisticated composite structure than subdividing a rope into two parts, even though the former lags behind the latter by approximately 1 year. But there appears to be a large jump between using a dyadic attentional pattern in fragmenting and using a triadic attentional pattern in systematic subdivision of a circular slab of clay. Between the ages of 4 and 4½ years, according to Piaget et al. (1960), children usually succeed in sharing out two equal parts, but they cannot share into three equal parts. Fragmenting a circular slab of modeling clay into three parts apparently requires the development of the operations that produce the initial number sequence.

The use of a triadic attentional pattern as a numerical pattern in the context of subdivision is exemplified in Protocol V.

Protocol V.

SES (6; 2) Starts off by cutting several series of small pieces and distributing these as he goes along. For the next cake he cuts off three large slices and leaves the remainder. Finally, he succeeds in cutting a third cake into three almost equal parts. (Piaget et al. 1960, p. 320)

The modification in SES's fragmenting activity was self-initiated, which solidly indicates a goal to share the cake among the three dolls. I believe his triadic attentional pattern was evoked and served as a template in fragmenting. That he succeeded in cutting the circular cake into three almost equal parts convinces me that he constituted the elements of his triadic attentional pattern as arithmetic units in the midst of his activity, if they were not already arithmetical units before he started.

Subdividing a circular cake: Five dolls. Another child, SIM (5; 9), who was also judged to be at the level IIB, succeeded in sharing a circular cake equally among three children, but proceeded as in Protocol VI for the case of five children.

Protocol VI.

What he does is to cut a series of small slices and deals them out as he goes along, leaving an unused remainder. When given another cake and told to finish it all, he cuts it into seven successive slices and distributes the first five only. Later, he finishes with six parts, but he never succeeds in dividing into five (Piaget et al. 1960, p. 323).

In this example, SIM initially focused on making fragments without coordinating the fragments with the whole. When he was asked to focus on the whole, he made enough fragments to exhaust the whole, but without coordinating the number of fragments with the whole. SIM definitely had a goal to share the cake among five children as indicated by distributing the first five of the seven slices. This goal could be made possible, I believe, by the activation of a numerical composite, five, and the projection of units of the numerical composite into the continuous unit, the cake. In actual fragmenting activity, SIM surely lost his sense of the numerosity of the composite unit, five; his goal was to make slices, where I emphasize an awareness of plurality rather than of specific numerosity. Initially, he may have intended to make five pieces, but in actual fragmenting, he focused on simply making pieces, an activity which was set in motion by his goal to make five pieces.

SIM's behavior is what I expect from a child who uses a *numerical composite*, as opposed to a *composite unit*, in fragmenting. A numerical composite is the result of generating unit elements, but because there is yet no unit containing the elements, that unit is not yet an object of reflection for the child and the child is not aware of the composite as one thing. Rather, the child operates at the level of the unit elements.

In a numerical composite, the child can produce the five cake pieces in visualized imagination, but cannot yet "hold them at a distance" to operate on these cake pieces by coordinating their size and shape with the whole they are to be part of. Similarly, when the child focuses on the whole of the continuous unitary item, the cake, he loses sight of the number of pieces, which is exactly what happened when SIM was asked to finish all of the cake (as indicated by cutting the cake into seven pieces and distributing five).

To reconstitute a numerical composite as a composite unit, the child must take the generated unit elements, "hold them at a distance," and reflect on them in order to make them inputs for the unitizing operation. Once the numerical composite has been unitized and becomes the elements of a composite unit, the child can simultaneously be aware of the five unit elements and the newly formed composite unit. In this sense, the child can now operate simultaneously with these two kinds of units. Once the unit elements of a composite are items of a reflection, the child can begin to mentally coordinate the size of the pieces, representing the unit elements, with the size of the whole. However, this may still require a bit of experimentation at first. An indication of mental as well as physical experimentation in cutting a cake into five pieces is contained in Protocol VII.

Protocol VII.
ROS (6; 8) Is asked to divide the cake into fifths. He begins by constructing a series of successive parts which together account for all the cake but which number seven instead of five. He therefore tries twofold dichotomy but soon rejects that hypothesis. Eventually, the cake is divided into five approximately equal pieces, but these are parallel as were his thirds (Piaget et al. 1960, p. 325).

ROS was judged by the authors to be at level IIIA. Of children at this level, Piaget et al. (1960) commented that, "While every one of these subjects is eventually successful in dividing into five or six equal fractions, they all begin with a certain amount of experimentation although this is no longer necessary to them when trisecting. Only at level IIIB do I find division into fifths or sixths carried out with the same assurance as trisection" (p. 326).

Given that ROS eventually made parallel cuts and approximately equal pieces of the circular region indicates that he mentally estimated where he should cut the cake. This ability to make mental estimates indicates that ROS mentally fragmented the cake and then united those fragments into a connected but segmented unit. He was clearly aware that the whole needed to be exhausted, and he coordinated the number and size of the pieces with that requirement as indicated by his starting over after making seven pieces. To do this, he would need to be aware that the pieces together comprised the whole. For this reason, I believe that ROS had constructed the uniting operation and could produce composite units using that operation. Consequently, I infer that ROS used his composite unit, five, as a template for fragmenting.

The basic difference in fragmenting into three pieces and into five pieces for children like SIM is that, due to the child's advanced triadic attentional pattern, the child can be aware of the cooccurrence of three individual unit items, while children generally do not have such an advanced quintic attentional pattern. The cooccurrence of the unit items can provide the child with a sense of a composite whole, even if the child has not constructed a composite unit of three. This allows the child to attack the problem similarly to the child who has constructed a composite unit. Even so, I would expect trial and error to be involved in fragmenting into three pieces, because the child cannot yet mentally experiment with fragmenting the cake. Remember that when three is constructed as a numerical composite, the child can use this structure to generate an image of three items, but when three is constructed as a composite unit, the child can also set the image of three "at a distance" and reflect on it. In the former case, the child is "in" the re-presentation whereas in the latter case, the child is "outside" of the re-presentation and so can set it at a distance and look at it. On the one hand, this allows children like ROS to fragment the cake into three equal slices without first experimenting. Five, on the other hand, is a large enough number of pieces that even children like ROS, who have constructed a composite unit of five, use a combination of mental and physical experimentation initially.

Number Sequences and Subdividing a Line

The distinctions between levels IIB, IIIA, and IIIB as articulated by Piaget et al. (1960) can be interpreted in terms of the three number sequence types. But before drawing the parallels between the levels and the number sequence types, I discuss tasks that Piaget et al. refer to as subdividing a line, because children's performance on these tasks are even more compatible with the number sequence types than the subdivision of cake tasks.

The authors devised a sequence of six experimental situations, the first two of which I discuss. The first situation concerned locating a point b_2 on a_2c_2 in Fig. 4.3 so that a_2b_2 would be just as long as a_1b_1. In the second part of this situation, the child was asked to start at c_2 rather than a_2.

In the second situation, illustrated in Fig. 4.4, the top segment was broken into the same two parts as Fig. 4.3. However, the endpoints of the top and bottom segments were no longer lined up.

Lack of subdivision. Piaget et al. (1960) observed that children in his stages I and IIA immediately solved the first part of the subdivision task of Fig. 4.3, but not the second subdivision task illustrated in Fig. 4.4. Of these children, the authors

Fig. 4.3. The first subdivision of a line task.

Fig. 4.4. The second subdivision of a line task.

stated, "They do not know how to transfer the distance a_1b_1 to the other string a_2c_2 and they simply put b_2 opposite b_1 without worrying about the inequality of the intervals so obtained (Piaget et al. 1960, p. 130). In my model, for the children to transfer the interval a_1b_1 to the other string, they would need to have at least constructed the concept of an interval as an arithmetical unit, which can occur at the level of the initial number sequence.[14]

Intuitive subdivision. The next step in children's progress identified by Piaget et al. (1960) is an intuitive solution of situation two; that is, children simply looked at the two lines and made visual estimates. These children do not measure, even though they were provided with a measuring instrument that was longer than the intervals to be measured. Making a visual estimate is a clear indication that the children were aware of the intervals and that this awareness is the result of projecting units into the segments. Piaget et al. confirm this when commenting on the fact that these children take the point of departure into account as well as the point of arrival: "This means that they are aware of the interval involved…That awareness leads to a second feature of their progress in that they begin to subdivide the line and relate the segments to its overall length" (p. 140). This major advancement is the result of the construction of extensive quantity and is compatible with how SIM in Protocol VI tried to fragment a circular cake into five equal pieces.

Measuring without unit iteration. Piaget et al. (1960) subdivided operational stage III into two parts that echo the construction of two number sequences beyond the initial number sequence. In substage IIIA, when the children had a measuring stick longer than the distance a_1b_1, success was assured, but when it is shorter, the children used auxiliary pieces of material to measure. This behavior is compatible with the construction of the tacitly nested number sequence, as I explain later.

Protocol VIII.
RAY (7; 10) Illustrates level IIIA. He gradually moves b_2 beyond b_1, saying: "I think it's right now." (Note: The ruler is shorter than a_1b_1). As the ruler is too short, he measures a_2b_2 by using his hand as well as the ruler, then in measuring a_1b_1 he uses a strip of paper to prolong the ruler, checking a_2b_2 in the same manner, and so achieving an accurate reproduction of the distance given (Piaget et al. pp. 142–143).

According to Piaget et al., for RAY, segment a_1b_1 would not symbolize the operation of iterating the segment to constitute a segment of length possibly equal to segment a_1c_1. However, the authors regarded RAY as quite capable of mentally fragmenting segment a_1c_1 into equal subsegments.

> *In the present experimental setup, AC is broken into the two portions AB and BC by the bead. Even here, subjects at levels I and IIA overlook the whole, AC, and fail to regard the interval AB as a part. But it is quite another matter to break up AB into a number of abstract segments, as given by successive momentary positions of a ruler or a strip of paper.*
> (Piaget et al. 1960, pp. 145–146)

From this comment, I infer that the authors regarded RAY as being able to break up segment a_1c_1 into a number of abstract segments as given by the segment a_1b_1.

---

[14] See discussion of the initial number sequence in Chap. 3.

I do agree with the authors that there is "nothing natural in subdividing a length where the parts are not perceptually given" (Piaget et al. 1960, p. 145). But RAY was wholly capable of making such an abstract subdivision. Like the children studied by Hunting and Sharpley (1991), who could use their dyadic and triadic attentional patterns in fragmenting, RAY could use his attentional pattern for making composite units as a template for fragmenting an unmarked segment. I believe that RAY could assemble and use composite units of unspecified numerosity as a template for fragmenting prior to action. In this, it is possible to appreciate the compatibility between how RAY and how ROS in Protocol VII experienced a composite unit.

If, as the authors claim, RAY did subdivide a_2b_2 as given by his ruler, he was yet to constitute that subsegment of a_2b_2 as being iterable. If it was iterable, he could have used it to reconstitute a_2b_2 as a partitioned segment by iterating the subsegment, and this would have been manifest by his iterating the ruler to measure a_2b_2. So, I find his use of his hand and a strip of paper to prolong the ruler as consistent with the interpretation that although he did subdivide a_2b_2 per the ruler, the subsegments were not identical subsegments. Rather, they were distinctly different subsegments that implied different perceptual material.

An important property of the tacitly nested number sequence that has not been mentioned is that because this number sequence is constructed by reprocessing the initial number sequence, children can, while operating with one number sequence, present another number sequence apart from the one with which they are operating and coordinate their use. What this means in Protocol VIII is that RAY could mentally project an indefinite number of units into segment a_1c_1 as determined by a_1b_1, and then form the goal of projecting segments of the same length into a_2c_2. This would be sufficient for RAY to spontaneously measure along a_2c_2 to find where b_2 should be placed.

Measuring with unit iteration. At substage IIIB, the concept of an iterable unit enables the children to apply iterative stepwise movements to a ruler shorter than the length to be measured.

Protocol IX.

BED (8: 7) Illustrates level IIIB. "If you go there (a_1b_1) then I must go here too (a_2b_2)." He measures the distance by hand and then by ruler, using several applications (Piaget et al., pp. 142–143).

The distinction in the two ways of measuring between RAY and BED is parallel to the distinction that was made between the tacitly and the explicitly nested number sequences. In fact, the advancement in measurement displayed by BED in measuring a_2b_2 is quite compatible with how Jason (Protocol V, Chap. 3) used his units of one as iterable during the latter part of his second grade in school. In this, it is possible to appreciate the difference between the conceptual structures implied by the measuring behavior of BED and RAY.

In the case of BED, I infer that segment a_1c_1 symbolized a composite unit of subsegments, which could be produced by iterating a_1b_1, and which, when joined together, were possibly of length equal to a_1c_1. In the case of RAY, I infer that segment

a_1b_1 was one of a number of different segments that could be made all of length equal to a_1b_1. This is a subtle distinction, but it is a critical one. For BED, there would be no felt necessity to perform the fragmenting actions to fragment the whole segment a_1c_1. For RAY, segment a_1b_1 was only one among other subsegments of a_1c_1, and units would need to be at least mentally projected into the whole segment in order to be "in" the segment.

Partitioning and Iterating

I can now understand how partitioning and iterating can be essential aspects of the same psychological structure. The unit structure diagrammed in Fig. 3.10 of Chap. 3 symbolizes a composite unit structure. This unit structure explains why a child can have a sense of simultaneity in the case of "seven" or any other such number word. If the number word "seven" activates the composite unit structure diagrammed in Fig. 3.10 of Chap. 3, the unit structure enables the child to experience seven as a unitary item rather than as a composite unit containing seven unit items. The child knows that seven unit items can be produced and is aware of a specific numerosity, but the child is not compelled to produce seven counted items to experience them. A sense of simultaneous cooccurrence of the seven unit items is thus produced as a byproduct of the symbolizing function of the iterable unit item of Fig. 3.10 of Chap. 3. The child can be also aware of sequentially producing the seven counted items of Fig. 3.10 of Chap. 3 for the same reason it can be aware of their simultaneous cooccurrence.

This sense of the simultaneous cooccurrence of seven unit items is involved when using the template of Fig. 3.10 of Chap. 3 as a partitioning structure. However, alone it is not sufficient because the child must disunite the unit items in partitioning. Heuristically, I think of disuniting what has been united as removing the boundaries (the unitariness) from the elements of the containing unit. But this understanding of disuniting essentially destroys the composite unit structure during the process of disuniting. Children who have constructed composite units can reunite the equal parts produced by fragmenting, but reuniting these fragments has to be carried out if only mentally to establish the fragments as elements of a composite unit (a connected but segmented unit).

In the case of the child who has constructed an iterable unit of one, I model the operation of disuniting in a way that preserves the composite unit structure. In the establishment of a composite unit at the experiential level, such a child produces an awareness [which could involve a figurative image] of the composite unit structure to be used in partitioning. Once the composite unit is used in partitioning, the unit items of the partition appear to the child to cooccur. This awareness of the cooccurrence of the unit items is made possible by the uniting operation, and it is essential for the items to be disunited.

The operation that produces a "removal" of the boundaries of the composite unit structure is reprocessing the unit items of the composite unit structure using the iterable unit.[15] The operation of reprocessing is already available to the child because that is precisely how the iterable unit was constructed. Disuniting, then, means that the child focuses his or her attention on the elements of the composite unit structure rather than on the unitary structure as one thing. In this, the child remains "above" the elements and can move back and forth between the elements and the unit structure that contains them. Experientially, the child retains a sense of the parts produced in partitioning as belonging to the original unit without needing to mentally reunite those parts. A part remains a part of the original unit as well as a unit in its own right.

It is important to note that focusing attention on the elements of the composite unit structure does not remove the boundaries. Rather, the unit structure of the composite unit is constituted as background and the elements as foreground. This is a critical aspect of using a composite unit that has been produced by an iterable unit in partitioning. If the elements of the composite unit are applied to a continuous unit in such a way that equal parts of the continuous unit are produced, this activity would be framed by the unit structure of the composite unit. The parts could be mentally reunited to establish a connected but segmented unit in that case where the child brings the background unit structure into the foreground, but the child understands that this can be done and therefore does not need to perform the operation to establish an awareness of the result.

But more is possible. Each part of the disunited composite unit is an instantiation of a unit item that was produced using the iterable unit. As such, each part maintains its iterable quality. What this means is that the partitioning child can use any part of the partition in iteration to produce a continuous but segmented unit of the same size as the original unpartitioned continuous unit.

So I do not refer to a fragmentation of a continuous unit as an equipartitioning unless, first, the operating child intends to fragment the continuous unit into equal-sized parts and, second, unless the operating child can use any one of these equal-sized parts in iteration to produce a connected but segmented unit of the same size as the original unit.

Levels of Fragmenting

I have identified four levels of the fragmenting scheme that correspond to the construction of number sequences. First, there is the primitive fragmentation of a continuous unit into two parts made possible by the dyadic attentional pattern. At this level, children can begin to share a rope, a string, or some other kind of linear object into two parts using a figurative dyadic attentional pattern. Such sharing activity, especially in the case of planar regions, opens the possibility for reprocessing

---

[15] I assume that the operations of partitioning at the experiential level are simply those operations that such a child can use at the re-presentational level.

the produced fragments using the same pattern that was used in fragmenting. This, in turn, produces a numerical pattern, two, and children can become sensitive to the equality of the parts. Social practices involved in sharing, such as children always wanting a "bigger half," can obscure this precocious use of dyadic patterns in sharing, but it does not exclude it. When children understand that an item is to be shared equally, the rather artificial sharing situations used by Hunting and Sharpley (1991) and by Piaget et al. (1960) do indicate that a majority of children from the age of four years can indeed make two equal shares.

Although sharing into three parts in the case of planar regions requires the development of the operations that produce the initial number sequence, the study by Hunting and Sharpley (1991) does indicate an earlier emergence of sharing a linear object among three dolls. So I would expect children who can establish figurative quantity to be able to share a linear continuous unit into two or three parts, and I would expect them to be able, in the process of sharing, to reconstitute the involved patterns as numerical patterns in an attempt to make fair shares. However, I would not expect these children to share linear continuous units into five or more parts regardless of the equality of the units.

In the second level of fragmenting, there is the use of numerical composites to project units into continuous units. This scheme is characterized by fragmenting continuous units into three parts after some trial and error and by attempting to fragment a continuous unit into five or more parts. In the latter case, there is a lack of coordination of the number and size of the parts with using the whole of the continuous unit. I believe that this coordination might be achieved after some trial and error, but that the child would need to actually carry out the fragmenting activity. There would be no a priori necessity to share the whole of the continuous unit into so many equal parts.

In the third level of fragmenting, there is the use of composite units to project units into continuous units. Here, sharing a continuous whole into three parts entails an a priori necessity to share the whole of the continuous unit into three equal parts. This certainty is made possible by the simultaneous awareness of the unit items and of the composite unit to which they belong, along with the simultaneous cooccurrence of the three unit items. In other words, the three unit items seem to cooccur, and the child can present them in that way while remaining aware of their unitary wholeness. I believe that the elements of any other composite unit that has been established as a numerical pattern could be used in the same way. But, for those composite units whose elements do not occur in a pattern, the child uses mental and physical experimentation to partition a continuous unit using those composite units.

In the fourth level, there is the use of composite units whose elements are iterable to project units into a continuous unit. Here, the child understands that any one of the units can be used to reconstitute a connected but segmented unit equivalent to the original unit. It is here that I would say that the child has constructed an equipartitioning scheme.

There is a fifth level involving partitioning n items among m children exemplified in a study of children's partitioning behavior conducted by Lamon (1996). She presented 11 partitioning tasks to 346 students distributed throughout grades four to eight. Three of the tasks were to share 4 pepperoni pizzas among 3 children, to share 4 chocolate chip cookies among 3 children, and to share 4 oatmeal cookies among 6 children. To partition the n items among the m children, a child would have to distribute the operation of projecting a composite unit of numerosity m across the n items. The construction of this distribution operation should have been a distinct possibility for almost all of the children in Lamon's sample. For example, in the case where the child intends to share four pizzas among three children, if the child's concept of three is constructed in such a way that the number word "three" refers to a unit that can be iterated three times, the child would intend to share all of the pizza equally among three children prior to actually making the shares. That is, it would be the child's goal to split all of the pizza into three equal parts. Upon experimenting and finding that there is no easy or practical way of cutting the four pizzas into three equal pieces by making just two cuts, the child might then search for another way of cutting the pizza. The child might restructure the pizza into a unit containing three pizzas and another unit containing one pizza with the intention of giving one whole pizza to each child and then splitting the remaining pizza into three equal shares. If the restructuring is the result of productive thinking rather than an accidental restructuring, the child would also understand that partitioning each pizza of the original unit containing all four pizzas into three pieces and then combining the parts would produce three equal shares of all of the pizza. This, of course, constitutes a distribution of the operation of partitioning over parts of the original unitary whole to produce a partitioning of the whole. This would seem to be a crucial operation in the construction of fraction schemes.

Of the 123 children in the fourth and fifth grades, Lamon (1996) reports that the following percentage of children displayed incomplete, incomprehensible, or invalid strategies: approximately 50% of the children in the case of the pepperoni pizzas; approximately 48% of the children in the case of the chocolate chip cookies; and approximately 57% of the children in the case of the oatmeal cookies. The difficulty in sharing n items among m children was not because there were more items to share than there were children, because the greatest percentage of children who experienced difficulty in sharing occurred in the case of sharing four oatmeal cookies among six children.

These percentages are alarmingly high when considering that the children in the sample were at least 9 years of age, an age where I would expect over 75% of the children would have constructed the explicitly nested number sequence and thus iterable units of one. Hence I suspect that these children lack the distribution operation. In support of that theory, Lamon reported the successful children operating precisely in a way that would indicate use of the distribution operation.

Final Comments

The operations symbolized by children's number sequences play an instrumental role in their construction of connected number sequences and fragmenting schemes. This finding unifies children's quantitative operations across the discrete and the continuous.

I have indicated that the composite structures children use to establish situations of their number sequences serve as templates for fragmenting unmarked continuous units into fragments. Sharing a continuous unit among so many dolls provided an excellent context for the natural use of discrete structures as templates for fragmenting. The particular case of sharing a circular cake fairly among three dolls proved to discriminate children who had constructed the initial number sequence from children who had constructed only the figurative number sequence. It may seem strange that this apparently simple task discriminated so well between the prenumerical and numerical children. But there is a good reason for its discrimination. Cutting a circular region into three approximately equal parts requires a concentrated effort regardless of whether the child makes parallel cuts or locates the center of the region and makes radial cuts. To be successful, at some point in the fragmenting process, the child is obliged to make mental estimates of where to cut and mentally compare the parts that would be made if the cake was actually cut. This involves a coordination of the number and size of the parts with exhausting the whole of the cake. Certainly, operations that make this possible emerge at the level of the tacitly nested number sequence. Then children can use the numerical composite, three, as if it involved the uniting operation because they can be aware of three parts simultaneously without uniting them into a composite whole. However, due to the special or early production of a triadic figurative pattern, it can be used as a pseudo composite unit before the construction of the tacitly nested number sequence.

Although it is possible to find children who have constructed three as a numerical pattern without having constructed the initial number sequence (Steffe and Cobb 1988), the latter construction follows on soon after the former, so I would expect subdivision of a circular region into three approximately equal parts to indicate the construction of the initial number sequence. Because of their nature, situations involving sharing a continuous unit into so many equal parts involve the use of discrete structures to project units into the continuous unit. However, in the tasks that involved the subdivision of a line, the only requirement was that the child place b_2 at a place on a_2c_2 in such a way that a_2b_2 was of length equal to a_1b_1. In that the tasks did not involve a specific numerosity like three or four, they provided a good test of whether the operations involved in constructing number sequences were also involved in fragmenting. It was a surprise that the operations involved in subdividing a line were even more compatible with the operations involved in constructing number sequences than were the sharing tasks. It was a surprise because the children's discrete structures were not activated by the utterance of a number word or by the presence of so many dolls. Yet, making visual estimates to find where b_2 should be placed and the corresponding two levels of measuring behavior were easily interpretable in terms of the three levels of number sequences.

The finding that child's quantitative operations emerge in both the continuous and the discrete case in the same time frame, and in quite similar ways, provides solid support for the reorganization hypothesis. However, the quantitative operations that are constructed in the continuous case are not accompanied by the concomitant establishment of a natural language notational system in a way that is analogous to the discrete case. When attempts are made to establish a notational system for continuous quantitative operations in the children's mathematics education, a fractional language notational system is emphasized, reserving development of the number sequence notational system to discrete quantity. I believe that this practice serves to separate the children's construction of fractional schemes from their number sequences. It places great demands on the continuous quantitative operations that are available to the children because it is very easy to go beyond the Stage IIIB quantitative operations in teaching fractions. It also serves to retard the elaboration and reorganization of the discrete quantitative operations. I believe that these practices only contribute to the separation of fractions and whole numbers in the mathematical experiences of children and may very well lead to whole number knowledge interfering with the construction of fractional schemes. Hence this interference, when it is present, is not due to the nature of learning, but to the nature of teaching.

Operational Subdivision and Partitioning

When numerical structures are used as templates for fragmenting, the items that are established (at least in the case of the explicitly nested number sequence) are construed as parts of the unit from which they originated. In fact, five of the seven aspects of operational subdivision identified by Piaget et al. (1960, pp. 309–311) are satisfied by the fragmenting operations of a composite unit at the stage of the tacitly nested number sequence. These are aspects 1–5, which I outline below. Six and seven await the construction of more advanced number sequences.

First, the continuous unit to which the composite unit is applied as a template for fragmenting is a "divisible whole, one which is composed of separable elements" (p. 309) because the elements of the composite unit comprise the "separable elements" when projected into the continuous unit. Second, there are a determinate number of parts in the case of a composite unit of specific numerosity. Children can also use a composite unit of indefinite numerosity in fragmenting, which provides them with even more possibilities than implied by Piaget et al. (1960). Moreover, when a composite unit is used as a template for fragmenting, the whole of the continuous unit is exhausted, and there is a coordination of the number and size of the parts with exhausting the whole of the continuous unit, which is the third aspect.

Establishing the relationship between the number of parts and the number of cuts is a possibility for children who have constructed the tacitly nested number sequence. In the case of a row of blocks, I indicated that children can coordinate a sequence of interiorized intervals and a sequence of interiorized blocks at the level of extensive quantity. Because the tacitly nested number sequence is one level above extensive quantity, the possibility is present for children to abstract this relationship, which is the fourth aspect.

Children are sensitive to the equality of the parts, the fifth aspect, even at the level of numerical composite, one level below the tacitly nested number sequence. Moreover, even though the children remove the uniting boundaries of the continuous unit in subdivision, these children can reunite the parts produced into a continuous but segmented unit that is equivalent to the original continuous unit. However, these are sequential operations and the children may regard the continuous but segmented unit as a result that is unrelated to the unit with which they began. It is only later, after the emergence of the explicitly nested number sequence, does the possibility open for the children to regard the whole as invariant and the sum of the parts to equal the original whole, which is the seventh aspect. Upon construction of the explicitly nested number sequence, children also construe the parts as units in their own right. But it is yet unknown if children consider them as units to be subdivided further, which is the sixth aspect.[16] It would seem as if children who have constructed the generalized number sequence can take a units of units of units as a given and use this unit structure in establishing a relation between any one of the subparts produced on the second subdivision and the original whole.

Children who use their number sequence in subdivision are also not restricted to subdividing into only a small number of parts. They can conceptualize the possibility of subdividing a continuous whole into as many parts as they know number words. Thus, they can establish meaning for fractions such as one one-hundred seventy-fifth, or any other fraction corresponding to a number word of their number sequence. This is especially the case for children who have constructed the explicitly nested number sequence. Finally, relationships among the number of parts and the size of the parts are within reach of these children.

Partitioning and Splitting

I now return to Confrey's idea of splitting and show how our analysis is compatible with how she defines splitting. In her definition, she commented that the "focus in splitting is on the one-to-many action" (Confrey 1994, p. 292). When a child has constructed the iterable unit of one, the child can focus on the continuous unit that is to be partitioned,[17] which is the unit item implied by "one" in "one-to-many," as well as on the number of parts into which the child intends to partition the continuous unit, which is the "many" in "one-to-many." This is made possible by the child being two levels removed from the elements of the composite unit that the child projects into the continuous unit. That is, the child can focus on the unit structure of the composite unit that is projected into the continuous unit in such a way that it both contains and partitions the continuous unit, and then focus on what is inside of the unit structure without destroying the unit structure. This produces an initial experience of one-to-many.

---

[16] This will be one of the major issues that are investigated in the case studies that follow.

[17] Here, I use "partitioning" in the sense in which Confrey uses "splitting." Hereafter, I use "fragmenting" and "partitioning" rather than "splitting" to maintain the distinctions between partitioning and the earlier forms of fragmenting that I have identified.

But the child can do more in partitioning. If the child focuses on the elements of the composite unit into which the continuous unit is to be partitioned, the child can move back again to the unit structure. Moreover, the child can unite any subcollection of elements together and disembed them from the composite unit and constitute it as a composite unit in its own right without destroying these elements in the containing composite unit formed by partitioning. This establishes the classical numerical part-to-whole operation that serves as a fundamental operation in the construction of fractional schemes.

Finally, the child can use any singular part of the original partitioning in iteration to establish a connected but segmented unit equivalent to the original continuous unit. The child can also use any singular part of the original partitioning in iteration to produce a composite unit of elements of numerosity less than the numerosity of the composite unit used in partitioning and then compare that composite unit with the original composite unit. In this way, a child can establish meaning for, say, four-sevenths as four of one-seventh. The child can also compare four-sevenths with seven-sevenths and understand four-sevenths as four units out of seven units.

These are all crucial operations in the construction of fractional schemes. Although I do not deemphasize partitioning operations, neither do I focus on units of units as an end in themselves, as Kieren (1994) suggests. Rather, I use units of units as templates for fragmenting actions. By acknowledging the distinctions among numerical composites, composite units, and composite units that contain an iterable unit, I am able to make distinctions in what else the child might be able to do after making the initial fragmentation. Making the fragmentation is essential, but it does not supply the mental operations that are necessary for the construction of fractional schemes.

A more detailed analysis of children's fractional schemes is carried out in the context of the analysis of the fractional schemes the children in the teaching experiment actually constructed. There are many more distinctions that I have yet to make. What I have tried to do in this chapter is to establish a developmental rationale for the reorganization hypothesis. It seems to have great advantages for school mathematics. But, rather than leave the hypothesis unexplored, its implications for the teaching of fractions is thoroughly investigated in the next four chapters.

Chapter 5
The Partitive and the Part-Whole Schemes

Leslie P. Steffe

The reorganization hypothesis – *children's fraction schemes can emerge as accommodations in their numerical counting schemes* – is untenable if counting is regarded only as activity. Focusing only on the activity of counting, however, does not begin to provide a full account. When using the phrase "the explicitly nested number sequence," I am referring to the *first part* of a numerical counting scheme, which consists of a number sequence, which is a sequence of abstract unit items containing the records of counting acts.[1] In this case, the activity of counting is interiorized activity and it is no exaggeration to say that it is contained in the first part of the counting scheme. This number sequence is an example of what Piaget (1964) meant when he commented that "there is a structure which integrates the stimulus but which at the same time sets off the response" when speaking of a stimulus from the point of view of the child. In this case, number words such as "seven" refer to a singular unit that can be iterated seven times to fill out a composite unit containing seven counted items. So, the first thing that had to be established to justify the reorganization hypothesis was to observe how ENS children might use their numerical concepts as templates for partitioning continuous unit items. Toward that end, I start with a discussion of two protocols extracted from two teaching episodes held on the 28th of April and the 1st of May with Jason and Patricia during their third grade in school. These episodes illustrate how the two children used their numerical concepts in constructing what I referred to as the equipartitioning scheme (cf. Chap. 4).

The Equipartitioning Scheme

Breaking a Stick into Two Equal Parts

Jason and Patricia's use of their numerical concepts in constructing a partitioning scheme occurred fortuitously in a situation that was not designed for that purpose. After drawing a stick in TIMA: Sticks, the children were asked to break the stick into

---

[1] See Chap. 3, especially the sections, "Numerical Patterns and the Initial Number Sequence," "The Tacitly Nested Number Sequence," and "The Explicitly Nested Number Sequence."

Fig. 5.1. Cutting a stick into two equal parts using visual estimation.

two substicks of equal length. Jason cut the stick into two pieces as shown in Fig. 5.1. The teacher asked, "How do you know that the two pieces are of the same size?"

Jason's suggestion was to "copy the biggest one and then copy them again." He then said, "no," shaking his head and the two children sat in silent concentration. The children had formed a goal of finding a way to test whether the two pieces were of equal size, but they seemed to have no action they could use to reach their goal. So, the teacher suggested to Patricia that she draw a shorter stick that would be easier to divide visually. After Patricia drew this stick, the teacher did not suggest the actions the children contributed that are reported in Protocol I. Rather, they arose independently from the children. In the Protocol, "T" stands for "Teacher," "P" for "Patricia," and "J" for "Jason."

Protocol I. Using a number sequence to break a stick into two equal pieces.
T: (After Patricia had drawn a stick about 1 dm in length.) Now, I want you to break that stick up into two pieces of the same size.
P: (Places her right index finger on the right endpoint of the stick, then places her right middle finger to the immediate left of her index finger. Continues on in this way, walking her two fingers along the stick in synchrony with uttering.) One, two, three, four, five. (She stops when she is about one-half of the way across the stick.)
J: (Places his right index finger where Patricia left off and uses his right thumb rather than his middle finger to begin walking along the stick. He changes to his left index finger rather than his right thumb after placing his thumb down once. Continues on in this way until he reaches the left endpoint of the stick.) Six, seven, eight, nine, ten. (Then.) There's five and five. (Smiles with satisfaction.)
P: (Smiles also.)

Patricia independently introduced the action of walking her fingers along the stick. Jason picked up counting where Patricia left off, which solidly indicates that he assembled meaning for Patricia's method of establishing equal pieces of the stick. It seemed that Patricia stopped counting at "five" because she reached a place that she regarded as one-half of the way across the stick. Patricia, as well as Jason, now had a way to at least justify where the stick should be cut. The pleased look on their faces indicated that they had achieved their goal.

Composite Units as Templates for Partitioning

Patricia established a blank stick as a situation of counting by projecting units into the stick so that she could imagine the stick broken into two equal-sized pieces, where each piece was in turn broken into an indefinite numerosity of pieces of the same size. The fact that Patricia counted indicates that she was aware of an unknown numerosity of pieces prior to counting. So, at least in the case of the two children, a very basic condition for the reorganization hypothesis to be viable had been established: The children used their composite units[2] as templates for partitioning a stick into equal and connected parts and I regarded the result – a connected number – as templates of a possible fraction scheme.

Based on the insight that the children independently used their number sequences, which involved iterable units, in partitioning, I hypothesized that the operations of partitioning and iterating were parts of the same psychological structure for the children. More specifically, I hypothesized that partitioning for ENS children included both operations of breaking a continuous unit into equal-sized parts and iterating any of the parts to reconstitute an equivalent whole. The hypothesis would be confirmed if any single part of a partitioning could be iterated to reconstitute a stick equivalent to the unpartitioned whole by iterating the part. In order to test my hypothesis, a task was designed to see if the children would iterate a part of the stick in judging whether it was one of the several equal parts.

Protocol II. Breaking off one of the four equal parts of a stick.

T: Let's say that the three of us are together and then there is Dr. Olive over there. Dr. Olive wants a piece of this candy (the stick), but we want to have fair shares. We want him to have a share just like our shares and we want all of our shares to be fair. I wonder if you could cut a piece of candy off from here (the stick) for Dr. Olive.
J: (Using MARKS,[3] makes three marks on the stick, visually estimating the place for the marks.)
P: How do you know they are even? There is a big piece right there.
J: I don't know. (Clears all marks and then makes a mark indicating one share. Before he can continue making marks, the teacher–researcher intervenes.)
T: Can you break that somehow? (The teacher–researcher asks this question to open the possibility of iterating.)
J: (Using BREAK, breaks the stick at the mark. He then makes three copies of the piece; aligns the copies end-to-end under the remaining piece of the stick starting from the left endpoint of the remaining piece as in Fig. 5.2.)
T: Why don't you make another copy (This suggestion was made to explore if Jason regarded the piece as belonging to the three copies as well as to the original stick.)?
J: (Makes another copy and then aligns it with the remaining part of the original stick. He now has the four copies aligned directly beneath the original stick which itself is cut once. The four pieces joined together were slightly longer than the original stick as in Fig. 5.3.)

Fig. 5.2. Jason testing if one piece is one of the four equal pieces.

Fig. 5.3. Jason's completed test.

---

[2] Because the children counted, the elements of their composite units were arithmetical units; and so it can be said that the children assimilated the situation using their number sequences.

[3] MARKS enables a child to position the cursor on a stick and, by clicking the mouse, place a hash mark at that place.

Jason independently copied the part he broke off from the stick three times in a test to find if the three copies together would constitute a stick of length equal to the remaining part. This way of operating was crucial because it was the basis of my inference that he anticipated producing the three copies prior to their production. This anticipation would require him to repeatedly use the operations involved in making a stick in visualizing a stick, which is essential in iterating the stick. This repeated use of his operation of making a stick in visualizing a stick was presaged by his comment "copy the biggest one and copy them again" preceding Protocol I.

Comparing the three copies with the remaining part of the original stick does indicate that Jason took the three joined copies as a term of comparison; that is, as a unit containing three units that he could compare with the unmarked part of the original stick. This opens the possibility that he could unite a current copy of the part with those he had previously made. The possibility is confirmed when he makes another copy and then aligns it with the remaining part of the original stick after the teacher suggested that he make another copy. Jason's way of operating in Protocol II was a modification of the operations that constituted his concept of four. In fact, the *result* of Jason's mathematical activity in Protocol II can be regarded as a connected number, four.

Patricia, in the same teaching episode, demonstrated that she too could operate in the way Jason operated in Protocol II. I call the scheme that Jason and Patricia constructed an equipartitioning scheme, which is the fourth level of fragmenting that I identified in Chap. 4. It is crucial to understand that the independently contributed language and actions of the children warranted imputing this scheme to the children.

Segmenting to Produce a Connected Number[4]

Due to scheduling difficulties, Jason was paired with another child, Laura, during their fourth and fifth grades. Laura was also judged to have constructed an ENS. Here I analyze the connected numbers that Laura constructed for comparison and contrast with those that Jason constructed.

Equisegmenting vs. Equipartitioning

In Protocol II, I assume that Jason estimated one share by using his composite unit, four, to mentally project four separated but connected units into the stick.

---

[4] I would like to thank Dr. Ron Tzur for making his transcriptions of the videotapes available for my use in viewing them.

5   The Partitive and the Part-Whole Schemes   79

This seemed to occur as one composite act rather than as four individual and sequential acts, which is why I regarded it as an act of partitioning. In contrast to Jason's partitioning, in the first teaching episode of their Fourth Grade, held on the 12th of October, Laura breaks a stick into three parts by sequential acts, which I refer to as

Protocol III. Sharing one stick equally among three people.
T: (Asks the children to share a stick among three people.)
J: (Marks a stick into three parts that were obviously not the same length.)
T: (To Laura.) Do you think they are equal?
L: (Shakes her head "no." She then erases the rightmost mark and reactivates MARKS. She then runs the cursor from the left endpoint to the mark in a uniform motion and continues on until reaching a place she thinks marks off a part equal to the first part and makes a mark. She then continues on running the cursor with the same uniform motion to the rightmost endpoint.)
J: (Right after Laura has finished sweeping.) That one is bigger.
L: (Erases the second mark.) This is going to take a long time! (Measures more slowly with her sweeping motion over the first part of the stick and continues along the stick until she makes another mark.)

segmenting. Laura uses one part of the stick three times in *segmenting* the stick rather than projecting three units at once to *partition* the stick.

I refer to Laura's scheme for sharing a stick into three equal parts as an *equisegmenting scheme*. I hypothesize that the experience of segmenting a continuous item, the stick, was being recorded in Laura's abstract unit items of her number concept, three, which is one way that I understand how a connected number is constructed. This is another way of saying that her segmenting the stick was being assimilated using her concept of three, and that her concept of three was being modified in the process of assimilation. So segmenting the stick was being assimilated using her concept of three, and her concept of three was being modified in the process of assimilation. Hence this assimilation was a generalizing assimilation.

An important distinction between Laura's equisegmenting scheme and Jason's equipartitioning scheme is that Laura did not pull the estimate out from the stick and iterate it, as Jason did. Recall that in Protocol II, Jason used his unit of one in disembedding the part he marked off to indicate one of the four equal parts whereas Laura engaged in actually making units of one on the original stick in Protocol III. She seemed to use an image of the first part of the stick in monitoring making the two next parts, and so she operated *as if* she had pulled the first part of the stick out of the stick and iterated it. However, her focus was on the motion involved in each length as opposed to the resulting units. As we will see in later protocols, she does not equate her sequential segmenting with an iteration of the first unit. This fundamental difference in the way the two children operated arose throughout the children's fourth and fifth grades as they constructed and used fraction schemes.

The Dual Emergence of Quantitative Operations

Based on Laura's actions in Protocol III, it is possible to give an account of her concept of length.[5] Running the cursor over the first part of the stick indicates that motion was a constitutive aspect. Because the motion was uniform, I infer also that a sense of duration of the motion was involved. Finally, because she visually compared the three parts, independently erased the second mark, and then made another estimate of where to place the second mark, I infer that the trace of the motion between its beginning and its end was a third constitutive aspect of length. In fact, she used the visual records of the motion from one hash mark to another hash mark as well as a sense of the duration of the motion in gauging where to put the second hash mark. I think of the stick concept as a template that includes records of moving from one site to another, the trace of the motion, the duration of movement, and occupied space. Laura's concept of the length of a stick consisted of the first three properties.

Following Piaget et al. (1960), I argued in Chap. 4 that children construct quantitative operations in the case of continuous quantity as well as discrete quantity. This dual emergence of quantitative operations is fundamental to the reorganization hypothesis because operations implied by the number sequence are specialized operations and cannot possibly be used to fully explain quantitative concepts like length. On the contrary, the reorganization hypothesis claims that children can use their discrete quantitative schemes to reconstitute continuous quantitative operations and that both the discrete schemes and the continuous quantitative operations are reorganized in that process. Clearly Laura brought a concept of length to the task at hand that developed before her attempt to use number concepts to segment an unmarked stick. Although her concept of quantity with respect to length seemed yet to involve an iterable length unit, her equisegmenting scheme is another initial confirmation of the reorganization hypothesis.

Making a Connected Number Sequence

In the above protocol, we focused on the children's construction of connected numbers when segmenting or partitioning a stick. We also encouraged the children to make connected number sequences in a way that stressed iteration of a unit stick without first partitioning the stick. Toward this end, we encouraged the children to make a connected number sequence[6] by constructing meaning for "so many times as long." Protocol IV came from a teaching episode that was held on the 14th of October of the children's fourth grade.

---

[5] Cf. Chap. 4 for a discussion of the concept of length.

[6] A connected number sequence can be thought of as a sequence of abstract unit items that contain records of counted segments joined end-to-end. Although it might be thought of as a number line, the term "number line" is associated with structures of the real number line that go well beyond a connected number sequence.

5 The Partitive and the Part-Whole Schemes

Protocol IV. Making sticks two, three, and four times as long as a unit stick.

T: What I want you to start off with today is to make a set of sticks starting with a small unit stick about a centimeter long. (Gauges a centimeter by holding up two fingers about 1 cm apart.)
J: (Quickly makes such a stick.)
T: Make a set of sticks starting with one that is twice as long, three times as long, up to... where do you want to go up to? You can do it the quickest way you can figure out using the (computer) buttons you got...can you do it Laura?
L: (Copies two sticks beneath the unit stick, then three sticks beneath those two, then four sticks beneath those three, then four sticks beneath those four. She then more neatly aligns the first four rows of sticks, leaving the last four sticks unaligned.)
T: Now, before you go any further, I want one stick that is twice as long as the unit stick. (Laura had not joined the sticks together.)
J: (Takes the mouse and starts to join the four sticks Laura had not aligned.)
T: (To Jason.) Maybe you didn't understand what I meant by a set of sticks.
L: (In explanation.) You join these together and that would be one, then you put them together and that would be twice as long. (The two sticks immediately below the unit stick.) And then three times as long...like that.
T: (To Jason.) I don't think that was what Laura meant, what you are doing....
J: (Joins all 14 sticks on the screen together.)
T: Ok. Break those apart, Laura, and do what you meant.
L: (Activates BREAK[7] and breaks the 14-stick[8] Jason made into its parts. She then drags the unit stick to the upper left hand corner of the screen and then drags another stick directly beneath it and joins that stick with another stick in the broken row of sticks. She repeats this, making a 3-stick.)
T: Can you explain to Jason what you are doing?
L: (Moves the unit stick with the cursor and then moves the 2-stick with the cursor.) That's twice as long. (Moves the 3-stick with the cursor.) That's three times longer than that one, then you can make another line of four, then five, and then have it more and more and more....

Laura's copying of two sticks beneath the unit stick, then three sticks beneath those two, etc., solidly indicates that "twice as long" and "three times longer" meant to iterate the unit stick two or three times. That is, Laura considered the stick that Jason made as an arithmetical unit item that she could iterate so many times – as an iterable unit of one. Of course, her numerical concepts do not supply all of the necessary meaning for twice or three times longer, but she operated as if she was operating using discrete units to produce specific numerosities. Presumably, she anticipated implementing her unit stick an unknown number of times because she said, "and then have it more and more and more."

When Laura joined the sticks together using Join, I infer that she used her uniting operation to mentally compound the sticks together into a composite unit. I base the inference on the language she used in explanation to Jason, "You join these together and that would be one, then you put them together and that would be twice as long (the two sticks immediately below the unit stick). And then three times as long...". What she meant by "that would be one" presumably was that it would be one of the several sticks that she made. But she also regarded it as a composite unit item in relation to the unit stick as indicated when she said, "that would be twice as long."

---

[7] When a word is in caps, it refers to an action in the computer program.
[8] The notation "14-stick" refers to a 14 part stick.

The two inferences concerning the operations of iteration and uniting provide the basis for referring to the composite unit sticks, which Laura made as connected numbers. In making these connected numbers, she did not first engage in partitioning the unit stick and reconstitute it as a connected number, as in Protocol III. Rather, she produced a connected number in much the same way that she would produce a composite unit of four. The only difference was in her use of JOIN to implement her uniting operation.

After Laura's explanation at the end of Protocol IV, Jason finally understood what the teacher intended. He joined four of the remaining copied unit sticks together and placed the 4-stick he made under the 3-stick Laura had made. There were only four copied unit sticks left, so he made another copy and joined the five copied unit sticks together and placed them under the 4-stick. To make the next stick, Jason copied the 5-stick and the unit stick and joined them together. After Jason made the 6-stick, the teacher asked Laura if she wanted to carry on and what the next stick would be. Laura said, "seven," so the teacher asked her to find a quick way to make the seven. Laura asked, "Can it be the same one he did?" asking if she could make it the same way. To make the 7-stick, Laura then copied the 6-stick and the unit stick and joined them together.

From her comments and action of making the 7-stick by joining a copy of the unit stick to the 6-stick, I infer that she re-presented Jason's actions and abstracted the process of producing a number by taking the previous number and adding one more stick. She operated analogously for the 8-stick and commented to the teacher that to make the next stick, she would add one more. She was obviously aware of what she was doing and we can say that for both Laura and Jason, a connected number was related to its successor by the relation of "one more" in a way quite analogous to how a whole number was related to its successor. For this reason, I infer that both children had constructed an *explicitly nested connected number sequence*.

In the case of the 9-stick, Jason changed from simply adding one more stick to the 8-stick and made copies of the 6-stick and the 3-stick and joined them together, confirming that the sticks he made were indeed connected numbers. Jason and Laura made the next five sticks in a similar way. The teacher asked the children to make the 15-stick, the final stick of the series, using only one kind of stick. Jason copied the 3-stick five times and joined them together and Laura said she was going to use the 5-stick. That is, they knew that five iterated three times or three iterated five times would produce 15. That the children assimilated the teacher's request involving the 15-stick using their units-coordinating schemes[9] and coordinated the 3-stick and

---

[9] To find the product of five and three, if a child mentally inserts the unit of three into each unit of five to produce five threes prior to actual activity, the involved scheme is referred to as a units-coordinating scheme (Steffe 1991). Based on the ease with which the children selected the 3-stick and the 5-stick, they seem to have abstracted "three times five is fifteen" and "five times three is fifteen" in their work on multiplication in their regular mathematics classrooms. For this reason, I made the judgment that they used their units-coordinating schemes in assimilation.

the 5-stick prior to actually iterating them are solid indicators that the assimilations using their units-coordinating schemes were generalizing assimilations.[10]

Jason did not initially understand the intention of the teacher, but operated very powerfully upon recognizing Laura's language and actions involved in making sticks twice as long and three times as long as the unit stick. There were no major modifications necessary in his numerical schemes for him to operate as he did. This is also a solid indication of generalizing assimilation. Laura operated smoothly throughout Protocol IV and as if she was operating with discrete items. So, if a generalizing assimilation was involved in her case, it was immediate.

An Attempt to Use Multiplying Schemes in the Construction of Composite Unit Fractions

Both children seemed aware of how many times they were going to iterate the sticks they selected to make the 15-stick, so during the teaching experiment we conjectured that the children could use their units-coordinating schemes in the production of composite unit fractions. For example, if Laura was aware that she could copy the 5-stick three times to make a 15-stick, then we conjectured that she should be able to establish the 5-stick as one-third of the 15-stick. In this case, we refer to one-third as a composite unit fraction. If the conjecture proved to be viable, then it would be possible for the children to establish one-third and five-fifteenths as commensurate.

Provoking the Children's use of Units-Coordinating Schemes

Five teaching episodes were held between those held on the 14th of October and the 2nd of December of the children's fourth grade year that were devoted to the children using their units-coordinating schemes to find what stick could be iterated a given number of times to produce a given stick (21st and 26th of October and the 9th, 11th, and 18th of November). To open the teaching episode held on the 21st of October, the children made all of the sticks from the 1-stick through the 10-stick and erased all of the marks on the sticks. The teacher then asked the children to make a 24-stick using the sticks of the graduated collection as a preliminary to a units-coordinating task he had planned. Jason selected the unmarked 3-stick and iterated it eight times to make a 24-stick$_8$ marked into eight parts.[11] Laura then

---

[10] An assimilation is generalizing if, first, the scheme is used in situations that contain sensory material that is novel for the scheme, and if there is an adjustment in the use of the scheme [cf. Steffe and Wiegel (1996) and Steffe and Thompson (2000)].

[11] The notation, "n-stick$_m$" is used to denote an m-part n-stick. In this case, an eight-part 24-stick.

selected the unmarked 6-stick and iterated it four times to make a 24-stick$_4$ marked into four parts. The teacher then asked the children to hide their eyes while he made a 24-stick$_3$ using the unmarked 8-stick. Each of the three parts of the 24-stick$_3$ was unmarked.

Protocol V. Iterating trial units in an attempt to produce the 24-stick$_3$.

T: Well, what we can do is to use MEASURE. (Measures the 24-stick$_3$ and "24" appears in the number box.) All right, we have a 24-stick. Now, can you find out what piece I used three times to get 24?
L: (Whispers.) 10-20-30 (Starts over and counts. She puts up a finger for each number word she utters.) And you only used it three times?
J: (Tries to figure it out mentally without overt indications of counting.)
T: Yeah, I used only one kind of stick three times, and made 24.
L: (After further activity.) I know!
J: I know!
T: All right, Jason, go ahead. Its your turn.
J: What are you saying? I don't get it all right. (Seems very confused.)
T: I am saying I used one of these sticks. (The unit stick through the 10-stick.) Three times to get the twenty-four stick. Can you tell me which one I used?
L: (Whispers.) Oh, I know, I know, I know.
J: (Nods, "yes," but seems uncertain that he can answer the question.)
T: (To Jason.) You can? Ok, let Laura try it, and then we'll go back to you.
L: Eight!
T: How did you figure that out?
L: I just went (Pointing to the fingers of her left hand.) 8, 16, 24.
T: (To Jason.) All right, what did you have?
J: I just added every one of them, and none of them....
T: Three times?
J: (Nods his head "yes.")

The teacher's intention was for the children, after the 24-stick$_3$ was measured, to select a trial stick and then iterate it to find if the trial stick worked. After Laura found that ten did not work, she continued on selecting trial numbers and iterating them three times to find if 24 would be the result. She found that eight would work: "I just went (pointing to the fingers of her left hand) 8, 16, 24." Her way of proceeding is a solid indicator that she was aware of the number of iterations, in this case, three, and that she used her units-coordinating scheme to find the desired stick. Her counting indicates that her units-coordinating scheme was reversible because, given a result of the scheme (twenty-four) and the number of iterations (three), she established a composite unit that she used in iterating three times.

Her counting "8, 16, 24" rather than copying sticks opens the question of whether her countable items were, in fact, images of sticks or whether the sticks were merely symbolized by her counting acts. If her countable items were images of sticks, this might open the possibility that she would regard an 8-stick as one of the three units of 24 and, hence, as one-third of 24. My argument will be that she indeed operated using images of sticks, and that her image of a stick was an *operative numerical image* in that she could use the images in numerical operating.[12]

In contrast to Laura, encouraging the children to make an estimate of a stick that could be repeated three times to make the 24-stick$_3$ did not seem to activate visualizing activity on Jason's part. After the teacher measured the 24-stick$_3$, Jason did try to find a number from one through ten that could be iterated three times to produce 24. His comments ("What are you saying? I don't get it all right." and "I just added every one of them, and none of them....") indicate that he did engage in numerical operating. But there was no basis to infer that he made operative numerical images of sticks and operated on those images.

In an attempt to generate more insight into how the children operated, the teacher asked the children to pose situations to one another. When one of the children was posing a situation, the other child was to close his or her eyes. Laura started by making four copies of the 10-stick and joining them together. After asking her to clear all the marks, the teacher asked Laura how many times she used the stick. Laura said "four" and, with the help of the teacher, Jason measured the 40-stick$_4$ and quickly said that Laura used the 10-stick. After the teacher asked him if it could be something else, Jason emphatically shook his head, "no," and said, "Well, it can be something else but, umm, it will be more than ten, I mean, more than four." This was an insightful comment and it indicates that Jason was aware that any stick from the unit stick through the 10-stick was a possibility, and also that if one of the other sticks were used, it would take more than four to make the 40-stick. That is, he was aware that the shorter the stick, the more times it would need to be iterated to make the 40-stick. He was also aware that none of the sticks shorter than the 10-stick would work because, after the teacher asked him if it could be something else, Jason emphatically shook his head, "no." In fact, his comment immediately above was given as a justification for why it could not be another stick. So, the indication is solid that he assimilated the situation using his units-coordinating scheme and thereby interiorized his figurative stick images, forming operative numerical images.

At this point, there was no corroboration of the hypothesis that Laura's images of sticks were operative, numerical images. Now that we know that Jason's images were operative and numerical, we can ask whether Jason had reason to attribute his way of operating to Laura. Immediately after the situation of Protocol V, Jason copied the 6-stick seven times, joined the seven copies together, and erased all marks on the resulting 42-stick$_7$. He expected that Laura could solve the task and, when coupled with how she proceeded in Protocol VI, there is reason to tentatively infer that her images of sticks were operative, numerical images. After telling Laura that he used a stick seven times, Laura proceeded as in Protocol VI.

---

[12] In the case of an operative, numerical image, the image would be a re-presentation of an interiorized stick concept, such as a transparent segment of thread, using her concept of the connected number, eight.

Protocol VI. Laura finding what stick used seven times makes a 42-stick.

L: (After measuring the 42-stick$_7$.) Forty-two. Umm, let me see, I know my multiplication... Oh gosh....
T: He used it seven times to get a 42 stick.
L: (Hesitantly, to the teacher.) Six?
T: Ask him. (Jason.)
L: (Looks at Jason.)
J: (Nods his head "yes.")
T: (To Laura.) How did you know? Tell him how did you know. I mean, that was....
L: I just remembered doing it in math today...we were doing six times seven is forty-two!

After Laura measured the 42-stick$_7$, she interpreted the task as multiplicative as indicated by her saying, "Forty-two. Umm, let me see, I know my multiplication... Oh gosh...." That the situation evoked multiplication facts in Laura is an indicator that her images of sticks were operative, numerical images, but I would want to see her reason strategically using her units-coordination scheme as Jason did before attributing operative, numerical images of sticks to her.

An Attempt to Engender the Construction of Composite Unit Fractions

As noted at the beginning of this section, the primary purpose of asking the children to use their units-coordinating schemes in the context of connected numbers was to establish schemes that might serve in the construction of composite unit fractions. After preliminary work with one-half in the teaching episodes held on the 9th and 11th of November, the teacher concentrated on one-third in the episode held on the 18th of November. The intention was for the children to make explicit that the number of times a stick is repeated to make another stick corresponds precisely to the fractional part it is of the other stick.

After the children had made unmarked sticks through the 10-stick, the teacher asked the children to make the 30-stick in the fastest way possible as a preliminary to finding the stick that is one-third of thirty.

Protocol VII. Making one-third of thirty.

L: (Makes a copy of an unmarked 15-stick that was already made and uses REPEAT to make the 30-stick.)
T: Don't erase the marks. We are going to leave the marks there. (Indicates that the children are to use FILL to make a part for each child and Laura fills the first part and Jason the second.)
T: So you used the fifteen to make the thirty. What part of the whole stick is the 15-stick?
L: One-half.
J: Half.
T: One-half! What makes it one-half? Why do we call it one-half? What did you do to build the 30?

L: Repeat.
T: Repeat the fifteen…
L: Two times.
T: Twice. You had to have the fifteen twice. That makes it half. Right? If you need to repeat it twice that makes it half. Which stick would be one-third of thirty? (The teacher then asks the children to build a 30-stick to be shared among the three of them.)
J: (After a slight pause, copies the 10-stick and uses REPEAT to make the 30-stick. He then uses FILL to color the three parts with different colors.)
T: Which one did you use?
J: Ten.
L: Ten.
T: Ten! How many times did you use it?
J: Three.
L: Three.
T: What made you think that ten would be one-third of thirty?
J: Three times ten is thirty. Three tens is thirty.
T: Did you understand what he said?
L: No.
T: Can you explain to her what you said?
L: I know what he said but I don't agree with what he said.
T: What did you say Jason? Tell her!
J: Three ten's is thirty. Three times ten is thirty.

The children's immediate response to the teacher's question, "What part of the whole stick is the 15-stick?" when coupled with Laura's repeating of the 15-stick, indicate that their meaning of "one-half" included repeating the 15-stick twice to produce the 30-stick. Although Jason's reply "half" may not warrant this inference, his subsequent actions and language after the teacher asked him "Which stick would be one-third of the thirty?" certainly do imply a similar meaning for "one-half."

It is very important that Jason independently selected the 10-stick and then made the 30-stick. Moreover, his explanation for why ten is one-third of thirty – "Because three times ten is thirty!" – indicates that he regarded the resulting 30-stick$_3$ as a result of using his reversible units-coordinating scheme in the context of connected numbers. Laura's comment that she did not agree with Jason opens the possibility that she was not aware of the relation between the number of times a stick is repeated to make another stick and the fractional part the first is of the second in the case of more than two repetitions. So, I look to the next teaching episode for further investigation of the children basing the construction of composite unit fractions on their units-coordinating scheme.

Conflating Units When Finding Fractional Parts of a 24-Stick

On the 2nd of December, the teacher asked a fraction question in the context of the children using their units-coordinating schemes in a further exploration of the children's conception of a composite unit fraction. As usual, the children had started by making all the sticks from the 1-stick through the 10-stick and erasing all of the marks on the sticks. There was now a graduated collection of ten unmarked sticks on the screen. The teacher then asked the children to make a 24-stick using the sticks of the graduated collection as a preliminary to a fraction task that he had planned. Jason selected the unmarked 3-stick and iterated it eight times to make a 24-stick marked into eight parts. Laura then selected the unmarked 6-stick and iterated it four times to make a 24-stick marked into four parts. The teacher then posed his fraction task. After asking the children to hide their eyes, the teacher made a 24-stick$_8$ using the unmarked 3-stick.

Protocol VIII. Finding the fractional part a 3-stick is of a 24-stick.

T: I used one of the sticks. Which one did I use and what fraction is it of the 24-stick?
L: It is either the two or the three.
J: Three. It's the three. (Laura agrees.)
L: And the fraction is three-eighths!
J: Three-eighths. (In agreement.)
T: (To Laura.) You said two or three and (To Jason.) you said three. How did you find out?
J: Same....
L: They look the same. (Referring to the unmarked 3-stick and to the eight parts of the 24-stick.)
T: And how did you find out? (To Jason.)
J: I went 3, 6, 9, 12 and ended up to 24. And I know it would be a three because I used a three. (He had made a 24-stick$_8$ earlier.)

The children focused on the numerosity of the 3-stick when saying "three-eighths" rather than on the stick as one composite unit even though the 24-stick was marked into eight equal parts and it was in full view of the children. So, the teacher continued to investigate whether this would recur in other cases.

Protocol VIII. (Cont).

T: Jason, close your eyes. Laura, pick any other way to make the 24-stick. Any way you like.
L: (After approximately 14 s during which she subvocally uttered number words while looking at her hands resting underneath the table, she makes a copy of the 2-stick and repeats it 11 times to make a 22-stick. She then makes a correction by copying the unmarked 2-stick and joining the copy to the 22-stick.)
T: Ok, Jason! That's the problem. You need to say which stick she used and what fractional part it is of the 24-stick. And you need to verify it.
J: (Uncovers his eyes. Copies the 2-stick into the RULER and measures Laura's stick. "12" appears in the number box.) She used the two.
T: And? What fractional part is it?
J: (Sits quietly without answering.)
T: Remember what we call it? What does this number ("12") tell you about the name of the fraction?
J: How many times you click!
T: So, what would you call the fraction?
J: Two-twelfth!

T: One-twelfth.
J: One-twelfth. (Almost simultaneously with the teacher.)
T: Laura, you close your eyes. (To Jason.) You are going to build a 24-stick but do not use the two or the three.
J: (After a short pause, makes a copy of the 6-stick and repeats it four times.)
T: (After asking Jason to think of what part the 6-stick is of the 24-stick.) All right Laura, here is the problem. (Points to Jason's 24-stick$_4$.) Find which stick it is and what fractional part it is.
L: (After visual inspection.) He used the six stick four times.
T: How did you know it?
L: I looked at this stick (The 6-stick.), and then I looked at it. (Jason's 24-stick$_4$.)
T: Oh! Like before with the two and three and you tried to see which one?
L: (Nods her head "yes.")
T: (Points to the first part of Jason's 24-stick$_4$.) And what do you call the 6-stick here? What part of the 24-stick is it?
L: Six-fourths.
T: Six-fourths. How do you verify? (To Jason.) Is this the name of the part she used?
J: The name of the part?
T: (Asks Jason what part the 6-stick is of the 24-stick.) What fraction name do we give it?
L: Six-fourths.

Laura knew that Jason repeated the 6-stick four times and that it produced a 24-stick. However, whether she was aware that iterating the 6-stick four times produced a composite unit containing four composite units of six elements each is problematic. The same question applies to Jason's thinking, because he said "Two-twelfth!" in the immediately preceding task. Both children seemed yet to establish the results of operating as a composite unit containing four component composite units of six individual units, say, and disembed one of the four component units of six from the composite unit containing these four component units and make a one-to-four comparison that implied a six to twenty-four comparison.

In a subsequent task, Jason repeated the 1-stick 24 times to make a 24-stick. He did not erase the marks, and Laura almost immediately said that he used the 1-stick 24 times and that it is one-twenty-fourth. Laura saying "one-twenty-fourth" does indicate that she was aware that the number of times that Jason used the 1-stick was necessarily equal to the number of parts in the stick produced. Moreover, she seemed to be aware of a comparison between one part and the twenty-four parts. When the stick iterated was a unit stick so that there were only two levels of units involved, not three, she operated as we hoped she would operate when the stick iterated was a composite unit stick.

Operating on Three Levels of Units

Indication of the status of the children's part-to-whole operations was provided in the task of the following protocol. Before the beginning of the following protocol,

Laura copied the 12-stick and repeated it to make the 24-stick and the teacher asked her to prove to Jason that she used the 12-stick.

Protocol VIII. (Second Cont).
T: How would you prove to him that you used the twelve? (Speaking to Laura.)
L: That I put the twelve into the MEASURE and I measure it and it came out to be two times.
T: And it is the 24-stick, so?
L: Measure it two times.
T: It needs to be the twelve?
J: It has to be.
T: It has to be? If you use the twelve twice you get twenty-four. Why does it have to be?
J: Because there is no other way, only twelve plus twelve is twenty-four.

When asked what fraction the 12-stick was of the 24-stick, Jason said "one-twoth"! In that he knew that Laura had used the 12-stick, he obviously focused on the 12-stick as one stick and made a one-to-two comparison. So, in the case of two 12-sticks embedded in a 24-stick, Jason could reason as if he took a unit of units as material of operating. His operations included disembedding a component unit of twelve units from the composite unit containing two units of twelve. That is, Jason could produce a composite unit containing two component composite units each of numerosity twelve and disembed one of the two component composite units of twelve and make a one-to-two comparison. But when more than two component composite units were involved, he compared the numerosity of one of the component composite units to the numerosity of all of the composite units.

Laura's "proof" that she used the 12-stick does indicate that she was aware of the two 12-sticks as units belonging to the 24-stick, but yet as units apart from the 24-stick. But saying, "I measure it and it came out to be two times" by itself would not be very convincing. However, in an earlier teaching episode, she said that measuring an unmarked 10-stick using an unmarked 5-stick would be two and explained as follows: "Because the five goes in one time and then another time (touches the tips of her two forefingers together simulating the placement of the fives in the 10-stick) and then five plus five is ten." This explanation convinces me that, like Jason, she regarded the 5-stick as a unit apart from the 10-stick and that she regarded the 10-stick as a composite unit comprising two 5-sticks.

Necessary Errors

I interpret the children's answers of "three-eighths" rather than "one-eighth," "two-twelfth" rather than "one-twelfth," and "six-fourths" rather than "one-fourth" as *necessary errors* rather than as errors due to a simple misinterpretation of the situation. An error (from the observer's perspective) is *necessary* if it occurs as the result of the functioning of a child's current schemes. In the previous section, I argued that both Jason and Laura did disembed a numerical part from a numerical whole containing two of these parts and conceive of the whole as consisting of its two parts. For example, Jason said that twelve is "one-twoth" of twenty-four. So, in that case where Jason conceived of twelve as a composite unit, and of twenty-four as a unit made up of two such units, he made an appropriate one-to-two comparison. However, this kind of reasoning seemed to be confined to cases where there were only two equal composite parts. This kind of reasoning would occur more generally if the children could use three levels of units generally in assembling the situations of their units-coordinating schemes.

Units-coordinating scheme is a multiplication scheme that gets its name from the coordination of, to the observer, two composite units of units where one composite unit is inserted into each unit item of the other composite unit. This is possible for children who have constructed the explicitly nested number sequence because "four," say, refers to an arithmetical unit item that can be iterated four times to produce a unit containing four arithmetical unit items. The child does not insert a complex of four unit items into the unit items of, say, the number six. Rather, the child inserts a single entity that could be iterated four times to produce the extension of the number, four, into the unit items of the number, six. The result is a unit containing six unit items each of which implies four unit items. In this way, the child has an awareness of six fours, or four six times.

From what I have explained, it might seem that a child would necessarily produce a composite unit containing six component composite units, each of which contains four individual units by means of the units-coordination. But there is no necessary transformation produced of the two-level unit structure with which the child started. This is apparent when Jason as well as Laura said that the fraction produced by iterating the 3-stick eight times was three-eighths in Protocol VIII and Jason counted by three eight times in verification that the 3-stick was the one which the teacher used. Saying "three-eighths" indicates that the children were aware that there were eight units each of which contained three individual units. But this was not enough for the children to judge that the 3-stick was one-eighth of the 24-stick. What is needed is for the children to take the eight units of three as a composite unit using the uniting operation because to make a part-to-whole comparison between one of the eight units and the unit comprising them means that the child has already taken the eight units of three as the elements of a containing unit. Uniting the eight units together into a composite unit is an act of abstraction that distances the child from the eight units and permits the child to regard the eight units of three as if they were eight singleton units while maintaining their composite quality.

That the children said "three-eighths" rather than "one-eighth," "two-twelfth" rather than "one-twelfth," and "six-fourths" rather than "one-fourth" could be interpreted simply as their whole number knowledge temporarily interfering with their making the appropriate unit comparisons. However, to say that the numerosity of the 3-stick interfered with the children establishing the 3-stick as one-eighth of the 24-stick does not take seriously the constraints imposed by their current units-coordinating scheme. This scheme is essential for the student to assimilate the situation multiplicatively, but the part-to-whole comparisons that they made were constrained by the structure of the results produced by the scheme, which was an experiential sequence of composite units rather than a unit containing that sequence that could be taken as input for further operating. My argument that this is not simply a temporary "error" is based on the consistency of their "errors" and by Jason's appropriate one-to-two comparison in the case of using a 12-stick to make a 24-stick. In that special case, Jason treated the two 12-sticks as two unit sticks apart from the 24-stick as well as comprising the 24-stick, and this permitted him to compare a 12-stick to a 24-stick in a one-to-two comparison. So, when appropriate operations were available to him, he used them not only in assimilating situations, but also in solving them.

Laura's Simultaneous Partitioning Scheme

The experiment to explore whether the children could use their units-coordinating schemes in the establishment of composite unit fractions essentially failed in that the children, although they could find what stick repeated, say, six times makes the 24-stick, conceived of it as four-sixths of the 24-stick rather than one-sixth. Because I construed this as a necessary error, we shifted from our attempts to bring forth the units-coordinating scheme as a cognitive mechanism for the construction of composite unit fractions, and we instead focused on developing the children's partitioning and segmenting schemes involving only two levels of units. This protocol from the teaching episode held on the 7th of December illustrates how the children could use their numerical concepts through ten in partitioning and segmenting.[13]

Protocol IX. Drawing a stick that is one-tenth of another stick.
T: Can each one of you draw one-tenth of that stick? The one who wins will be the one that will be closer.
L: (Draws her estimate.) Right there!
J: (Looks at the screen for some time and draws his estimate.)
L: That's the same!
J: No it isn't, no it isn't!
L: Ok! I will go first here! (She repeats the stick ten times and it is too long.)
J: (Even though his estimate is longer than Laura's, he still repeats it ten times to check.) Oh gosh! (Both children giggle.)
T: You want to try one more?
L: I want to try it one more time!
T: One-tenth, all right!
L: One-tenth, one-tenth! Ok! This is my color! Ok that was too long… ok! That long! (Draws her estimate.)
J: (Draws his estimate, both children laugh.)
L: (Repeats her estimate while counting out loud, and the estimate is very accurate.) Just about!
T: Very close!! Let's see Jason. That's very nice!
T: (Speaking to Jason.) What do you think yours is, too short or too long?

---

[13] Note that the children do not use MARKS to put a mark on the stick, but draw a new stick.

L: Too short!
J: (Repeats his estimate which is shorter than Laura's and produces a shorter stick than the original stick. The children giggle.)
T: All right a little bit too short!

The children became deeply engaged in the task and expressed pleasure at making an estimate by drawing a stick and then testing their estimates by using REPEAT. The initial estimates of both children were closer to one-eighth and one-seventh of the unit stick (Laura and Jason, respectively), and Laura's second estimate was uncannily accurate. Thus, the children used the iterative aspect of the connected number, one, to test their estimates by iterating it ten times and comparing the result against the original stick. Recall that in Protocol III, Laura segmented her stick by transposing a unit from one site to another on a given stick when the unit that was being transposed was a part of the stick being segmented. Unlike Jason, she did not pull out her estimate and iterate it. In Protocol IX, there was no necessity for the children to disembed their estimate from the original stick, so I still could not impute an equipartitioning scheme to Laura. I could not infer that the parts of the stick she produced by projecting parts into the stick were identical parts rather than equivalent parts. So, I interpret her actions of iterating the stick she drew as actions of segmenting the original stick. For this reason, I attribute an *equisegmenting scheme* to her.

That Laura made such an uncannily accurate estimate on her second trial may have been fortuitous. So, in the next teaching episode, held on the 8th of February of their fourth grade year, the teacher posed a task involving sharing a stick into eight equal parts. Other than serving as a check of Laura's as well as Jason's estimates, the teacher wanted to explore whether Laura's use of iteration in Protocol IX was specific to the estimation task. Sharing tasks emphasize partitioning rather than segmenting, so in a task that Laura construed as sharing, given her uncannily accurate estimates, she may have no reason to iterate in order to verify her estimates of an equal-sized part. The initial task was, after Jason drew a segment the same length as a Snickers candy bar, to share it equally among eight people. The teacher imposed the constraint that they could make only one mark.

Protocol X. Sharing a candy bar among eight people by making only one mark.

T: (Counts all the persons in the room aloud: 1, 2, 3, 4, 5, 6, 7, 8.) Your first task is to share this candy bar among these people. But use only marks. Remember you can move marks. But mark only the share of one person and use that to create all the shares. Go ahead.
J: (To Laura.) Go ahead.
L: (Takes the mouse and activates MARKS.) But we can use a lot of marks to....
T: Use one mark, if it will not come out as a fair share then you can use another, but try to make it as close as you can in the beginning.
L: (Activates MARKS again and tries to estimate where to put the first mark. She makes an uncannily accurate estimate. The mark she makes on the stick is apparently one-eighth of the unit stick, but she is yet to verify her estimate.)

T: You know, we can still play with the screen. Remember PULL PARTS[14] and REPEAT? (Encouraging Laura to use these computer actions.)
L: Ok, there's....
T: You remember, PULL PARTS and REPEAT.
L: So, can I make another mark?
T: No, no, just one mark. Now see if it's a fair share.
L: (Seems confused and looks for a button to use.)
T: Do you want to pull the part first? (Again encouraging Laura to construct an equipartitioning scheme.)
L: Ok. (Activates PULL PARTS and pulls the greater of the two parts from the marked stick. She sets the cursor over the smaller piece.) Do I do this piece too? (The marked piece that she estimated as the share of one person.)
T: Which one do you want to use to check to see if it's one-eighth?
L: Umm, this one? (Points to the 7/8-stick she pulled out from the marked unit stick.)
T: Now how can you tell that this (Pointing to the 7/8-stick.) is exactly one-eighth – one-eighth of the candy bar? This is the candy bar. (Opening his hand over the length of the original stick.)
L: I don't know.
T: (To Jason.) Jason, do you have an idea?
J: (Nods "yes" and takes the mouse, drags the 7/8-stick to the top of the screen, then pulls out Laura's estimate from the marked stick.)
T: Can you tell Laura what you are going to do?
J: I'm gonna'...pull one of these. (Points to the marked part of the unit stick.) And put it under there and see if....
L: (Enters Jason's talk, nodding "yes.") And repeat.
T: Ok.
J: (Repeats the 1/8-stick eight times until it reaches the end of the original stick. The resulting 8/8-stick seems to be exactly the same length as the unit stick.)
T: Wow, Wow, Wow! Laura you made it so quickly!! One, two, three, four, five, six, seven and eight. I don't believe it! Isn't that great! It's really good! (The children then mark the unmarked original stick using the 8/8-stick as a template.)

The accuracy of Laura's estimate should not be regarded as fortuitous. Her estimate of one-tenth in the 7th of December was also uncannily accurate on her second trial and there were other occasions where she made similar accurate estimates. Her comment, "But we can use a lot of marks to..." should be considered as indicating that she visualized marks on the stick so that eight parts would be formed. She could then accurately gauge the length of one of the parts.

Although Laura made an uncannily accurate estimate of one-eighth of the stick, she did not independently use PULL PARTS and use REPEAT to verify the estimate. Pulling the 7/8-stick from the marked unit stick, when coupled with her choice of the 7/8-stick after the teacher asked her which of the two parts she wanted to use to check to see if its one-eighth, indicates that she intended to continue on marking the 7/8-stick.[15] Laura definitely could use her number concepts up to ten as templates

---

[14] Using Pull Parts, a child can activate that action button by clicking on it and then click on one or more parts of a stick. The child can then deactivate the action button and drag copies of the parts out of the stick while leaving the stick intact.

[15] Unfortunately, the teacher asked her how she could tell if the 7/8-stick is exactly one-eighth, and his question closed off any further actions she may have taken with the 7/8-stick.

5   The Partitive and the Part-Whole Schemes   95

for partitioning blank sticks in the true sense of a partitioning. In fact, in Protocol IX, it is plausible that she used the composite unit, ten, as a partitioning template in making her estimate. In that case, she would *simultaneously* project the units of her composite unit into the blank stick and experience the parts as co-occurring. In Protocol IX, this partitioning activity seemed to close off her need to verify the part she marked off by pulling the part from the original stick and iterating it eight times to make a test stick. The operation of iteration unquestionably was available to her as indicated by Protocol IX. But in that case her estimate was not a part of the stick of which she was estimating a part and so she did not need to disembed the estimate, whereas in Protocol X, her estimate was a part of the original stick. Although iterating, partitioning, and disembedding were operations of her number sequence, she seemed to only use partitioning in Protocol X. That is, she used her connected number concept, eight, to project units into the stick. Jason, on the other hand, disembedded the part of the stick Laura had made and iterated that part eight times in an attempt to find if it was indeed one-eighth of the unit stick. In that Laura made such uncannily accurate estimates when partitioning a blank stick into up to ten parts, I refer to the scheme she used as a *simultaneous partitioning scheme* to distinguish it from Jason's equipartitioning scheme. The apparent difference in the two children seemed to reside in Jason's construction of the equipartitioning scheme, which involves simultaneous partitioning along with the ability to disembed and iterate a partitioned part, whereas Laura had constructed only the equisegmenting scheme, which involves sequential segmenting without disembedding. Her uncanny ability to make accurate estimates may have served to suppress her use of the disembedding and iterating operations because she apparently felt no need to verify her estimates. Nevertheless, her images of sticks seemed to be operative, numerical images because she could use them in numerical operating.

An Attempt to Bring Forth Laura's Use of Iteration to Find Fractional Parts

Based on Laura's drawing, an estimate of one-tenth of a stick and then iterating the estimate ten times to test its accuracy in the 7th of December teaching episode, I formed the hypothesis that she had indeed constructed the iterative operation involving a singleton length unit that was based on her explicitly nested connected number sequence, but not on an equipartitioning scheme. This hypothesis is especially plausible because in the teaching episode on the 9th of November, she said "You could use the one (1-stick) and use it thirty times!" when it was her goal to find which of the 1-stick through the 10-stick could be used to make the 30-stick and how many times it needed to be used. The comment provides solid indication that she imagined the operation of iterating the 1-stick thirty times to produce the 30-stick. It also indicates that she visualized the 30-stick, not as a blank stick, but rather as a marked stick prior to imagining iterating the 1-stick, which is an essential inference that must be made before making the claim that the 1-stick was an iterable unit for her.

In the case of the equisegmenting scheme that was imputed to her in the 7th of December teaching episode, it was her goal to draw one-tenth of a stick and then iterate the drawn stick ten times to find whether the result was the same length as the original stick. In the 8th of February teaching episode, on the other hand, her initial goal was, starting with a blank stick, to share a stick into eight equal parts rather than draw an estimate of one-tenth of the stick. In the sharing case, iteration was not activated. Rather, she used her connected number concept, eight, as a template to partition the blank stick and this led to imputing a simultaneous partitioning scheme to her.

In the teaching episode that was held on the 10th of February, the teacher encouraged the children to engage in cognitive play in TIMA: Sticks. Cognitive play is a necessarily pleasurable, largely unguided investigation of a mathematical situation including tools, in this case. The primary reason for doing so was to encourage Laura to independently use PARTS rather than MARKS to partition a stick into so many equal parts. The difference in these two computer actions is that the former is used to simultaneously partition a stick into so many parts whereas the latter is used to mark off one part at a time. Up to this point in the teaching episodes using TIMA: Sticks, Laura always used MARKS and gauged where to place the marks in acts of segmenting. Jason used PARTS[16] rather than MARKS, so the hypothesis was formed that a child's independent use of PARTS indicated an awareness of the operation of partitioning a stick simultaneously into so many parts and the independent use of MARKS indicated the operation of partitioning a stick by using sequential acts of segmenting.

In the 10th of February teaching episode, Jason playfully drew a stick spanning the screen and used PARTS to mark it into 99 parts. Laura followed by drawing a stick approximately 4 cm in length and used PARTS to mark it into 43 parts. Following these two activities, Jason drew a stick approximately 1 cm in length and again marked it into 99 parts. Laura then drew a stick even shorter than Jason's and marked it into 20 parts. Jason then drew a stick spanning the screen and again marked it into 99 parts and Laura tried to use PARTS to mark a copy of an unmarked stick into one part, which leaves the stick unmarked. These tests of the extreme values of PARTS provide solid indication of awareness of a relation between the number of parts and the size of the parts. Laura verbalized this awareness when she marked a stick into 20 parts. It was also manifest in an attempt by Jason to mark what amounted to a stick fragment he drew into 99 parts. Marking a short stick into a rather large number of very short parts fascinated both children, especially Jason. In fact, Jason subsequently drew a sequence of eight sticks, where each succeeding stick was shorter that the preceding, and then marked each stick into 99 parts without comment.

---

[16] After activating Parts, any number through "99" may be selected as the number of parts to be made. Then positioning the cursor anywhere on a stick and clicking the mouse partitions the stick into that number of parts. If "14" is selected, for example, 13 equally spaced hash-marks appear on the stick.

5 The Partitive and the Part-Whole Schemes   97

The current teaching episode held on the 17th of February followed upon the children's cognitive play using PARTS. The goal of the teacher in the current teaching episode held on the 17th of February was to bring forth iterating as a means to an end in the context of Laura using her simultaneous partitioning scheme. After a preliminary cognitive play session again using PARTS, the teacher asked Laura to share a stick that Jason drew among seven people.

Protocol XI. Laura sharing a stick among seven people.
T: Draw a stick please. (To Jason.) Laura, if you were to share this stick among, say, seven people, can you show me your share?
L: (Starting at the left end of the stick using MARKS, Laura quickly places the first mark on the stick and then continues on making marks across the stick, gauging each placement so that the marks stand approximately evenly across the stick as shown. She then adjusts the second mark using MOVE MARKS.)
J: And now the other way. (Speaking to the teacher while grabbing the mouse and activating REPEAT.)
T: Just let her take her share. (Speaking to Jason.) Take your share out. Take only your share out. Ok?
L: (Laura uses PULLPARTS to pull the first piece.)
T: (To Jason.) Copy the stick. (The 7-stick.) And then erase the marks.
J: (Jason accidentally erases the marks before he copies the stick.)
T: (Asks Laura to copy the stick.)
L: (Makes a copy of the stick and moves the part she pulled to the left endpoint of the original stick.)
T: Do you think your share is a fair one? Can you use that one for all of the seven people?
L: Mm hmm. (Meaning "yes.") (Aligns the left endpoints of the blank 7-stick and the part she pulled out. She then makes a mark on the blank stick at the endpoint of the pulled part and then moves the cursor along the blank stick at intervals approximately equal in length to the part she marked off and makes five more marks. The last part is shorter than those preceding, which are all very close in length.) No, that one is...a little off.

Fig. 5.4. The results of Laura marking a stick into seven parts.

Because Laura had previously mentally partitioned a blank stick simultaneously into eight parts and because she had used PARTS in cognitive play activity in the 10th of February teaching episode as well as in the current teaching episode, the teacher fully expected her to use PARTS to mark the stick in Protocol XI into seven parts. However, she instead used MARKS and used the result to segment the original stick in a test to find if the part she marked off was a fair share. This was a surprise especially because she had just used PARTS to mark sticks into 20, 43, 99, and other such numbers of parts (Fig. 5.4).

In retrospect, it was fortunate that she did use MARKS rather than PARTS because of the operations indicated by her use of MARKS. In her initial marking of the stick, Laura used the first (and left most) part of the stick she made with the first mark as a segmenting unit in making the remaining marks. Again, she was uncannily accurate, which indicates that she visually projected seven equal parts into the stick prior to making her first estimate. Sequentially making marks carried the force of mentally sliding her initial estimate along the blank stick because she had already mentally partitioned the stick and was aware of what the results of sequentially making marks would look like. So, there is reason to believe that the operation of segmenting was indeed a constitutive operation integrally involved in her sharing of a stick as well as in simultaneous partitioning.

In short, iterating involves explicitly pulling a part from a stick and iterating that part to establish a stick to compare with the original in a test of whether the pulled part is a fair share, whereas segmenting involves using a part of a stick within the stick as a template to establish other parts of the stick. So, in segmenting, the child operates within the stick whereas, in iterating, the child operates outside of the stick to establish another stick that can be compared with the given stick. The units that are established by iterating are projected into the original stick. Apparently, Laura's goal when using PARTS was not a sharing goal. Rather, her goal seemed to be to partition the stick into a rather large number of parts without actually intending to share the stick among so many people.[17] So, there is good reason to believe that iterating a stick as Jason did is an abstraction of the segmenting operation in which Laura had engaged.

Jason's Partitive and Laura's Part-Whole Fraction Schemes

Lack of the Splitting Operation

Given Jason's ability to iterate a part of a stick to produce a stick that could be compared against the unit stick in a test of whether the part would be a fair share, the possibility arose that his equipartitioning scheme was a multiplicative scheme. His concepts of number words were multiplicative in that seventy, say, referred to a unit that could be iterated seventy times to produce a composite unit that contained seventy unit items. That is, for Jason, seventy was conceived of as one seventy times. Analogously, Jason's equipartitioning scheme would be a multiplicative scheme if he could make a stick so that a given stick was, say, five times longer than the stick to be made. To conceive of such a stick a priori, the child would need to posit a hypothetical stick and understand that iterating the hypothetical stick five times would produce a stick identical to the original stick. This would also entail partitioning the original stick into five parts of equal length with the understanding that one of the parts could be iterated five times to produce the original.

---

[17] A sharing goal involves an intention to actually break the stick apart. Laura's use of PARTS seemed to not involve such an intention.

5 The Partitive and the Part-Whole Schemes

Protocol XII. Making a stick so that a given stick is five times longer than the stick to be made.

T: Ok! Let's draw a stick....(Draws a stick approximately of length 2 in.) Can you show me a stick that is five times longer than that stick?

L: (Activates COPY and makes a copy of the teacher's stick. She then activates REPEAT and makes a 6-stick instead of a 5-stick because she interprets "five times longer" as "five more.")

T: (To Jason.) Can you show me a stick so this one (Points to an existing stick in the screen that Laura had drawn just before the teacher drew the 2-in. stick.) is five times longer than the stick you show me?

J: WHAT?

T: Ok! This is my stick. (Pointing to the existing stick.) I want you to make a stick such that mine is five times longer than yours.

J: Five times longer?

T: Yes. Mine will be five times longer than yours.

J: (Makes a copy of the stick, activates PARTS, dials to "10," and clicks on the copy, marking it into ten equal pieces. He then breaks the marked stick using BREAK and joins the first five pieces back together. He then drags the five extra pieces into the TRASH.)

T: Ok. Mine is five times longer than yours? Can you show me that?

J: Mmm Mmm. (Yes.) (Repeats the 5/10-stick he made and places the resulting 10/10-stick in the middle of the screen. He then places a copy of the unmarked original stick directly above it and then places the 5/10-stick above the unmarked original stick aligned at the left endpoints as shown in Fig. 5.5.)

T: How many times did you repeat when you did it?

J: Two, one time.

T: So when you repeat it...is this five times longer?

J: This one is (Pointing to the top 5/10-stick.) but that one no, this is ten (Pointing to the bottom 10-stick.).

Fig. 5.5. Jason's attempt to make a stick so that a given stick is five times longer than the new stick.

It is quite significant that Jason did not confuse the question the teacher asked him – "Can you show me a stick so this one is five times longer than the stick you show me?" – with the question the teacher asked Laura – "Can you show me a stick that is five times longer than this one?" To solve the problem as the teacher intended, it would have been necessary for Jason to posit a *hypothetical stick* such that repeating that stick five times would be the same length as the teacher's stick. This would require the operations of partitioning and iterating be implemented simultaneously rather than sequentially. That is, he would need to not only posit a hypothetical stick, but also posit the hypothetical stick as one of the five equal parts (of the teacher's stick) that had been already iterated five times and see the results of iterating as constituting the teacher's stick. This is a *composition* of partitioning and iterating and I refer to it as a *splitting operation*.[18] Since operations involving

---

[18] See Sáenz-Ludlow (1994) for an analysis of the constructive power of a child, Michael, whom I infer had constructed this operation.

two as a quantity often develop precociously, it is interesting that in this case, the children also could not make a stick such that a given stick was twice as long as the stick to be made.

Jason's Partitive Unit Fraction Scheme[19]

Jason's choice of ten parts in Protocol XII does indicate that he was aware that the stick he was to make had to be shorter than the teacher's stick. It also indicates that he was aware that the stick he was to make was a part of (or embedded in) the teacher's stick. Choosing ten as a partition of the original stick and then using a unit of this partition to make a stick five times longer than the unit stick (a 5-stick) also indicates an intuitive awareness that the stick he was to make of necessity needed to be iterated five times. His choice of ten enabled him to establish a stick so he could give meaning to "five times longer" in that the 5-stick was a 5-part stick. A sense of embeddedness and a sense of iterability are essential in the construction of the composition of partitioning and iterating. I regard Jason's attempted solution as creative mathematical activity as well as indicating that the splitting operation was within possibility for him. But he was yet to construct the composition of partitioning and iterating, which is crucial in establishing a unit fraction because of the whole-part relationship a unit fraction implies. Unit fraction language had meaning for both children, but it was not based on the composition of partitioning and iterating.

In the following continuation of Protocol XII, it is possible to understand what "one-tenth" meant for both Jason and Laura.

Protocol XII. (Cont).
T: What do you mean when you say ten?
J: Well...it's the same size as...yours. (Pointing to the copy of the original stick.)
T: So, what is this (Pointing to one of the pieces of the 10-stick produced by Jason.), what would you call it? How much is it...pull that part....
J: One-half! (Referring to the 5-stick's relation to the 10-stick.)
T: Pull that part. Pull one small part of your parts.
J: (Jason pulls one of the parts of his 10-stick.)
T: Ok! This is your part for the time being.
J: One-tenth.
T: So this part is one-tenth of mine?
J: Uh-huh. (Yes.)
T: What do you say? He says that this piece is one-tenth of mine. Is that ok? (Speaking to Laura.)
L: Yeah.
T: Why do you think it is one-tenth? (Speaking to Jason.)
J: Because this is one out of the ten little pieces. (Holds his left thumb and forefinger about 1 cm apart indicating a little piece.)

---

[19] See also Tzur (1999) for a discussion of this scheme.

5 The Partitive and the Part-Whole Schemes

T: Ah! I see. Now if you repeat this one ten times....
J: Which one?
L: This one. (Laura is following the dialog and knows what the teacher wants Jason to do.)
T: What would you get?
J: Which one do I repeat?
T: Yours.
J: Mine?
J&L: Ten times?
L: Ten-tenths!
T: Ten-tenths! What do you say? (Speaking to Jason.)
J: Ten-tenths?
T: Do you want to do it?
J: Repeat that one?
T: Your piece, yes.
J: (Jason repeats the 1/10-stick ten times.)
T: So how much is ten-tenths, Laura?
J: The whole stick. (Note that Jason is aware of the whole stick.)
L: Ten of those little sticks.
T: Did you hear what Jason said? That ten-tenths is the whole stick? Do you agree with it?
L: I guess so.
T: Why?
L: I don't know.
T: Why do you say that ten-tenths is the whole stick?
J: Because it is ten little pieces and it is a how long the whole stick is. So one whole stick is ten pieces of those little ones.
T: Did you get it? (Speaking to Laura.)
L: Yes.

In Jason's explanation for why the teacher said "one-tenth," the language "out of" is a key to understanding his meaning. In one *out of* ten parts, the child conceptually disembeds one part from the ten parts while leaving the part in the ten parts as well. So, the part conceptually belongs to the ten parts while being, at the same time, conceptually separated from the ten parts as an entity independent of the ten parts. In one of the ten parts, the part is distinguished within the ten parts without being disembedded from the ten parts. In other words, his language indicates that he regarded one little piece as one unit part out of the ten unit parts from which it originated as well as one unit part within the ten unit parts. Consequently, iterating the one-tenth part produced a stick identical to the original stick and, hence, the ten parts constituted the length of the original stick. So, he used one-tenth as if it were an *iterable fraction unit* that was on a par with his iterable unit of one.

I call the scheme he used to establish one-tenth a *partitive unit fraction scheme*[20] to emphasize that the dominant purpose of the scheme was to partition the connected number, one, into so many equal parts, take one out of those parts, and establish a

---

[20] A partitive fraction scheme extends the partitive unit fraction scheme in that it can be used to produce proper fractions.

one-to-many relation between the part and the partitioned whole. The iterative aspect of the scheme served in justifying or verifying whether a unit part of the connected number, one, was one of so many equal parts. It also served in producing a partitioned stick that was the length of the connected number, one.

It may seem as if Jason had constructed the splitting operation when he said, "Because it is ten little pieces and it is how long the whole stick is. So one whole stick is ten pieces of those little ones." in explaining why ten-tenths was the whole stick. Based on this comment, there is no doubt that he understood that the ten pieces of "those little ones" comprised the whole stick. He also regarded the length of the stick as ten little pieces. So it would indeed seem that he had constructed the splitting operation. However, there is little indication that he had constructed a multiplicative relation between the whole unpartitioned stick and one of its hypothetical parts prior to actually partitioning, which is essential in the construction of a unit fraction. If one-tenth had been constructed as a unit fraction, Jason would be aware of the whole stick as a unit stick and of a hypothetical part of the unit stick such that the unit stick consisted of ten iterations of the hypothetical part.

I emphasize "hypothetical" because a splitter produces an image of *some* stick (a hypothetical stick) and mentally sets it in relation to the unit stick in such a way that iterating the hypothetical stick produces the unit stick, prior to any observable action. Instead of producing a hypothetical stick and considering it as defining a partition of the unit stick in the beginning of Protocol XII, Jason instead engaged in the operation of partitioning the unit stick into ten parts to establish a target number of parts. That is, Jason did not take partitioning as a given, i.e., he actually had to engage in partitioning in order to produce something he could consider as a part of the unit stick. This observation countermands the hypothesis that an independent use of PARTS indicates that the partitioning child is aware of the *operation* of partitioning as well as the results of partitioning that I made earlier on concerning Laura's use of MARKS rather than PARTS in contrast to Jason's independent use of PARTS.

After Jason actually established one-tenth of the unit stick, it is consequential that both he and Laura knew that if one of the ten little pieces was iterated ten times, the result would be ten-tenths of the stick. This is basic, and I regard it as essential for a scheme to be called a unit fraction scheme. Although there is some doubt whether Laura knew that ten-tenths constituted the whole stick in the continuation of Protocol XII, she was the first to say "ten-tenths" for the result of repeating one-tenth ten times. Still, she did not seem to explicitly realize that "ten-tenths" referred to the whole stick partitioned into ten equal parts and that it was the length of the stick.

Laura's Independent Use of Parts

Whether Laura had constructed a partitive unit fraction scheme that was on a par with Jason's is at issue even though she said that ten-tenths is the result of iterating one-tenth ten times. Based on Protocol IX, Laura could draw one-tenth of a stick and then iterate it ten times and compare the results with the original stick to verify whether the part was one-tenth of the whole stick. But when the estimated part was an actual part of the original stick, Laura did not pull the part and iterate it to find whether the estimated part was a fair share. There was nothing in the continuation of Protocol XII that would serve as an indicator that she could disembed a unit fractional part from a stick and iterate it when it was her goal to verify that the part was a fair share. So, we decided to provoke her use of PARTS instead of MARKS to partition a stick in an attempt to bring forth an equipartitioning scheme. Even though the hypothesis that an independent use of PARTS can be used as an indicator of the child's awareness of the program of operations constituted by partitioning has been countermanded (cf. Protocol XI), Jason independently used PARTS and he also had constructed the equipartitioning scheme. He was definitely aware that he could mark off one of so many equal parts of a stick and that he could iterate the stick in a test to find if it was a fair share prior to engaging in the activity.[21] That is, his equipartitioning scheme was an anticipatory scheme and his use of PARTS seemed to stand in for that scheme. So, it was well-worth exploring whether Laura's independent use of PARTS indicated a change in her simultaneous partitioning scheme.

In the current teaching episode held on the 22nd of February, PARTS was deactivated and was not available to the children for their use. The idea was that Laura might experience the microworld she created using DRAW, FILL, COPY, MARKS, REPEAT, MEASURE, and PULL PARTS (but not PARTS) as different than the microworld she created in the past when PARTS was available. This idea was based on Laura making sense of Jason's independent use of PARTS in the past and on her use of PARTS in cognitive play.

The first task the teacher presented to the children was to work together to share a stick first among two people, then three, and so on. A stick had been placed into the RULER so copies could be made of that stick and the children understood that they were to share copies of that stick.

Protocol XIII. Laura independently asking to use PARTS.
L: (Activates MARKS and makes a mark on the stick to the right of the center of the stick.)
T: Fill the two parts differently.
L: (Using FILL, colors the leftmost part of the stick blue.)
J: (Grabs the mouse and fills the other part yellow. He then clicks on PULL PARTS and pulls the left most part from the original stick. He then aligns it directly beneath the right most part with left endpoints coinciding.[22] He shakes his head "no" and uses MOVE MARKS in an attempt to center the mark on the original stick. As he moved the mark to his left, he also moved the pulled part to the left keeping left endpoints aligned. He then gives the mouse to Laura.)

---

[21] However, he still did have to engage in the activity of partitioning in order to produce a part of the whole that he could reason with. Hence partitioning is anticipatory but not taken as given.

[22] At this point, there was only one blue stick beneath the original stick in Fig. 5.6 and it was beneath the yellow part of the original stick with left endpoints coinciding.

L: (Makes a copy of the pulled part that now lies directly beneath the right most part of the top stick in Fig. 5.6 and aligns its left endpoint with the left endpoint of the original stick. The two sticks overlap as shown in Fig. 5.6. She then makes a copy of the topmost stick in Fig. 5.6 upon the teacher's direction and places it directly beneath the overlapping sticks as in Fig. 5.7.)
T: What could you do to make the sticks even?
L: (Moves the mark on her copy of the topmost stick to the midpoint of the two marks as shown in Fig. 5.7. She then stops and Jason takes the mouse.)
J: (Pulls the rightmost part from the bottom stick and repeats it to find if it is equal to the leftmost part. The resulting stick is very close in length to the stick immediately above it.)
T: (Asks the children to make shares for three people so they get equal parts, all three of them.)
J: (CLEARS the screen and then makes a copy of the stick in the RULER. He then proceeds to use MARKS to make three equal parts. He is uncannily accurate.)
L: (While Jason is making marks.) I know we could do it, but we don't have PARTS!
T: (To Laura.) You want to have PARTS?
L: (Nods her head "yes.")
T: (To Jason.) You want to have PARTS?
J: (Nods his head "yes.")
T: (To Laura.) How would you do it using PARTS?
L: Like if you want to do it in four pieces, you put the number on four and turn PARTS on and then put them on the line.
J: (Reconfigures TIMA: Sticks using the CONFIGURE menu to include PARTS.)
T: Now PARTS will work.
L: (Takes the mouse, dials PARTS to "3," and clicks on a copy of the stick in the RULER.)

Fig. 5.6. Laura's attempt to make two equal shares of a stick.

Fig. 5.7. Laura's attempt to mark a stick into two equal shares.

Laura's comment "I know we could do it, but we don't have PARTS!" was a surprise. It is interesting that it occurred in the context of Jason using MARKS to mark the stick into three equal parts. This suggests that she attributed the use of PARTS to Jason in that her experience of Jason making equal shares in the past involved the use of PARTS. Her explanation of how to use parts immediately preceding Jason's reconfiguration of TIMA: Sticks to include PARTS, when coupled with her actually using PARTS after he reconfigured TIMA: Sticks, together corroborate the inference that she indeed constituted the use of PARTS as her way of making equal shares.

Her use of "we" – "I know *we* could do it…" does indicate that her goal to share the stick into three equal parts was a social goal, and further, that how they made fair shares was a social activity in a way that is similar to how her goal was a social goal.[23] This explanation of goals fits with Laura's comment about using PARTS. Her observation of Jason using PARTS in the two teaching episodes preceding this one was apparently assimilated by her as Jason's way of making shares of sticks but not necessarily as her way of operating because she used MARKS rather than PARTS to make shares. Nevertheless, Jason's use of PARTS had to be meaningful for her or else she would not have spontaneously said, "I know we could do it, but we don't have PARTS!" in Protocol XIII. The question now arose in the teaching episode whether Laura's use of PARTS was on a par with Jason's uses of PARTS. That is, did it imply an equipartitioning scheme?

The goal of the teaching episode had been established at the outset by the teacher when he asked the children to work together to share a stick among two people, then three, and so on. This goal apparently permeated each specific goal of the children to share a stick in Protocol XIII because Laura's hypothetical example that she gave in the continuation of Protocol XIII involved four people. Her example indicates an awareness of a general goal, so it is legitimate to say that the task the teacher presented was an *open-ended task* for the children. The continuation of Protocol XIII picks up the children's activity immediately after Laura made the 3-stick using PARTS.

Protocol XIII. (First Cont)
T: (To Jason.) Go ahead.
J: (Drags the 3-stick to the bottom of the screen and then makes a copy of the stick in the ruler and places it directly beneath the 3-stick. Using PARTS, he makes two equal parts of this stick. He then makes another copy of the stick in the ruler and places it directly over the 3-stick and uses PARTS to make four equal parts of this stick. He then gives the mouse to Laura.)
L: (Makes a copy of the stick and places it directly over the 4-stick and uses PARTS to make five equal parts of the stick.)
T: I want both of you to prove that each of these five people get the same share. (Points to the 5-stick.)
J: (Moves the 5-stick to the top of the screen and gives the mouse to Laura.)
L: (Activates PULL PARTS and pulls the left most part from the 5-stick. She then activates REPEAT and makes another 5-stick directly beneath the original by iterating the part five times.)
T: Why is this a proof that it is a fair share?

Fig. 5.8. Using PARTS in partitioning.

---

[23] For a goal to be a social goal, the child must be able to infer, based on the language and actions of another child, that the other child does indeed have intentions. The case of social activity is quite similar to that of a social goal in that a child might assimilate the language and actions of the other child that constitutes a mathematical activity and then reenact the assimilated activity in constituting the activity as a personal activity.

L: Because I pulled a part from the top and copied it, it was the same as the first one.
T: (Copies a stick from the RULER and places it at the top of the screen.) Can you, Laura, tell me what would be one-seventh of that stick?
L: One-seventh – (Dials PARTS to "7" and clicks on the stick.)
T: Show me the stick that would be one-seventh.
L: Oh! (Activates PULL PARTS and pulls a copy of the left most part of the stick.)
T: Can you prove to me that it is one-seventh of the whole stick?
L: (Activates REPEAT and makes a 7-stick directly beneath the original.)

The way in which Laura "proved" that the part she pulled from the 7/7-stick was one-seventh of the whole stick opens the possibility that her use of PARTS was on a par with Jason's use of PARTS. She was explicitly aware that she could use each part of the five-part stick she made to make another 5-stick because she pulled a part from the stick and then proceeded to make another 5-stick which was, in her words, "the same as the first one." She also explained what she did in reply to the teacher's query concerning why using REPEAT constituted a proof – "Because I pulled a part from the top, and copied it and it was the same as the first one." Although she used the first part of the 5-stick, her use of the indefinite "a" does indicate an awareness that a part other than the first part could have been used as well. Using PARTS to partition a stick and REPEAT to prove that each part was a fair share were not unrelated operations. However, her use of REPEAT seemed restricted to the situation where she was asked to make "proofs" as indicated in the second continuation of Protocol XIII (W stands for a witness).

Protocol XIII. (Second Cont)
W: I want Laura to find one-fifth. (The children had completed making the 6-stick to the 10-stick in Fig. 5.8.)
L: (Counts from the bottom up and selects the 5-stick.)
T: He wants the stick that is one-fifth; the part that is one-fifth.
L: (Browses the cursor along the 5-stick.) This one that is right here.
T: Is this one the one you are going to use? Ok, pull the part that is one-fifth.
L: (Moves the 5-stick to the upper part of the screen.)
T: He wants to see one part that will be the one-fifth.
L: (Moves the 5-stick back to its place and pulls the first of the parts out.)
W: Why is that one-fifth, Laura?
L: Because in PARTS we had put five and putted on the stick. (Meaning she clicked on the stick.)

Laura's understanding of one-fifth is indicated by her comment, "Because in PARTS we had put five and putted on the stick." She apparently thought of "one-fifth" as a plurality of five parts rather than as one single part out of five parts. Moreover, when the witness asked, "Why is that one-fifth, Laura?" rather than ask her to *prove* that it was one-fifth, Laura resorted to explaining that she partitioned the stick into five parts rather than explaining how she could repeat a pulled part five times. So, in retrospect, when Laura made one-tenth and one-eighth of a stick in the section on her simultaneous partitioning scheme, these fraction number words may well have meant "ten" and "eight," respectively.

Laura's Part-Whole Fraction Scheme

Another insight into the nature of Laura's fraction scheme occurred in the teaching episode held on the 3rd of March. The children were posing tasks to each other, and Laura posed the task in Protocol XIV.

Protocol XIV. An incisive question.
L: I am thinking of a stick that has twenty pieces.
J: (Partitions the stick into twenty parts using PARTS, pulls the first part out of the stick using PULL PARTS and then repeats it twenty times using REPEAT.)
T: (As Jason repeats the part.) What was the name of the part that you thought of?
L: The piece?
T: How much of the stick?
L: One-twentieth.
T: So you think the piece is one-twentieth of the whole stick, the red stick?
L: (Nods her head "yes.")
T: I see. And what did Jason do?
L: He repeated it to make it the same amount.

Immediately before Protocol XIV, Laura operated similarly to the way Jason operated in Protocol XIV in the case where Jason posed the task, "I am thinking of a stick that is one-seventh of that one." When she posed her own task, however, she said "I am thinking of a stick that has twenty pieces." rather than "I am thinking of a stick that is one-twentieth of that stick." The actions of pulling a part from the stick and iterating it to "make it the same amount" apparently still were not an integral part of her simultaneous partitioning scheme. She knew what to do to make a stick that is one-twentieth of a given stick after partitioning the stick into twenty pieces (pull parts), and to prove that the stick is one-twentieth of the given stick (repeat), but these actions were apparently separated from the operation of partitioning a stick into twenty pieces. On the several occasions that were observed up to this point where she used PULL PARTS and REPEAT, it was necessary for the teacher to provoke these actions through questions or comments or for her to believe that Jason was going to execute them. So, the proposition that she had constructed a fraction scheme is contraindicated by the way in which she posed the task in Protocol XIV and by her concept of one-fifth in the second continuation of Protocol XIII.

Protocol XV. Using sharing language in an attempt to provoke an accommodation in Laura's simultaneous partitioning scheme.
T: I am thinking of a stick that is the share of one person in a party. Six came to the party.
J: (Takes the mouse and starts to dial PARTS.)
T: Let Laura do it.
L: (Dials PARTS to "6" and clicks on the stick. She then activated the menu ERASE and activated ERASE A LINE rather than ERASE A MARK. She then clicked on the stick and it disappeared.) Ohh!! (In surprise.)
T: (Laughing.) All right, do it again. (While he is reposing the task, Jason makes five copies of the stick in the RULER.)
L: (Partitions a copied stick into six parts using PARTS and, this time, she activates ERASE A MARK. She then erases each of the last four marks from the fifth to the second, leaving the first.) There's one piece left.
T: Can you show me the piece out of the "candy"?
L: (Pulls the part using PULL PARTS.)
T: How much would be the share of this person out of the whole red stick?
L: He will get one-sixth.

The change in the teacher's language from fraction language to sharing language – "I am thinking of a stick that is the share of one person in a party. Six came to the party." – did orient Laura to erase the fifth to the second mark, leaving only the first mark to mark off one part of the stick. However, these actions should be regarded as indicating that the share for one person was one *of* the six parts, not one *out of* the six parts.[24] Consequently, there is no indication of an accommodation in Laura's simultaneous partitioning scheme. But, in three tasks after the one in Protocol XV Laura began to establish a part-whole fraction scheme.

Protocol XVI. A slight modification in Laura's sharing language.
L: I'm thinking of a stick that, um, um, this is a birthday cake, and if they cut it up into five pieces, how much would one – how much would everybody get?
T: Everybody?
L: How many? (Holds her right index and thumb about 1 in. apart.)
J: How big a piece you mean?
L: Yeah.
T: I see Jason agrees with you.
L: Yeah....
J: (Partitions the stick into five parts and pulls one part out.)
L: Yeah, that's right!
T: Can you tell Laura how much is the share this person gets from the cake?
J: One-fifth. (We can see Laura muttering "one-fifth" to herself almost simultaneously.)
L: Yeah.
T: (Speaking to Laura.) That's what you thought of?
L: Yeah. (Nodding her head.)

---

[24] Recall that in one out of the six parts, the child conceptually disembeds one part from the six parts while leaving the part in the six parts. While in one of the six parts, the part is distinguished within the six parts without being disembedded from the six parts.

T: Now I will ask a question – the same thing – people came to the party, the birthday party (changes the people who came to the party to the children's class) but we had all your class. How many kids are in your class? Do you remember?
L: Twenty-two.
J: Twenty-one–twenty-two.
T: And all these kids came to the party! And I want you to show me the share of…eleven people.
L: (Dials PARTS to "11" and clicks on the stick.)
T: Will that be enough for the whole class?
L: No.
J: Out of eleven?
T: Eleven kids – but all the twenty-two came in.
L: OK! (Looks puzzled.) (Erases all marks and then dials to "22" in PARTS and clicks on the stick. She then activates PULL PARTS and clicks on the first part of the stick. She then clicks on the second part, and in the process of pulling the second part, Jason comments "repeat." So, Laura drags the second pulled part into the trash and repeats the first pulled part eleven times, making an 11-stick.) There's eleven people…that can share something.

In the context of sharing a birthday cake, Laura's question "How much would everybody get?" does indicate an awareness of all of the people sharing the cake. But it also indicates an awareness of how much of the birthday cake each person gets. This awareness is indicated by Jason's question "How big a piece you mean?" because that was his interpretation of Laura's question. In that Laura agreed with him, the task Laura posed to Jason constitutes a shift from focusing on the number of pieces of cake to focusing on how much cake each person gets relative to the whole of the cake as well. This is a crucial shift in modifying her simultaneous partitioning scheme into a fraction scheme. This shift is also indicated by Laura's attempt to verify what she meant by "everybody" when she held her right index finger and thumb about an inch apart as she said "How many." Both meanings – all of the people and the amount of each person relative to the whole – are indicated by her clarification.

At this point in Protocol XVI, the teacher was encouraged by this shift and posed the more complex sharing question where he asked Laura to show him the share of eleven of the twenty-two people in the children's classroom. In that Laura initially used PARTS to partition the stick into eleven rather than twenty two parts does indicate a lack of mentally disembedding a connected number eleven from the connected number twenty-two prior to partitioning the stick into eleven parts, and establishing a part-whole relation between the two. Nevertheless, the teacher's question, "Will that be enough for the whole class?" served to reorient her. She promptly exclaimed, "No!" and proceeded to repartition the stick into twenty-two parts and then began to pull out eleven of them. This attempt is a solid indicator that she was in the process of establishing a part-whole fraction scheme where the operations of partitioning a whole into a specific number of parts and disembedding several of these parts from the partitioned whole were assimilating operations of the scheme.[25]

---

[25] Comparing the disembedded parts to the partitioned whole is not indicated in Protocol XVI.

The operations that Jason performed actually using sticks in TIMA: Sticks apparently stood in for operations he could also perform on his re-presentations of his more prosaic situations. His clarifying comment "How big a piece you mean?" is an indication that he could operate equally well in his re-presentations of his ordinary situations and in the situations he created in TIMA: Sticks. Corroboration of this proposition is found when he suggested to Laura to repeat the part she pulled from the 22-stick she made using PARTS.

Establishing Fractional Meaning for Multiple Parts of a Stick

Based on the way both Laura and Jason produced the share of eleven people in Protocol XVI, we proceeded to explore their production of fraction language as a consequence of the functioning of their fraction schemes. The task of Protocol XVII was extracted from the same teaching episode as Protocols XIV through XVI, held on the 3rd of March.

Protocol XVII. Meanings of three-twenty-fourths and six-twenty-fourths.
T: The birthday party was going on when I came into the class. What will be the share that three people get? (There were twenty-four people at the party sharing a birthday cake.) On the cake, show me the piece that, let's say, three of us will get.
J: (After partitioning the stick into twenty-four parts, pulls the first part from the stick and repeats it to make a 3/24-stick.)
L: That's what I was going to do.
T: All right, Laura, if that was what you were going to do, tell me how much the share of three of us together will be out of the whole cake.
L: Three twenty-fourths!
J: (Nods "yes.")
T: So, now, let's say another three people came in. Can you show me the part that will be?
J: Three more?
T: Three more. So we have altogether six. We still have 24 but we want to see the share of six of us.
L: Six of us. (Activates REPEAT and clicks on the 3/24-stick.) Oh! There.
T: That's it?
L: That's it.
T: Explain.
L: Ok. I repeated it, and now there are six people that could get their…. Ok (Points to each piece with the cursor.) 1, 2, 3, 4, 5, 6.
J: (With Laura.) Their share.
T: All right. So how much is this out of the whole cake?
J&L: Six-twenty-fourths.

It is quite important that Jason did not simply partition the stick into twenty-four parts and then fill three of them with a color. Rather, he used the operations of his partitive unit fraction scheme when, after partitioning the stick into twenty-four parts, he pulled one part out and then iterated it three times to produce the share of three people. Nonetheless, whether he regarded the 3/24-stick as three-twenty-fourths *because* it was three times the 1/24-stick is ambiguous up to this point in my analysis. In later teaching episodes, however, I will discuss contraindications. My current interpretation is that Jason's meaning for "three-twenty-fourths" came out of the 3/24-stick as a part of the 24/24-stick, as opposed to coming out of the 3/24-stick as three times the 1/24-stick.

Laura does say, "That's what I was going to do," in referring to Jason's iterative actions. She also knew that the stick Jason made for the share of three of the twenty-four people was three-twenty-fourths of the stick, so I infer that she assimilated Jason's actions using her partitioning and disembedding operations. However, in that it was Jason rather than she who initiated the actions, her comments do not indicate that disembedding and iterating were coordinated operations of her partitioning scheme. It is also important to note that the teacher asked, "All right. So how much is this out of the whole cake?" before either of Jason or Laura said, "Six-twenty-fourths." Moreover, Laura produced what to the observer is a 6/24-stick in order to produce the share of six people rather than to produce six-twenty-fourths of the stick. She also counted the parts of the 6/24-stick to verify that it indeed contained six parts. So, although Laura replicated the 3/24-stick to make the 6/24-stick, whether she was explicitly aware that the 6/24-stick was indeed six-twenty-fourths of the 24/24-stick because it was twice three-twenty-fourths is at issue. She knew that six was twice three, but the 6/24-stick was called "six-twenty-fourths" because it was six out of twenty-four pieces, not because it was twice three-twenty-fourths. This may have been the case for the 3/24-stick as well. That is, Laura constituted the 6/24-stick and the 3/24-stick as fractional parts of the original stick by comparing them to the original stick rather than reasoning that the 6/24-stick was twice the 3/24-stick or that the 3/24-stick was three times the 1/24-stick.

Neither Jason nor Laura had constituted a connected number (a 3/24-stick) as a *fractional number*, which takes its fractional meaning from the fractional part of which it is a multiple. The construction of fractional numbers apparently requires the construction of the splitting operation. The child must first be aware of the connected number (the 3/24-stick) as a composite unit item containing (three) equal units, where the novelty is that the composite unit item could be produced by iterating any one of the three unit items it contains (a whole-to-part relation). In this case, the child would be explicitly aware of the multiplicative relation between the connected number as a composite unit item and any one of its parts. For example, three-twenty-fourths is three times one-twenty-fourth and one-twenty-fourth iterated three times is three-twenty-fourths. This opens the possibility of three-twenty-fourths being considered as a fractional number because its fractional meaning would no longer be directly dependent on its relation to the whole of which it is a part. The relation to the whole would be inferential in that it could be established by means of reasoning of the sort, "This stick is three-twenty-fourths of the whole stick because it is three times one-twenty-fourth of the whole stick." Neither Jason nor Laura independently generated this sort of a reason for why they called the 3/24-stick "three-twenty-fourths" or the 6/24-stick "six-twenty-fourths."

A Recurring Internal Constraint in the Construction of Fraction Operations

The children's construction of fraction operations were both enabled and constrained by their current operations. In the discussion of Protocol VIII and its continuations, my claim was that the children could not take a unit of units of units as material for further operating, and this lack constrained the construction of fraction operations involving a composite unit fraction. The teaching episode containing Protocol VIII had occurred on the 2nd of December and the current teaching episode was held on the 3rd of March of the same school year, so it is interesting to note that the children had still not made progress in producing a composite unit fraction commensurate with six-twenty-fourths. Producing such a composite unit fraction would also constitute an indication that six-twenty-fourths was a fractional number.

Protocol XVII. (Cont)
T: Can you find another name for that piece (the 6/24-stick)?
J: Ah, let's see....(Makes three copies of the 6/24-stick and aligns them end-to-end with the original 6/24-stick directly beneath the 24-stick.) Ah, four-twenty-fourths.
T: It was six-twenty-fourths. And now it's four....
J: Ah, see, there's ah (Moves the last 6/24-stick back and forth.)...Ah, if it's, see, that can do it...
L: Six times four is twenty-four.
T: Six-twenty-fourths. (Apparently, he did not hear what Laura said.)
L: No.
T: (To Jason.) What did you think of?
J: That's six-twenty-fourths.
T: (Asks Jason to erase the marks from one of the 6/24-sticks. He then points to that stick and to the 24/24-stick.) How much is this stick of the whole stick?
J: Four-twenty-fourths.
L: One-twenty-fourth.
T: One-twenty-fourth?
J: One-twenty-fourth, as you put three more there is four-twenty-fourths.

I consider Jason's answer of four-twenty-fourths and Laura's answer of one-twenty-fourth as contraindication that they used the four 6/24-sticks that Jason established to unitize each stick, disembed one of these four singleton units from the four singleton units, and make a one-to-four comparison. These operations would entail operating on a unit that contains four composite units of six length units each; that is, a unit of units of units. The unit structure that Jason made by copying the 6/24-stick three times and aligning those copies along with the original 6/24-stick end-to-end did involve two levels of units – the four composite units containing the four 6/24-sticks and the twenty-four length units of the 24-stick as indicated by his answer of "four-twenty-fourths." This answer corroborates my earlier inference that Jason was operating at two levels of units but not three levels of units. Even after Jason erased the six marks on each 6/24-stick he still compared the four composite units each of six length units with the twenty-four length units of one. At this point, Laura seemed to interpret a blank 6/24-stick as a single length unit as indicated by her answer, "one-twenty-fourth."

Continued Absence of Fractional Numbers

That both children were yet to construct fractional numbers became even more apparent in the teaching episode held on the 31st of March. After spending approximately 6 min of the teaching episode reaching agreement with the children on how to use REPEAT to make a stick so many times longer than a given stick, the teacher asked the children to pose "I am thinking of a stick" situations to each other using elevenths.

Protocol XVIII. Making a stick twice as long as a 3/11-stick.
L: (Draws a stick.) I am thinking of three-elevenths of this stick.
J: (Dials PARTS to "11" and clicks on the stick. He then fills the first three parts.)
L: (Nods her head.) Yeah.
T: (After Laura had agreed.) Now, you want to pull out?
J: (Using PULL PARTS, pulls the first part out of the stick and then using REPEAT, makes a 3/11-stick by repeating the 1/11-stick.)
T: Ok. All right. (Indicates to Jason that he is to pose a situation.)
J: (Draws a stick.) I am thinking of a stick that is six-twelfths. (The teacher asks him to use elevenths.) – That is six-elevenths.
L: (Uses PARTS to partition a copy of the unit stick into eleven parts. She then uses FILL to color the first six parts. She activates PULL PARTS and clicks on each one of these filled parts and pulls them as an intact stick from the 11/11-stick.)
T: Is that correct? (To Jason.)
J: Mm-hmm. (Yes.)
T: All right. Now, I am thinking of a stick that is twice as long as the three-elevenths. (Pointing to the 3/11-stick that Jason had made.)
J: (Makes a copy of the 3/11-stick and joins it to the original 3/11-stick.)
T: (Points to the 6/11-stick that Jason just made.) How much is that one of the original stick? (Points to the endpoints of the original stick with his left thumb and forefinger.)
J: Six-elevenths.
T: How did you know that?
J: Because three plus three is six, and there is eleven of them.
L: (While Jason is explaining, Laura moves the 6/11-stick that Jason just made and places it so that its right end point is aligned above the right end-point of the 6/11 stick she made from the stick Jason drew when he posed his problem.)
T: (To Laura.) What do you say? Is it six-elevenths, and if it is, how do you explain it?
L: Which one? This one?
T: The one I thought of. I thought of something that was twice as long as that one.
L: I think it would be five.
T: Five-elevenths?
L: Five-elevenths.
T: Or five what?
L: (Louder.) Five-elevenths!
T: Why?
L: Because two more than three would be five.

Jason pulling a 1/11-part from the 11/11-stick and repeating it to make a 3/11-stick is, again, solid indication that he regarded the 3/11-stick as three times the 1/11-stick. However, the status of the 3/11-stick that Jason produced is clarified when he justified why two 3/11-sticks was six-elevenths. There, he resorted to a part-whole explanation rather than explain that six-elevenths is six times one-eleventh. A stick was not six-elevenths because it was six times one-eleventh. Rather, it was six-elevenths because it was six out of eleven parts. Hence I infer that he was working with a partitive fraction scheme,[26] but was yet to construct fractional numbers, which would require the construction of the splitting operation.

Laura's way of operating corroborates the inference that her fraction scheme was a part-whole fraction scheme. To make a 6/11-stick using the 11/11-stick, she first filled six parts using FILL and then pulled the six parts from the 11/11-stick as an intact 6/11-stick rather than make a 6/11-stick by repeating a 1/11-stick. In addition, the difference in the fraction schemes of the two children was underlined after the teacher asked them how much the stick Jason had produced (as twice as long as the 3/11-stick) was of the original 11/11-stick. Jason reasoned that it was six-elevenths "Because three plus three is six, and there is eleven of them." During the first 6 min of the teaching episode, it had become evident that Laura repeated the stick twice using REPEAT (this produces a 3-part stick rather than a 2-part stick) to make a stick twice as long as a unit stick. The teacher had attempted to alter her use of REPEAT and she had used it appropriately. But when Laura said that Jason's stick would be five-elevenths "Because two more than three would be five." her comment should be interpreted to mean that, for her, "twice as long" still meant to make two more unit parts beyond the three parts that were already present. This way of operating corroborates that a fraction unit was not iterative for her because, in the case where she was asked to make a fraction twice as long as a unit stick, she did not include the unit stick with the two units she produced to make a stick twice as long as the given fraction unit. So, it is further corroboration that she was yet to construct a partitive fraction scheme.

An Attempt to Use Units-Coordinating to Produce Improper Fractions

In a preceding teaching episode held on the 10th of March, the children used PARTS to partition a stick into eight equal parts, then pulled a 1/8 part from the stick, made five copies and joined those copies to the original pulled part to make six-eighths. This was done in the context of the children producing a fraction language that developed naturally out of their current schemes. The present teaching episode was a continuation of the March 10 teaching episode with the exception that the teacher introduced an intervention that he hypothesized would result in the children producing meaning for conventional improper fraction language. In the current teaching episode, Laura eventually posed the task, "I am thinking of a stick that is eleven-elevenths of that stick!" But neither child posed a task involving an improper fraction, so the teacher intervened and posed the task of Protocol XIX.

---

[26] Recall that a partitive fraction scheme extends the partitive unit fraction scheme in that it can be used to produce proper fractions.

5 The Partitive and the Part-Whole Schemes

Protocol XIX. Using repeat to produce sticks longer than the unit stick.

T: Now, I am thinking of a stick that is twice as long as this six-elevenths. Let Laura do it, you did it last time.
L: (Makes a copy of the 6/11-stick and then repeats the copy once using REPEAT, making a 12/11-stick. She then drags this stick to the end of the 6/11-stick.)
T: Now, I have a question. (Points to the 11/11-stick.) Is that the original one we started with? (Both children indicate "yes.") How much is this one of the....
J: (Interrupting the teacher.) Twice as long as the green one. (The 6/11-stick was green.)
T: The original one?
J: Oh! How much is it? It is ...
L: There is only one left over from this one. (The original stick.)
J: There is eleven, there are twelve pieces and people come to the party and they take eleven, so there is one more on.
T: So, how much is it?
J: So it is eleven, twelve-elevenths!!
T: Twelve-elevenths. (To Laura.) What do you think?
L: I don't know.
T: How did you figure it out? (To Jason.)
J: (Pointing to the 12/11-stick.) There's six, and six plus six is twelve, and there's eleven here. (Pointing to the 11/11-stick.)
T: What do you say? (To Laura.)
L: Yes.
T: (To Jason.) You know what? Make a stick three times as long as this six-elevenths (Points to the 6/11-stick.) And you (Speaking to Laura.) will tell me how much it is.
J: (Makes a copy of the 6/11-stick and uses repeat to make an 18/11-stick.)
T: So, it is three times as long as the six-elevenths. So, Laura, how much is of that one?
L: Bah, bah, bah – eighteen-elevenths!!
T: How did you know that?!
L: Three times six is eighteen.
T: I see! Now you used what Jason explained to you before, to do the same thing with the three?
L: Yes.
T: All right! What if I would ask five times as long as the six-elevenths?
J: Thirty.
L: Yeah, thirty-elevenths.

It was never a goal of the children to produce a fraction greater than the whole before the teacher intervened. So the teacher decided, given that the children were working at the upper boundary of what their fraction schemes made possible, to provoke the children to embed their units-coordinating schemes in their fraction schemes. After the teacher seized upon the moment and asked Laura to make a stick that was twice as long as the 6/11-stick, Laura actually produced the stick by making a copy of the 6/11-stick and then repeating the copy once, making a 12/11-stick. Laura immediately answered that there was only one left over after the teacher asked the children "How much is it?" which indicates that she compared the original 11/11-stick with the 12/11-stick. But, her saying that she did not know if it was twelve-elevenths indicates that she was aware of a twelve-part stick, but each part had lost their status as 1/11 of the original stick. Her choice to say "eighteen-elevenths" after Jason had made a stick three times as long as the 6/11-stick rather than simply "eighteen" apparently was because both the teacher and Jason had used "elevenths" in referring to the 12/11-stick.

In fact, she hesitated before saying "elevenths" when she said "eighteen-elevenths." Nevertheless, she seemed to have learned an appropriate way of acting and speaking in the event that she was asked to iterate a fractional part of a stick. Her ability to do so (she said that "three times six is eighteen" in justifying why she said "eighteen-elevenths") was based on her ability to use a composite unit in iterating, which was inherited from her units-coordinating scheme for whole numbers.

Jason's comment, "So there are twelve pieces and people come to the party and they take eleven, so there is one more" also indicates a comparison between the original stick and the 12/11-stick Laura made. In that he chose to speak in terms of people coming to a party, it seems that his way of thinking in the context of TIMA: Sticks was also his way of thinking about his more or less everyday situations. However, his comment indicates that his thinking is quite similar to Laura's in this situation. So, there seemed to be a lacuna in his reasoning in that he did not regard each of the twelve people as having a part of the original – only eleven.

Nevertheless, his comment "twelve-elevenths" does constitute an independent and creative production of fraction language that was based on the operations he used to produce it – "There's six pieces and six pieces so there's twelve, and there's eleven here." He seems to lack a reversal of the direction of his part-to-whole comparisons, in which, not only does the size of the part receive meaning from the size of the whole, but the size of the whole receives meaning from its iterative relationship to the unit part. He could disembed, from a whole, a part that itself consisted of parts (e.g., six parts from eleven parts). He could also inject the part into the whole after he actually disembedded it from the whole: "Six-elevenths" meant "six parts out of eleven equal parts" *and* indicated how much the six parts were of the eleven parts. But the meaning of "twelve-elevenths" needs to transcend this part-to-whole meaning. This involves what I have called splitting operations because the child has to take the 11/11-stick as a unit containing hypothetical parts each of which can be iterated eleven times to produce the whole. In this way of thinking, a unit fraction (a hypothetical unit part of the 11/11-stick) becomes a fractional number *freed from its containing whole* and available for use in the construction of a 12/11-stick. The multiplicative relationship between the whole and the iteration of the 1/11-stick is indirectly maintained in the construction of the improper (or proper) fraction. This allows the child to inject the whole (the 11/11-stick) into the 12/11-stick with the potential to disembed the whole from what was formerly only considered as a part (in other situations like six-elevenths). Thus, he would be able to restructure what was formerly a part (12/11-stick) into a composite unit containing the original whole unit (the 11/11-stick) and another unit (the 1/11-stick). Upon the emergence of the splitting operation, I regard the partitive fraction scheme as an *iterative fraction scheme* that can be used to produce improper fractions.

A Test of the Iterative Fraction Scheme

This realization that Jason was lacking reversibility of the part-whole relationship was realized only in retrospective analysis. At the time, the children's spontaneous use of improper fraction language led to a working hypothesis that the two children had in fact constructed the iterative fraction scheme. This hypothesis was tested on the spot.

Jason used his partitive fraction scheme in an attempt to make fourteen-eighths and, as a consequence, he made eight-fourteenths instead. Provoking the children's

Protocol XX. Failure to structure fourteen-eighths as eight-eighths and six-eighths.

T: I am thinking of a stick that is fourteen-eighths of that stick. (Points to an 8/8-stick.)
J: (Erases all marks on the 8/8-stick, partitions the resulting blank stick into 14 parts using PARTS.) Fourteen-eighths?
T: Fourteen-eighths.
J: (Fills the first six parts of the 14/14-stick he made using FILL. He then uses PULLPARTS to pull parts from the right-hand side of the stick, eventually making an 8/14-stick.)
T: (To Laura.) How much is it?
L: Eight-fourteenths.
T: I asked about fourteen-eighths and you said this is eight-fourteenths. Why?
L: Because there is fourteen little marks and eight in them.
T: That's eight-fourteenths, she said. (Indicates to Jason that he is to make fourteen-eighths and challenges the children to make it. However, both sit quietly, so the teacher makes a copy of the 14/14-stick and erases marks upon a suggestion by a witness.) Can you think of another way to make fourteen-eighths? (Asks each child to explain a way to make fourteen-eighths to the other child.)
L: (Points to the blank stick.) But that one is eight out of fourteen.
T: (Asks the children if that would make fourteen-eighths, attempting to be nonevaluative.)
J&L: (Sit quietly.)
T: I am thinking of a stick that is seven-eighths of this one. (The blank stick.)
J: (Partitions the stick into eight parts, colors the first seven, activates PULL PARTS and pulls the first filled part from the stick and makes a 7/8-stick using REPEAT.)
T: Now, I am thinking of a stick that is twice as long as this stick. (Pointing to the 7/8-stick).
L: (After several attempts, repeats a 7/8-stick she made into a 14/8-stick.)
T: Is this twice as long, Jason?
J: Yes it is.
T: How much is it?
J: No it is not. (Mutters.) (Says that Laura should have used sevenths and not eighths.)
T: (The teacher clears Laura's marks and asks again if either of them know how much the 14/8-stick is of the original one.)
J&L: I don't know.

use of their units-coordinating scheme to make a stick twice as long as the 6/11-stick to make a 12/11-stick in Protocol XIX did not induce an accommodation in their fraction schemes that would enable them to produce a 14/8-stick. Had the children reenacted the operations they used to make a 12/11-stick to make a 14/8-stick in the original problem of Protocol XX, this would have been solid indication that they had made an accommodation in their fraction schemes in the production of improper fraction language. But there is no indication of such an accommodation.

It is especially revealing that, after the teacher returned to the situation he had used to bring forth the production of improper fraction language in Protocol XIX, neither Jason nor Laura could say that the stick that was twice the 7/8-stick was a 14/8-stick. So, the advance they seemed to make in Protocol XIX was a temporary advancement that appeared to be based on the teacher's directives and their use of their units-coordinating schemes.

Discussion of the Case Study

In the formulation of the reorganization hypothesis, I did not assume that children use their number sequences in the production of continuous units. Rather, my assumption was that children had already constructed continuous units alongside the discrete units of their number sequences. This assumption finds support in the analysis of the development of the unit of length given by Piaget et al. (1960) that I discussed in Chap. 4.

> Unlike the unit of number, that of length is not the beginning stage but the final stage in the achievement of operational thinking. This is because the notion of a metric unit involves an arbitrary disintegration of a continuous whole. Hence, although the operations of measurement exactly parallel those involved in the child's construction of number, the elaboration of the former is far slower and unit iteration is, as it were, the coping [capping] stone to its construction. (p. 149)

The Construction of Connected Numbers and the Connected Number Sequence

Knowing that a unit of length for Piaget et al. (1960) was an iterable unit, and that Jason and Laura had constructed an iterable unit of one in their ENS, I had reason to believe that the operations of partitioning – an arbitrary disintegration of a continuous whole – and iteration would be available to them in the context of continuous units. In fact, Piaget et al. found that three-quarters of the children they studied from 7 years 6 months to 8 years 6 months had attained operational conservation of length, which "entails the complete coordination of operations of subdivision and order or change of position" (1960, p. 114). Hence, I assumed at the outset of the teaching experiment that these operations were available to Jason and Laura, as 9-year-olds. I certainly did not set out with the assumption that we needed to induce these operations in them through their use of their number concepts to make fair shares. Rather, I assumed that these operations would emerge in the context of making fair shares of sticks in TIMA: Sticks and that they would be available to the children as they learned to use their numerical concepts as templates for partitioning unmarked sticks into so many equal parts.

Jason cut a stick into two equal parts using visual estimation in Protocol II and then constructed the equipartitioning scheme. That is, he estimated a partitioning of a whole into four equal parts and checked his estimate through iterating. Hence, the assumption was justified for him. Jason's independent iteration of the part, which he broke off three times in an attempt to find if the part was a fair share, had its origin in the operations that he constructed in the case of continuous quantity, and it would be a misconstrual of the reorganization hypothesis to claim that his iterating the part to find if it reconstituted the whole had its sole origin in his numerical concept, four, even though four was a multiplicative concept for Jason.[27] Nevertheless, I do argue that his number concept, four, was inextricably involved in his construction of the

---

[27] Four being a multiplicative concept means that four is conceived of as four times one of its units.

equipartitioning scheme and that he produced a connected number, four, as a result of iterating the part he broke off.

Laura's attempt to segment a stick into three equal parts in Protocol III is wholly compatible with Piaget et al.'s (1960) analysis of measuring without unit iteration that I discussed in Protocol VIII of Chap. 4. Laura's lack of iteration of length units corroborates Piaget et al.'s (1960) finding that unit iteration in the continuous case lags behind unit iteration in the discrete case because she could iterate discrete units of one. So, my assumption that iterable arithmetical units would imply iterable length units was unviable. It would be possible to interpret Laura's lack of construction of iterable length units throughout the duration of her fourth grade as countermanding the reorganization hypothesis. However, Laura's use of her concept of three as a guide in segmenting activity in Protocol III served as corroboration that her numerical schemes played an integral part in developing fraction schemes. Moreover, in the very next teaching episode, Protocol IV, Laura produced an explicitly nested connected number sequence by iterating a stick that she used as if it were a discrete unit of one. Laura's connected number sequence seemed to be an advancement over her segmenting operations in which she engaged in Protocol III because she used the operations of her explicitly nested number sequence in producing the sequence. However, this explicitly nested connected number sequence did not produce an iterable length unit and she treated the segments she joined together as if they were discrete unit items. The reason for this state of affairs is that the explicitly nested number sequence she constructed was not constructed by means of first partitioning a segment into equal-sized parts, disembedding a part from the partitioned segment, and iterating the part to produce other connected numbers that could be then compared with the original segment in a part-to-whole or a whole-to-part comparison. The explicitly nested number sequence that she did construct was unrelated to her simultaneous partitioning scheme and Laura's segmenting operations on a stick were not reorganized as iterative operations.

In Protocol IX, when Laura drew an estimate for one-tenth of a stick, she did iterate the estimate to produce a connected number ten that she compared with the original stick. In fact, I attributed an *equisegmenting scheme* to her where the operation of the scheme was iterating the part used in segmenting. But this scheme did not include the operation of disembedding a part from the whole stick to be used as an estimate. For Laura, partitioning, disembedding, and iterating did not seem to be parts of the same psychological structure in the case of length units as they were for Jason and she did not construct an equipartitioning scheme throughout fourth grade. The consequences of this will be extensively explored in Chap. 6 using teaching episodes involving her and Jason during fifth grade.

On the Construction of the Part-Whole and Partitive Fraction Schemes

Laura's Part-Whole Fraction Scheme

The similarity in the construction of quantitative operations in the continuous and discrete cases should not be taken to indicate that children's construction of connected numbers by means of partitioning emerges spontaneously. It was very striking that the connected numbers that Jason and Laura produced by using their numerical concepts in partitioning were constrained by their continuous quantitative operations. Laura used her numerical concepts in partitioning, but the connected numbers she produced were constrained by her inability to mentally disembed parts of a partitioned stick. Her lack of disembedding restricted Laura to simultaneous partitioning and segmenting. These latter two operations were involved in her use of MARKS rather than PARTS in TIMA: Sticks to mark off equal shares of a stick (cf. Protocols X and XI). Because Jason's preference was to use PARTS rather than MARKS to partition sticks, I hypothesized that an independent use of PARTS indicates the equipartitioning scheme. Although it is an indication, it is not a prima facie indication because in Protocol XIII, Laura independently chose to use PARTS to eliminate a perturbation she experienced when attempting to partition a stick into two equal parts without also constructing the operations of disembedding and iterating for length units. That is, it is possible to provoke the use of PARTS for children who have constructed only simultaneous partitioning.

Although I did not mention it in the discussion of Protocol XIII and its continuations, Laura's use of PARTS did involve an abstraction. It is no coincidence that her independent use of PARTS occurred in the context of attempting to mark a stick into two equal parts using MARKS. In her attempts to mark the stick into two parts, Laura unitized the two parts and took them together as a composite unit. Making a connected number, two, in this way is an act of reflective abstraction because the act of unitizing the sensory items strips them of their sensory material to create abstract unit slots. I use the word "slot" because, although the current sensory items can occupy these slots, other figurative or sensory items can be assimilated into the slot structure as well. In this case, I consider the involved sensory items to be the segments that Laura established using MARKS and the sensory material that is stripped away to be the length of the segments. The presence of sensory material that constitutes length would now provide an assimilating situation for the projection of composite slot structures into segments. Simply stated, Laura eliminated the need to use segmenting when making a partition of a stick.

Laura had yet another step to take in establishing her part-whole fraction scheme. She still focused on the numerosity of five parts in the second continuation of Protocol XIII rather than on a unit containing the five parts. For one of the five parts to have a fractional meaning, the five parts have to be taken together as a composite unit containing the five parts, the part has to be disembedded from the composite unit while leaving the composite unit intact, and the disembedded part has to be compared to the composite unit. All of these operations were available to Laura in the case of discrete quantity, so when the teacher-researcher in Protocols XV and XVI used sticks as a quantitative item to be shared among so many people, Laura assimilated the sticks using her numerical concepts and operated on them as if they were discrete quantities. This permitted Laura to partition a stick into twenty-two parts, take the twenty-two parts as a composite unit, disembed eleven of the twenty-two parts from the composite unit containing the twenty-two parts, and compare the composite unit containing the eleven parts with the composite unit containing the twenty-two parts (cf. Protocol XVI). So, after Laura had established the disembedding operation for connected numbers, she produced her part-whole fraction scheme.

Jason's Partitive Fraction Scheme

The operations of the equipartitioning scheme include mentally partitioning a continuous quantity while maintaining the resulting parts of the partition as elements of the abstract composite unit (numerical concept) used in partitioning. The operations also include disembedding a part of the partition from the partition and iterating it so many times to establish a connected number to compare with the original. Jason modified his equipartitioning scheme in his construction of the partitive fraction scheme, which I consider as the first genuine fraction scheme. It was a genuine fraction scheme because, after Jason partitioned a stick into so many parts, he was aware that he could iterate any of the parts to produce a stick of length equal to the original. He was also aware that if a stick was called "one-fifth," he could iterate it five times to produce the partitioned stick of which it was a part. Furthermore, if he made an estimate of one-tenth, for example, of a stick, he was aware that if the estimate was accurate he could iterate it ten times to produce a partitioned stick of length equal to the original. So, his fraction unit symbolized a composite unit that contained a sequence of units each identical to the fraction unit; this identity relation between each part of the partition is essential for a unit to be constituted as iterable. However, the iterability of the fraction unit was inherited from the iterability of the unit of one in the discrete case.

The continuation of Protocol XII indicated that Jason had begun to construct operations involved in measurement because he regarded ten-tenths as how long the whole stick was after iterating one-tenth of the stick ten times. But he had yet to construct the splitting operation and the iterative fraction scheme, among other measurement operations, that are based on a unit of units of units. Jason's lack of construction of the iterative fraction scheme serves as contraindication that his operation of iteration of partitive unit fractions was a multiplicative operation. This issue will be investigated further in Chap. 6.

The Splitting Operation

The constructive path for producing the iterative fraction scheme is much more demanding than generalizing assimilation, in which the assimilating structure of an existing scheme is modified. Developing the iterative fraction scheme involves constructing a new operation, the splitting operation, that is qualitatively different from the equipartitioning scheme. In the equipartitioning scheme, partitioning and iterating are operations that are more or less sequentially performed, whereas in the splitting operation, the child's awareness of a multiplicative relation between a whole and one of its hypothetical parts is produced by the *composition* of partitioning and iterating. In other words, they are realized simultaneously. It is this multiplicative relation that transforms partitive unit fractions like what Jason produced into genuine unit fractions.

My current conjecture is that interiorizing the operations of the equipartitioning scheme produces the composition of partitioning and iterating. Such an interiorization produces vertical learning,[28] and I think of it as a metamorphosis of the scheme containing the operations. A metamorphic accommodation is much like the strong form of Piaget's (1980) reflective abstraction:

> Logical-mathematical abstraction...will be called "reflective" because it proceeds from the subject's actions and operations...we have two interdependent but distinct processes: that of projection onto a higher plane of what is taken from the lower level, hence a "reflecting," and that of "reflection" as a reorganization on the new plane. (p. 27)

When projecting and reorganizing operations are already available for a given scheme, it is sometimes difficult to distinguish between a generalizing assimilation and a reflective abstraction. However, when these operations are not yet available for a given scheme, they must be assembled in experiential situations, and the projection from one level to the next, allowing the student to put the action of the scheme "out there" and reflect upon them, may be a protracted process. In these cases, I consider the operation of projection to be set in motion by the interiorization of actions or operations carried out at the experiential level by means of reprocessing completed actions or operations in the service of a local goal.[29] As neither Jason nor Laura constructed the iterative fraction scheme, it was not possible to engage in a retrospective analysis of their case study in search for these acts of interiorization that might have appeared to be temporary modifications in their ways of operating that preceded the reorganization. Jason and Laura's apparent production of improper fraction language in Protocol XIX is the kind of phenomenon I would look for in a student who is on the brink of reorganization, although, in this case, neither student was.

Acknowledgment I would like to thank the editors of the Journal of Mathematical Behavior for granting permission to publish parts of an earlier version of this chapter in this book.

---

[28] "Vertical learning" refers to the reorganization of schemes at a level that is judged to be higher than the preceding level. New ways of operating are introduced that are not present at the preceding level.

[29] See Steffe (1994a) for a model of the interiorization of acts of counting that producethe initial number sequence.

Chapter 6
The Unit Composition and the Commensurate Schemes

Leslie P. Steffe

By the end of his fourth grade year, Jason had constructed the partitive fraction scheme and Laura had constructed the part-whole fraction scheme, and the children had used these schemes to produce fractional parts of a fractional whole. But the children could not use them to produce fractional amounts that exceeded the fractional whole. Nor could the children produce composite unit fractions commensurate[1] with fractional parts of connected numbers such as five-fifteenths as one-third. Further, the children could not produce what I have called fractional numbers. On the basis of these constraints in their constructive activity, a major goal of the teaching episodes with Jason and Laura during fifth grade was to explore whether the children could produce composite unit fractions commensurate with a fractional part of a connected number they had just established. In retrospect, it might have been more appropriate to begin with the goal of bringing forth the splitting operation in Jason and Laura. But at the time of the teaching experiment, I had not yet constructed the concept of the splitting operation, so we relied on our conceptual analysis of fractional equivalence and improper fractions to guide us. We hypothesized that the operational bases of producing a class of equivalent fractions and producing a class of fractions with a constant denominator, while not identical, were on a par with one another in terms of their level of abstraction.

I formed the hypothesis that the differences in the partitioning schemes of the two children would be reflected in the schemes the children constructed to produce fractions commensurate with a given fraction. This hypothesis is related to the hypothesis that the equipartitioning scheme can be reorganized into the *splitting operation* whereas the equisegmenting scheme cannot be directly reorganized into the splitting operation. Recall from Chap. 5 that in equipartitioning, partitioning and iterating are yet to be composed and are enacted sequentially. The results of a mental partition produce a situation that the child uses to estimate an *actual part* of the whole stick. The child then iterates this estimated part to produce a

---

[1] I use "commensurate" rather than "equal" to indicate that the relation is a result of operating: children operate to produce a composite unit containing three composite units each of numerosity five, disembed one of the three parts and establish it as one-third.

partitioned whole that can be used to test whether the estimated part is a fair share. This is quite different than splitting a stick. In splitting, the child produces a *hypothetical stick* that is both separate from and a part of the given stick. This stick is both a result of a possible partitioning and an input for a possible iteration. Given Jason's equipartitioning scheme, I hypothesize that the splitting operation should emerge in his case. I am uncertain about whether the splitting operation will emerge in Laura's case. The lack of iteration in her simultaneous partitioning scheme could prove to be an internal constraint that leads to necessary errors on her part.

The Unit Fraction Composition Scheme

The initial teaching episode with Jason and Laura during their fifth grade occurred on the 25th of October. There were two primary goals of the teaching episode. The first was for the children to reestablish their use of the possible actions of TIMA: Sticks after the summer vacation, and the second was for the children to reestablish their use of their fraction schemes using TIMA: Sticks. In the first 20 min of the teaching episode, the children reestablished the use of COPY, PARTS, DRAW, and LABEL. While they were in the context of showing the teacher all they knew about three-fourths, an entirely unplanned event occurred as a result of the social interactions of the three participants.

Protocol I. Making a fraction of a fraction.
J: (Makes a copy of a 4/4-stick, which he had been using and colors three parts of it. He then pulls these three colored parts out of the 4/4-stick, releases the mouse, and sits back in his chair.)
T: OK, so now you have three-fourths. So now I want to see another way to deal with PULL PARTS and do the same three-fourths.
L: You can make it smaller!
T: Go ahead. I don't see what you mean so let's see.
L: (Takes the mouse and pulls one part from the 4/4-stick.)
T: Now, you can use that one to make three-fourths.
L: (Dials PARTS to "4" and clicks on the pulled part. Following this, she colors three of the four parts and uses PULL PARTS to pull them from the stick.)
T: Wait, wait, wait, wait. Now I want to ask you a question because what you did was so nice! Can you give a name, a fraction name, can you tell me how much this is out of the whole (indicates the unmarked original whole in the RULER)?
L: Three-tenths.
J: (Puts a hand under his chin and thinks) three-sixteenths.
T: Because you have different answers and you are a team, you want to give me one and explain to each other until you get to a solution.
L: Uh-oh! Well, we had ... I don't know...
J: (Points at the 3/16-stick Laura made.) See, if we would have had it in that (Points to each part of the 3/4-stick he made by pulling parts.) Four, four, four, and four – sixteen. But you colored three, so it is three-sixteenths!
L: Oh! I thought you meant the thing we first started with was a ten. (Presumably referring to the 4/4-stick as a tenths stick.)

After Laura pulled one part from the 4/4-stick, the teacher intervened and told Laura to use that part to make three-fourths. His expectation was that Laura would use REPEAT or COPY to make three-fourths of the unit stick. It was a complete surprise to the teacher that she used the ¼-stick to make three-fourths of the one-fourth stick. In retrospect, her making three-fourths of one-fourth instead of iterating the ¼-stick was a confirmation of her part-whole fraction scheme because she did see the ¼-stick as something that could generate three-fourths of the stick.

Jason's Unit Fraction Composition Scheme

When the teacher asked the children to give a fraction name for how much Laura's bar was out of the whole, Laura's answer of "three-tenths" and her explanation: "I thought you meant the thing we first started with was a ten" had no observable basis in the context of the teaching episode, as the children had not made tenths. Jason's answer of "three-sixteenths" and his explanation of his answer was also a complete surprise. It was as if he had constructed new operations in his partitive fraction scheme over the summer vacation.

Jason's comment, "See, if we would have had it in that (points to each part of the 3/4-stick he made by pulling parts) four, four, four, and four – sixteen." indicates that he reversed the operations used in making three-fourths of one-fourth of the stick. Prior to reversing these operations, he would first need to relate the three parts that Laura pulled out of the 1/4-stick back to the original 1/4-stick that she started with to establish the goal of finding how much the three parts were of the original stick. Establishing this goal is a solid indicator of reversibility in his partitive fraction scheme.[2] However, reversibility in his partitive fraction scheme alone would not be sufficient to establish the fractional part of the original stick the three-fourths of one-fourth comprised. Reversing the operations in making three-fourths of one-fourth is also necessary, and it involves recursive partitioning. For a composition of two partitionings to be judged as recursive, there must be a good reason to believe that the child, given a partial result of the composition of the two partitionings, can produce the numerosity of the full result. But this is not all, because the child must also use the partial result of the second of the two partitions (the one that is not fully implemented) in the service of another goal. The importance of this latter judgment is that, to produce the numerosity, sixteen, Jason must have *intentionally chosen* to partition each of the remaining fourths of the original partition into fourths when his goal was to find the fractional part of the original stick the three parts comprised. This amounts to embedding a subscheme in the reversible partitive fraction scheme.

Given three-fourths of the one-fourth Laura made, the expected results of the reversible partitive fraction scheme is to produce a partition of the whole fraction stick whose elements are of the same size as the elements of the three-fourths of the

---

[2] Reversibility of a scheme entails taking the results of the scheme as input for producing a situation of the scheme.

one-fourth. Of course, the discrepancy between the whole stick not being partitioned into these elements and the expectation that it should constitute the perturbation that drives the search for a way to make the partition. But the reversible operations of the reversible partitive fraction scheme are not sufficient to eliminate this perturbation.[3] The only way to partition the whole stick is to distribute the partitioning of each fourth across the three remaining three-fourths as exemplified by Jason. So, the child is left in a search mode if he or she is yet to construct recursive partitioning.

Jason's modification of partitioning to constitute recursive partitioning is nothing other than a novel use of his units-coordinating scheme. His goal to find how much three-fourths of one-fourth would be of the whole stick, when coupled with the visible results of Laura's two acts of partitioning, evoked the productive act of distributing the operation of partitioning (units-coordinating) across the results of the first partition. In other words, Jason took the results of Laura partitioning the whole stick into four parts as input elements for further partitioning into four parts. Although it was Laura who made three-fourths of one-fourth in Protocol I, Jason's actions indicate that he regarded her actions as if they were his own. What this means is that he mentally performed the actions that he observed Laura carry out. So, when the teacher asked, "... can you tell me how much this is out of the whole?" Jason had already produced the three-fourths of one-fourth with Laura.

If the productive thinking in which Jason engaged proves to be more or less permanent, then there would be reason to think of it as a *unit fraction composition scheme*. The goal of this scheme is to find how much a fraction of a unit fraction is of a fractional whole, and the situation is the result of taking a fractional part out of a fractional part of the fractional whole, hence the name "composition." The activity of the scheme is the reverse of the operations that produced the fraction of a fraction, with the important addition of the subscheme, recursive partitioning. The result of the scheme is the fractional part of the whole constituted by the fraction of a fraction (Fig. 6.1).

Corroboration of Jason's Unit Fraction Composition Scheme

Given the results of Protocol I, the teacher now had two goals. The first was to explore whether the change that he experienced in Jason's way of operating with fractions would recur in other situations, and the second was to bring forth a unit fraction composition scheme in Laura as a result of interacting mathematically with Jason.

---

[3] To recreate a fractional whole when given, say, a 3/5-stick where there are no marks on the stick, the child who has constructed reversibility in her partitive fraction scheme can split the stick into three parts and use one of the parts in iteration to recreate the fractional whole.

6   The Unit Composition and the Commensurate Schemes   127

Protocol I. (Cont) Corroboration of Jason's unit fraction composition scheme.
T: (To Laura) Show me more about three-fourths.
J: (While the teacher is talking to Laura, Jason pulls two adjacent parts from a copy of the original 4/4-stick he made, erases the mark separating the two adjacent parts, partitions this 1/2-stick into four parts using PARTS, and fills three of these four parts with a different color.)
L: (Looks at the results of Jason's activity.) Oh, I know another way.
T: (To Laura.) Wait, wait, wait, wait, wait, don't try. That's good, keep it.
J: I got two of these, erased that line, and I ... um ... put four pieces.
T: How much is this one (The 3/8-stick Jason made.) of this one (The unmarked unit stick in the RULER.)?
J: (After approximately 10 seconds) three-eighths.

Jason saying "three-eighths" provides corroboration that recursive partitioning did indeed recur in his independently executed productive activity of making three-fourths of one-half of a copy of the original 4/4-stick. Changing the situation from making three-fourths of one-fourth to making three-fourths of one-half does indicate that he could willfully generate situations of his unit fraction composition scheme. However, he did not seem to know in advance that he had made a 3/8-stick because it took him approximately 10 seconds to say "three-eighths" after the teacher asked, "How much is this one of this one?" Nevertheless, by changing the situation he seemed to have the confidence that he could produce a fraction number word for the stick he made, which is solid indication that his unit fraction composition scheme was an anticipatory scheme.

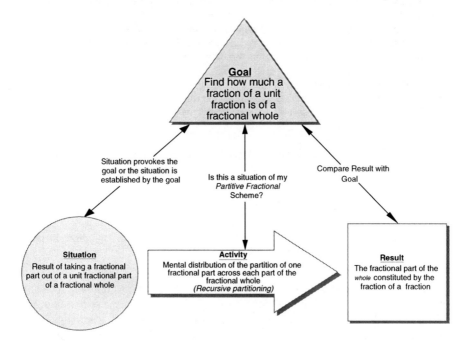

Fig. 6.1. Jason's unit fraction composition scheme.

Laura's Apparent Recursive Partitioning

Laura initially could not say how much the stick that Jason made was of the whole unit stick, so the teacher continued exploring how she might respond to the situation. In that she commented in Protocol I, "Oh! I thought you meant the thing we first started with was a ten." there is indication that she did in fact return to the original stick and make an estimate of how many little pieces would be in the whole unmarked stick. This would constitute a fulfillment of the expectation of her reversible part-whole scheme but without implementing its activity. We know from the teaching episodes with her during her fourth grade that she had a propensity for making such estimates.

Protocol II. An associative link between two schemes.

T: Ok (Looks at Laura in anticipation of her answer concerning the request to show him more about three-fourths.).

L: (Sits quietly for approximately 18 seconds) I don't know, ... am ... am ... am. I don't know. I guess that's right because that came out of there. (Pointing to the four-part 1/4-stick and then to the 4/4-stick.) Those two (Pointing to the four-part 1/2-stick Jason made as if it was still a 2/4-stick.) came out of ... (Pointing to the 4/4-stick.).

J: (Interrupting.) I got two out of this one (Pointing to two parts of the four-part 1/2-stick.), and then we have four of these (Pointing to the four parts of the 4/4-stick.), and two and two, ... (Meaning that there are two parts in each 1/4 of the 4/4-stick for a total of eight parts.) and there's four (Pointing at the parts of the 4/8-stick he made.).

T: (To Laura.) Because I am not sure that I understand Jason right, I want you to explain to me what he said.

L: These two (Points to each one-half of the 4/8-stick Jason made with two fingers, one finger on each half.) came out of here (Points to two one-fourths of the 4/4-stick with two fingers, one finger on each fourth.), and that would be two, four, sixteen. (She partitioned each part of the 4/4-stick into four parts.)

T: But Jason said three-eighths.

L: Oh well, ... Oh, well, ... Oh, OK. See, four and four (Pointing to the first two parts of the 4/4-stick.) and then four (Pointing to the 4/8-stick.), and then these three (Pointing at the 3/8-stick Jason made.) came out of here (The 4/8-stick.).

J: (Makes another explanation.)

L: OK. (Draws a new stick approximately the same length as a 1/8-stick, partitions it into four parts, and pulls three of the four parts in a reenactment of both her and Jason's making three out of four parts.)

J: Oh my God!!

T: Before you try to find out how much this three is of this one (The unmarked stick in the ruler), what is this three-fourths of?

L: (While the teacher is speaking, Laura nods her head toward the screen and subvocally utters number words.) That's four thirty-seconds!!

Laura's reenactment of making three-fourths of a part of the whole stick indicates that she made an analogy between her construction of three-fourths of one-fourth in Protocol I and Jason's construction of three-fourths of one-half. Although Laura initially could not say how much the stick that Jason made was of the whole stick, she rapidly progressed to explicitly counting by fours in distributively partitioning each of the four parts into four parts using her units-coordinating scheme: "See, Oh well, ... four and four ...". Moreover, after she drew a stick approximately the same length as the 1/8-stick, partitioned it into four parts and then pulled out three, her goal seemed to be to find how many parts would be there in the 8/8-stick if she were to partition each eighth into four parts rather than to find how much the three parts were of the 8/8-stick. This is indicated by her answer "four thirty-seconds" rather than "three thirty-seconds." Therefore, like Jason, she was able to determine the number of pieces resulting from a second partition that were contained in the whole.

Whether this distribution of partitioning each eighth into four parts constituted recursive partitioning as explained in Jason's case is not clear. If she was not simply reenacting Jason's actions, however, she may have embedded recursive partitioning through units-coordination in her reversible part-whole fraction scheme. The alternative is that, after making three-fourths of one-fourth and making three-fourths of one-half, the perturbation induced by her unfulfilled expectation of finding how much of the fractional whole she had made focused her on listening to Jason's explanation. She could recognize his explanation because she, too, had established a units-coordinating scheme. In this case, her choice of using her units-coordinating scheme would not be made independently, but as a result of observing Jason using his scheme. So, the relation she established between her units-coordinating scheme and her reversible part-whole fraction would be an association rather than an embedding.[4]

Rather than constructing a unit fraction composition scheme, she may have constructed an associative chain of schemes, where any scheme in the chain was triggered by the results of the scheme immediately preceding. If so, then Jason would be able to independently use his scheme, whereas Laura would be unable to independently use her scheme. To explore this possibility, the teacher asked the children to show him three-fourths of one-half in a teaching episode held on the 1st of November. Jason took the lead in both activity and explanation while Laura sat quietly, so there was no indication that Laura had done more than form an associative link between the two schemes.

Producing Composite Unit Fractions

On the basis of the new possibility that both children could partition a partition, the teacher explored whether the children could engage in the operations necessary to establish four-twelfths as one-third. He was not yet to make the distinction between recursive partitioning and simply partitioning a partition, nor had any other member of the project team. So, whether Laura might establish four-twelfths as one-third was more problematic than we envisioned when planning the teaching episode held on the 8th of November one week after Protocol II. The teacher's plan was to begin by asking the children to partition a stick into 12 parts using PARTS, but under the constraint that they could dial PARTS to any number less than 12, but not 12.

---

[4] By "embedding," I mean that the units-coordinating scheme is contained in the first part of the fraction composition scheme.

Protocol III. Laura's attempt to partition a stick into 12 parts without dialing PARTS to "12."

T: OK. Now, you know what you have to do first, right? We have twelve children come to the party. The RULER will be the store of birthday cakes, so you can always make a copy. Go ahead. Remember that you work as a team. ...
L: So, we are supposed to make the cake?
T: Copy a cake and divide it among twelve kids.
J: (Grabs the mouse and makes a copy of the stick in the RULER. He then dials PARTS to "12.")
T: Can you do that? ... Don't cheat on me!!
L: Can we use PARTS at all?
T: Sure, you can use PARTS, PULL PARTS, everything.
L: (Dials Parts to "11" and clicks on the stick. She then counts the parts, pointing to each part with the mouse cursor.) 1, 2, 3, 4, 5, 6....
J: (Interrupts Laura.) Eleven?
L: (Uses PULL PARTS to pull the last part out of her 11/11-stick and joins it to the end of her 11/11-stick.)
T: But we don't have that cake. You made a cake that we don't have. This is the cake (Pointing to the stick in the ruler.). We would like to add our cake, but we can't!

Laura's action of partitioning the stick into 11 parts and then pulling the last part out and joining it to the 11-part stick is certainly rational, as it served her in making a 12-part stick that could be shared equally among 12 children. But she did not engage in recursive partitioning. Not being allowed to dial PARTS to "12" served as a constraint in her use of PARTS, and this constraint was as close as we could come to establishing a situation that would induce recursive partitioning without stipulating how the children should operate. In contrast to Laura, recursive partitioning was evoked in Jason by the constraint.

Protocol IV. Jason's attempt to partition a stick into 12 parts without dialing PARTS to "12."

J: (Grabs the mouse while the teacher is talking.) Heyyy...! (Drags the stick Laura made into the TRASH and makes another copy of the stick in the RULER. He then dials PARTS to "3" and clicks on the ruler. He then dials PARTS to "4" and clicks on each of the three parts of the 3/3-stick, making a 12/12-stick.)
T: Can you explain to me as a child in this party why we would now have the same, an even piece for each one?
L: He had three pieces and he added four in each thing.
T: What makes it even? Tell me more because I am not sure I understand. Sounds good to me, but I'm not sure that.... Can you pull the piece of one kid out?
L: (Takes the mouse and pulls the first part of the 12/12-stick out from the whole stick.)
T: Can you tell me what fractional part of the cake is that one?
L: (Immediately.) One-twelfth!
J: (Following Laura.) One-twelfth.

Jason's decision to partition the stick first into three parts and then each part into four parts is corroboration of the inference that he had constructed recursive partitioning operations. His manner of operating was distinctly multiplicative in that he seemed to be aware simultaneously of the result, 12, of his partitioning activity, and of 12 partitioned into three composite units each of numerosity four.[5] He seemed to be aware not only of this structure, but also of the operations that produce it – first partition the stick into three parts and then each of these parts into four parts. In that both children said that one of the 12 parts Jason made was one-twelfth of the stick, he continued on with his plan to explore whether the children could produce one-third as commensurate with four-twelfths.

Protocol V. Explaining why one-third is commensurate with four-twelfths.

T: Can you pull out... four pieces?
J: (After dragging the 1/12-stick into the TRASH, pulls four parts from the stick using PULL PARTS.)
T: How much of the whole cake is the share of the four kids?
J: (Immediately.) Four-twelfths.
T: Can you explain that to me?
J: There's twelve pieces and there are four that's colored (Jason had colored the four pieces he pulled out.).
L: (Talking with Jason.) Twelve pieces and four colored.
T: OK, now we are coming to a problem. Can you measure it and see if what we have is four-twelfths?
L: (Takes the mouse and uses MEASURE to measure the 4/12-stick using the stick in the RULER. While she is measuring, she says "one-twelfth" as a guess of what will appear. Jason guesses "three-fourths." "1/3" appears in the NUMBER BOX.)
J&L: (Both are surprised that "1/3" appeared.) One-third?! (Almost simultaneously.)
J: (Almost immediately.) Oh, oh, I see that!! (Grabs the mouse.)
L: (As Jason grabs the mouse.) Oh, I do too because first he, first he (Pointing to the 12/12-stick excitedly.) First, he put it in three pieces and he made four in each thing (Pointing three times at the 12/12-stick from the left to the right in designation of "thing.") so four, that would be one of the four (Pointing back and forth between the 12/12-stick and the 4/12-stick), one of the four pieces, ah, that would be one of the three pieces (Making brackets with her hands as if there is something in between them.).
J: (As Laura finishes her explanation, activates REPEAT and repeats the 4/12-stick twice, making a 12/12-stick.)
T: OK!! All right, that's very nice, so you did it, you gave me a very good explanation. You gave me one (Points to Laura.), and you gave me another one (Points at Jason.). (To Laura.) Did you see what Jason did?
L: No.
T: When you were talking, explaining to me, and that was marvelous, he took the first piece, what did you do, Jason?
J: Repeated it.
T: How many times?
J: Two.
T: So, altogether, how many do you have?
J: Three. And that one was like, purple, and I put four, four, four, and that equals twelve.
T: (To Laura.) Which is exactly what you said, isn't it?
L: (Before the teacher addressed her, she was looking around the room, and gave no indication of understanding what Jason said other than a little head nod.)
J: (Continues on explaining.) And we colored, like we colored four, and its one-third.

---

[5] This may indicate that Jason had constructed operations that produce three levels of units as assimilating operations.

Given that we had had little indication of Laura's recursive partitioning, it was a surprise that she explained why "1/3" came up in the NUMBER BOX after she measured the 4/12-stick using the stick in the ruler. Her comments – "First, he put it in three pieces and he made four in each thing ... so four, that would be one of the four ... one of the four pieces, ah, that would be one of the three pieces." – do indicate that she established three composite units with four elements in each in her regeneration of Jason's partitioning activity and that she regarded the piece Jason made as one of those three composite units.

When "1/3" appeared in the NUMBER BOX, it was surprising to her and it is no exaggeration to say that she experienced perturbation. Her explanation was brought forth by this perturbation and served to eliminate it. In the process of elimination, her regeneration of Jason's partitioning activity solidly indicates that she could indeed perform the operations involved in partitioning a partition. Not only does it indicate performance of these operations, but it also indicates that she created a unit structure using the result of the repartitioning – a composite unit containing three units of four units each – and that this unit structure was within her awareness. So, it is possible that she made an accommodation in her concept of one-third as I explain below.

Her goal was to explain why the 4/12-stick could also be one-third of the 12/12-stick. We know that, for Laura, "one-third" meant to partition a stick into three equal parts and then disembed one part from the three parts and compare the part to the whole. I assume that "1/3" activated this scheme of operations and she used it in regenerating Jason's actions. Her saying, "First he put it in three pieces...." indicates that she regenerated Jason's act of partitioning the stick into three parts using the partitioning operations in her concept of one-third. Further, her comment "he made four in each thing (Pointing three times at the 12/12-stick from the left to the right in designation of 'thing'.)" indicates that she regenerated Jason's actions of distributing partitioning into four parts across the now given three abstracted unit items. This seemingly produced a novelty in her concept of one-third in that she could now regard the *composite* units of four elements of the 12/12-stick as a third, as indicated by her making brackets with her hands as if something was in between them. So, not only did she regenerate her experience of Jason making the second partition, but she also united each of the four elements produced into a composite unit, an operation that she was quite capable of performing. Finally, her comment "that would be one of the four (Pointing back and forth between the 12/12-stick and the 4/12-stick.), one of the four pieces, ah, that would be one of the three pieces" indicates that she disembedded one of the composite units from the three composite units and compared it with the others.

So, there were two novelties in Laura's use of her concept, one-third. First, she was able to assimilate the situation of composite units of four into her fraction scheme as if these composite units were units of one. Second, she used her concept of one-third in monitoring her activity as she inserted units of four into each of the three units she produced when regenerating Jason's partitioning the stick into three pieces. What monitoring means in this case is that she seemed explicitly aware of three composite units of four, and further that she was aware of focusing her attention on the four unit items within each of the three composite units and then shifting her attention from these four items to the three composite units. That she did indeed make checks is indicated by the self-correction: "that would be one of the four (Pointing back and forth between the 12/12-stick and the 4/12-stick.), one of the four pieces, ah, that would be one of the three pieces (Making brackets with her hands as if there is something in between them.)."

She was aware of two levels of units, and could deliberately shift her attention between the two levels as it suited her purpose. But this does not complete the account of her monitoring, because she used her concept of one-third to structure the situation as one-third, and in doing so, she made checks to be sure her operations with composite units were also operations for making one-third. That is, her actions of regenerating Jason's actions fed back into her concept of one-third in the process of regenerating them. If Laura's ability to monitor her making of one-third was a permanent modification, then she should be able to operate similarly with other unit fractions.

Jason's repeating of the 4/12-stick also indicates that he used his concept of one-third in reconstituting the 4/12-stick as one of the three composite units. This is the first time that Jason seemed to be explicitly aware that if a composite part can be repeated three times to reconstitute the whole, then the part is one-third of the whole. He seemed to be aware *before* he repeated the 4/12-stick to make the 12/12-stick that repeating the former three times could produce the latter. This is indicated by his actions of repeating the 4/12-stick as well as by his answer, "Three. And that one was like, purple, and I put four, four, four, and that equals twelve." after the teacher asked, "So, altogether, how many do you have?" He used this result in making his judgment that the 4/12-stick was indeed one-third, as indicated by his comment, "And we colored, like we colored four, and its one-third."

If Jason was aware that the 4/12-stick could be repeated three times to make the 12/12-stick before he repeated it, then this would indicate that he had established four as an iterable unit without needing to actually engage in the operations that produce this structure, i.e., he would need to take a unit of units of units as a given. If he had indeed constructed the ability to take the results of actually making a unit of units of units as a given, this would constitute a major advancement in his numerical operations. In retrospect, his use of recursive partitioning in the first continuation of Protocol IV is an indication that he could take the structure of a unit of units of units as a given prior to operating.

Laura's Reliance on Social Interaction When Explaining Commensurate Fractions

Both children could assemble the operations that were needed to explain why a 4/12-stick can be also a 1/3-stick given the computer readout "1/3" after measuring the 4/12-stick. The difference in the two children's explanations is that Jason was the one who independently partitioned the stick first into three parts, then each part into four parts, to produce a 12-part stick. Laura, on the other hand, mentally reorganized the perceptual material that was available to her into three composite units each of numerosity four in a reenactment of Jason's actions. Moreover, Jason's explanations of repeating the 4/12-stick three times seemed to carry little significance for Laura. There was no indication that she assimilated Jason's iterative actions and his explanation of them, which is compatible with the inference that she could not take a unit of units of units as a given prior to operating. Given the differences in the two children, the expectation was that Laura would be restricted to regenerating actions such as those carried out by Jason in her explanations, whereas Jason would be able to independently generate explanations. In Protocol VI, still from the November 8 teaching episode, we see the necessity of social interaction in Laura's explanations.

Protocol VI. The role of social interaction in Laura's explanations.
T: You start with four pieces. That's always the way that you start.
L: (Dials PARTS to "4" and clicks on a copy of the stick in the RULER.)
T: Now, Jason, think of a number of kids that came to the party, and ask Laura to make this.
J: Um, twenty-four.
L: Twenty-four (Nods her head "yes.").
J: No, sixty-four! (Playfully.)
T: No, just wait a little bit with the big numbers. We will make up to forty.
J: Thirty-nine.
L: OK. That's not an even number.
T: (To Jason.) You always have to be able to make it, OK?
J: Oh....
T: Think of something that you will be able to make and then ask her, OK?
L: Make it twenty-four.
J: Thirty-eight.
L: Thirty-eight? Thirty-eight? Ah, four and then thirty-eight would be...
J: (While L is thinking out loud, with his eyes upward he reconsiders his choice.) Thirty-six!
L: Thirty-six. How many fours in thirty-six? (Sits looking into space with her eyes fixed upward and with her hands crossed.)
J: I know it! I know it! I know it!
L: (Mumbling.) Times eight is thirty-two, and four times nine will be...thirty-six. Ha! (Takes the mouse.)
T: Before you make it, will you make different toppings, please (i.e., color the parts of the 4/4-stick different colors.)?
L: Oh, yeah. (While she fills the parts of the 4/4-stick with different colors, Jason sits and watches.)
L: OK. (Dials PARTS to "9" and partitions each part of the 4/4-stick into nine parts.) Nine, nine, nine, and nine.
T: (To Jason.) Is that what you thought of?
J: (Nods.)
T: So, how much is the share of one kid?
J: One ... thirty-sixth.
L: (As Jason is speaking, subvocally utters.) One thirty-sixth.
T: Is that right?
J: (Grabs the mouse.)
L: Four thirty...
J: (Uses REPEAT to make a 4/36-stick using the 1/36-stick. He then measures the 4/36-stick using MEASURE and "1/9" appears in the NUMBER BOX.)

T: Whoops!
L: (Sits and smiles weakly.)
J: One ninth!? Oh, four, four, four, four, four…(To Laura.) Oh, four times nine is thirty-six.
L: Yeah, I got it now. Four, and four, and four, and four, and four, and four, and four, and four, and four. (Pointing to the 36/36-stick from the left to the right nine times in synchrony with uttering each "four.") So, that's one (Pointing to the 4/36-stick Jason made.) out of all those fours!

After Jason finally decided that the cake was to be shared among 36 kids, Laura independently incremented one more beyond eight and four more beyond 32 to produce 36 as one four more than 32. Her reasoning demonstrates her ability to engage in independent and productive thought when using an anticipatory scheme, in this case her units-coordinating scheme. Her strategy, when coupled with her question "How many fours in thirty-six?" indicates an explicit intention to find out how many parts each of the four parts of the stick would need to be partitioned into to make 36 parts. Note that the wording of her question implies that the units she used in units coordination were interchangeable.

After "1/9" appeared in the NUMBER BOX, Jason experienced what appeared to be a strong perturbation, as he expected "4/36" to appear. He restructured the situation into "four times nine is thirty-six." and used this as an explanation to Laura for why "1/9" appeared. Laura, however, sat and smiled weakly, suggesting that she did not understand how "1/9" could appear. Jason's explanation "Oh, four times nine is thirty-six." was assimilated by Laura using her units coordinating scheme, as indicated by her comment, "Yeah, I got it now." and by her subsequent explanation of what she "got": "Four, and four, and four, and four, and four, and four, and four, and four, and four (Pointing to the 36/36-stick from the left to the right nine times in synchrony with uttering each "four".). So, that's one (Pointing to the 4/36-stick Jason made.) out of all those fours!"

This use of her units coordinating scheme differed from making a units coordination using discrete quantity. Her meaning of "four times nine" was the partitioning activity in which she engaged, and her meaning of "thirty-six" was the numerosity of the result of the partitioning. But the result was not simply 36 parts. Rather, it was nine composite units each containing four elements, and she seemed to be explicitly aware of this result as indicated by her saying, "So, that's one (Pointing to the 4/36-stick Jason made.) out of all those fours!" She definitely treated the fours as if they were singleton unit items. Notice, though, that her explanation again relies on making a part-whole comparison; unlike Jason, she does not establish four as an iterable unit.

Whether Laura would have independently produced her explanation for why "1/9" appeared in the NUMBER BOX without Jason's explanation is doubtful because she did not seem to use her production of nine as the number of parts into which each of the four parts of the original stick was partitioned in her formulation of an explanation. Instead she seemed stymied until Jason intervened with his explanation. Nevertheless, her recognition of Jason's explanation and her translation of it into a justification of her own does indicate that at that moment she did restructure the situation she produced as "one out of all those fours." Whether this recognition alone would permit her to explain why 4/36 is 1/9, without a prior explanation by Jason, is yet to be explored. In Jason's case, whether he could independently generate the goal to find another way to think of 4/36 without measuring is also yet to be explored. Still another issue is whether the children regarded 1/9 and 4/36 as commensurate fractions and, if so, whether they could transform the former to the latter. I explore these issues further in my analysis of the teaching episode that occurred a week later.

Further Investigation into the Children's Explanations and Productions

Protocol VII, held on the 15th of November, can be used to interpret whether Laura had learned to make explanations independently of Jason's explanations.

Protocol VII. Establishing three-fifteenths as commensurate with one-fifth.
T: Fifteen kids came to the party. Start with three cuts on the birthday cake.
L: (Takes the mouse and dials PARTS to "3" and clicks on a copy of the sick in the RULER. She then colors the parts each a different color, dials Parts to "5," and clicks on each of the three parts.)
T: (To Jason.) Do you agree with that?
J: (Nods.)
T: What would be the share for three people?
L: (Immediately.) Three-fifteenths!
J: (In agreement with Laura.) Three-fifteenths. (He then makes a 3/15-stick by repeating a 1/15-stick that he had pulled out.)
T: I say it would be one-fifth!
J: (Looks intently at the sticks in the computer screen.) I agree.
L: (Sits and looks intently at the sticks in the computer screen for approximately 15 more seconds.) I don't agree.
T: (Asks Jason whether it is three-fifteenths or one-fifth.)
J: It's one-fifth.
T: It's one-fifth?
L: I don't know.
T: (Asks the children to explain why it could be both.)
J: (Points to the 3/15-stick.) There's five of these... (Takes the mouse and pulls a copy from the 3/15-stick and then makes copies of this copy. He then aligns four copies with the 3/15-stick directly beneath the 15/15-stick.)
L: I get it!
J: Counts the copies of the 3/15-stick in the row, pointing to each in synchrony with uttering.) 1, 2, 3, 4, 5. (Explains to Laura that these sticks go five times into the original 15/15-stick.)
T: (To Laura.) You said it was three-fifteenths. (To Jason.) You said it was one-fifth. It's both?
J: It's kind of both!

Similar to the way she could make a partition of a four-part stick into 36 parts in Protocol VI, Laura made 15 cuts given three cuts by partitioning each part of the 3/3-stick into five parts using her units-coordinating scheme.[6] She did independently say, "three-fifteenths" for the share of three people. However, from that point on, she did not produce an explanation for why the 3/15-stick could also be a 1/5-stick independently of Jason's explanation. After Jason aligned five 3/15-sticks end-to-end beneath the 15/15-stick in explanation for why a 3/15-stick could also be called "one-fifth," only then did Laura say, "I get it!" She was not simply waiting for Jason to make an explanation. To the contrary, she said "I don't know," when pushed by the teacher, which implies that she genuinely did not know why the teacher referred to the 3/15-stick as "one-fifth." Had Laura been able to make an explanation independently of Jason's explanation, this would have constituted a solid indication that she could use her concept of one-fifth to restructure three-fifteenths because she had partitioned the stick first into three parts, and then each one of these parts into five parts rather than the other way around. Consequently, she had no immediate past experience she could use as input for using one-fifth to restructure three-fifteenths.

Laura was stymied, so the teacher posed a task where Laura could choose the number of fifteenths that she would use in explanation. Rather than pull out five parts as the teacher expected, Laura pulled out three parts, so the task served as an occasion to test whether Laura would independently recognize the three parts as one-fifth. Jason, however, concentrated on pulling out five parts, so the task served as an occasion for Jason's independent production of five-fifteenths as one-third.

Protocol VII. (Cont)

T: (Posing another task.) Take a number of parts from up here. (The original 15/15-stick.) Fifteenths. A number of fifteenths. Take it out so here. (The NUMBER BOX.) We will get another number.

L: (Fills the first three parts of a 15/15-stick and then measures them after pulling them out. "1/5" appears in the Number Box.) Oh! (Looking very disconcerted.)

J: I got it!! (Colors two more parts of the 15/15-stick and pulls them out, making a 5/15-stick. He then colors the remaining ten parts of the 15/15-stick with two different colors, five one color and five another color. He then starts to measure the 5/15-stick.)

T: (Taking the mouse from Jason and laughing.) Don't measure! Don't measure! What were you thinking about?

L: Three-fifteenths!

T: (To Jason.) Three-fifteenths?

J: One-third.

T: (Asks Jason to explain to Laura.)

J: (Points to the 5/15-stick.) There's five – three of these and that's just one.

L: So that's five-thirds!!

J: (The teacher further pursued why the 5/15-stick could be both five-fifteenths and one-third.) Yes, but there are three groups, and one is fit into the other three!

---

[6]This should not be confused with recursive partitioning.

After the teacher asked the children to take a number of fifteenths out of the 15/15-stick so that they would get a different number in the Number Box, Laura interpreted his request using the 3/15-stick she had just experienced. When she measured it and "1/5" appeared in the Number Box, she looked very disconcerted. "One-fifth" did not seem to be what she expected to appear, even though she had just said, "I get it!" in the first part of Protocol VII. Jason independently produced one-third as commensurate to the five-fifteenths that he made. His comment, "There's five, three of these and that's just one." indicates that he a priori made five-fifteenths because he knew that 15 could be structured into a composite unit containing three units of five. For this reason, I infer that he had constructed an equipartitioning scheme for connected numbers greater than one that was on a par with his equipartitioning scheme for the connected number, one.[7] The inference that his composite unit of five was an iterable unit is based on not only the above comment, but also on his actual iteration of a 4/12-stick in protocol V. Apparently, recursive partitioning and the unit fraction composition scheme are also based on the child's construction of a unit of unit of units as a structure whose units can be used as material in further operating. Laura's lack of an explanation in Protocol VII for why three-fifteenths also could be one-fifth, when coupled with both her disconcerted look in the continuation of Protocol VII when "1/5" appeared in the NUMBER BOX and her statement that a 5/15-stick was five-thirds, constitutes contraindication that she had constructed this unit structure as a structure that she could take as a given for further operating.

Producing Fractions Commensurate with One-Half

The difficulty Laura experienced in the teaching episode held on the 15th of November in independently explaining why a 3/15-stick could be also a 1/5-stick served as a constraint for the teacher. He realized that Laura did not engage in the operations necessary for generating an explanation regardless of his attempts to bring those operations forth. So, in the teaching episode held on the 22nd of November, he decided to change the situations of learning to explore those accommodations that Laura might make in her part-whole fraction scheme, which would enable her to transform a unit fractional part into a commensurate fractional part. Because of the intuitive nature of one-half, the teacher decided to start with one-half of a cake and ask the children to take a piece of the cake that would be the same size as one-half of the cake but which was a different fraction name. Protocol VIII begins with the children engaged in the first task of the teaching episode in which Laura had already partitioned a stick into two parts, pulled one part, and labeled it "1/2." Most of the sticks mentioned in Protocol VIII were copies of a stick in the RULER, which the children pretended was the uncut cake.

---

[7] This scheme is referred to as an equipartitioning scheme for connected numbers.

6 The Unit Composition and the Commensurate Schemes 139

Protocol VIII. A plurality of fractions commensurate with one-half.

T: What will be the next one that will be one-half of the cake, but it will be another fraction?
L: What am I supposed to do? (She then drags an unmarked copy of the stick in the RULER beneath the 1/2-stick she made. The 1/2-stick is below the 2/2-stick.)
T: Take a piece that will be the same size as the one-half, but it will be a different fraction. You remember that last time we had three-fifteenths that was also one-fifth?
L: (Moves the copy around with the cursor.)
J: I think I know how.
T: (To Laura.) Look and see what he is doing.
J: (Dials PARTS to "4" and clicks on the unmarked stick beneath the 1/2-stick. He then pulls two parts out of this 4/4-stick, making a 2/4-stick.)
T: That's beautiful!! How would you label it?
J: Two-fourths.
T: Go ahead, label it.
J: (Activates LABEL and labeled the 2/4-stick "2/4.")
T: So, now we have one-half and two-fourths. Another member of the family, please?
L: (Grabs the mouse and makes a copy of the stick in the RULER.) Do we only make one-half?
T: I didn't hear you.
J: I know four of them.
L: (Dials Parts to "5.") Do we have to put one-half right now (that is, begin with one-half.)?
T: The family we are after is the one-half family. (An evasive answer.)
J: I know more than six!
L: Can I put ten on here? (The dial of PARTS)
T: Go ahead, as long as you can take a piece that is like one-half, that's OK.
L: (Dials PARTS to "10", clicks on the stick she copied, drags the 2/4-stick Jason made to the vicinity of the 10/10-stick and erases the mark on it. She then drags the resulting 1/2-stick directly over the 10/10-stick with left endpoints aligned. She then looks at the teacher as if done.)
T: Can you pull out the part that you think is one-half?
L: Yes. (Using the 1/2-stick as a guide, she pulls a 5/10-stick from the 10/10-stick.)
T: So, how much is it of the big cake?
L: Five-tenths. (She then labels the stick "5/10" using LABEL.)
J: (As Laura is labeling the 5/10-stick.) I know a lot of them. There are more than six! I can do it for 100, 50, ...
T: (To Laura.) How did you find five-tenths?
L: Because I know that one-half of ten is five.

Jason's comment, "I know a lot of them. There are more than six! I can do it for 100, 50, ..." solidly indicates that he eliminated the necessity of actually carrying out a recursive partitioning to make fractions commensurate with one-half. He posited possible partitionings and focused on the numerosity of the parts of those partitionings. Initially, however, Laura did not know what to do – "What am I supposed to do?" – and generated five-tenths only after Jason made two-fourths. After Jason made two-fourths, she seemed to establish a goal but it certainly did not involve recursive partitioning. Rather, her goal seemed to be related to Jason's partitioning a copy of the unit stick into four parts. Her question, "Do we have to put one-half right now?" and her subsequent act of partitioning the stick into ten parts together indicate that her goal was not to recursively partition the 2/2-stick.

Rather, it was to find a number (ten) that she could partition into two equal parts – "I know that one-half of ten is five!" Laura abstracted no general way of operating. Rather, she selected a number of which she could find one-half. Laura did independently produce five-tenths after the teacher asked the children for "another member of the family." At this point, my hypothesis is that her search for other such numbers might yield such specific numbers, but she would abstract no general way of operating to produce the numbers, other than selecting those for which she knew she could find one-half.

Following Laura's lead, in the next task of the teaching episode, Jason made an 18/18-stick and then pulled nine parts of the stick out to make a stick commensurate with one-half. He gave the same type of explanation for why he chose 18 that Laura gave for why she chose ten. Laura then, quite strongly, asserted that she knew one, and chose 48. However, she did not know what one-half of 48 was and resorted to counting from one in order to segment 48 into two equal parts. After a long pause, Jason asserted he got it, but the teacher never asked him for his answer or how he arrived at it. Upon the teacher's suggestion, Laura dragged a 1/2-stick directly over the 48/48-stick and used it as a guide to pull out 24 parts. She then said that was twenty-four forty-eighths and used LABEL to label it "24/48." While she was completing her labeling activity, Jason said, "I know a big one!"

Protocol IX. Jason using recursive partitioning to make a 200/200-stick.

J: I know a big one! (Copies a stick from the RULER and activates PARTS.) How many does this one go to (PARTS dials only to "99.")?
T: Let's see if you remember.
J: (Groans.)
T: How many do you want?
J: Two-hundred!
T: Well, think of a way to make two-hundred! I saw both of you doing two-hundred and five-hundred!
J: (Dials Parts to "50" and clicks on the stick. He then dials Parts to "4" and clicks on each one of the fifty parts of the 50/50-stick.)
T: (To Laura.) What do you think he is going to do (While Jason is dialing "4.")?
L: I don't know…
T: (To Laura.) How many parts do you think he is going to have? Is it going to be two-hundred?
L: Yeah, because fifty times four is two-hundred.
T: Can you think of a way to make two-hundred that will not take you so long to do it?
L: (After a long pause.) I know a way to make one hundred. That will be fifty and fifty and fifty and fifty.
T: How much did you say?
L: Two-hundred.
T: Well, then try it.
J: That's what I did.
T: Yeah. But can you think of a way to make two-hundred instead of making all these fours again, again, and again?
J: I know how. (Makes a copy of the stick in the Ruler, dials Parts to "4" and clicks on the stick. He then dials Parts to "50" and clicks on the first two parts, but is interrupted by the teacher.)

6 The Unit Composition and the Commensurate Schemes 141

T: Stop here, because you have the half and if you are going to do that whole thing, you are going to have to click how many times?
J: Four!
T: After you finish, you will have two-hundred pieces. How many pieces would you take out in order to have a half of two-hundred?
J: A hundred.
T: Are you going to do that?
J: (Shakes his head "no.")
T: Can you think of a way to take one-hundred pieces?
J: (Cuts the stick at the midpoint.)
T: That's good. (Asks Jason to label it. Jason creates the label "100/200" using LABEL.)
T: (To Jason) You said earlier that you know of six, and then you said that there are a lot more than six. How many do you think there...?
J: There's like six hundred, something like that.
T: Why do you think there are so many?
J: There are many numbers that you can put ... No there are more than six hundred! One-hundred thousand or something like that.

Jason's choice of two-hundred and his partitioning the stick into 50 parts and then each of the 50 parts into four parts after he realized that he could not dial PARTS to "200" indicates the power recursive partitioning held for him. Had he first partitioned the stick into four parts and then each part into 50 parts, it would not have been as dramatic as the other way around because that could be based on knowing that fifty plus fifty is one hundred, so fifty plus fifty plus fifty plus fifty is two-hundred. In fact, Laura made two-hundred in just this way, but there was yet no indication that she engaged in recursive partitioning. The power of his abstraction is illustrated by his comments in the last part of Protocol IX. He even thought there were more than six hundred commensurate fractions – "One-hundred thousand or something like that."

For Jason, one-hundred thousand was a sort of unbounded number. Thus, his recognition that there are one-hundred thousand fractions commensurate with one-half indicates that he had established the existence of a rather unbounded plurality of such fractions. His number sequence was apparently evoked as he made particular partitions of the 2/2-stick because he no longer needed to carry out the operations of recursive partitioning to produce fractions commensurate with one-half. He was now operating symbolically, taking for granted the results of recursive partitioning, which is a crucial step for him to establish a class of fractions commensurate to one-half.[8] Although it is unclear whether Jason actually constructed this class, he was aware of a plurality of fractions commensurate with one-half, which is quite different than producing just a few such fractions.

---

[8] At this point, it is possible that I should use "equal" instead of "commensurate" to indicate Jason's abstraction. If he could do the same for other fractions, including nonunit fractions, I would be compelled to use the term "equal."

Producing Fractions Commensurate with One-Third

Laura engaged in independent mathematical activity in Protocol VIII and in the task following that protocol when it was her goal to produce numbers she could take one-half of. She could anticipate partitioning numbers into two parts, so the teacher chose to explore how she might operate in the case of one-third in the teaching episode held on the 29th of November. The teacher was also interested in how Jason would operate in the case of one-third. Would he establish a plurality of fractions commensurate to one-third, as he had for one-half?

Protocol X. Fractions commensurate with one-third.
T: Now, please start with a three-thirds stick. Make me a three-thirds stick.
L: (Makes a copy of a stick that is in the RULER and partitions it into three parts.)...
T: I want another fraction, another fraction that will be like one-third.
J&L: (Sit quietly for approximately twelve seconds.)
J: (Jason takes the mouse and dials Parts to "5" and clicks on each of the three parts, partitioning each into five equal parts. He then activates PULL PARTS and pulls out the first three parts. The teacher asked, "Can you pull out a third for me?" while he was pulling the three parts, but that seemed irrelevant in Jason's activity.)
T: Is that piece one-third of the whole stick? (The stick in the RULER.)
L: No, it is three-fifteenths!
J: (Drags the 3/15-stick into the TRASH. He then activates PULL PARTS and sits quietly for about fifteen seconds, then pulls five parts out of the 15/15-stick he made.)
T: Is that one-third?
L: Because there are three whole pieces there (Points three times in succession at the 15/15-stick from the left to the right.) and there is one there (Points to the 5/15-stick Jason pulled out.).
T: Five-fifteenths. And before that you said one-third. One-third or five-fifteenths of what?
J&L: Of the whole cake. (Laura actually made a copy of the stick in the RULER and partitioned it into three equal parts using PARTS. She then said, "That's a cake," when the teacher asked her what it was.)

Both Jason's and Laura's language and actions in Protocol X were remarkable, but for quite different reasons. After the teacher finally said, "I want another fraction, another fraction that will be like one-third." Jason and Laura sat for approximately 12 seconds deep in thought. It is not surprising that it was Jason who then recursively partitioned the 3/3-stick by partitioning each part into five parts. When coupled with the 12 seconds he sat deep in thought, partitioning each third into fifths was indeed a creative act on a par with his production of a unit fraction composition scheme in Protocol I.

His pulling of three rather than five parts to make one-third indicates that he was indeed engaged in rather strenuous thought because it indicates that he did not restructure one-third as one of three units containing five units. Rather, he regressed and used his concept of one-third that involved three individual parts. So, he was in the process of constructing a scheme for producing fractions commensurate to given unit fraction. That he could use his concept of one-third to establish one out of three units containing five units is indicated by his self-correction. Even though the teacher asking, "Is that one-third?" and Laura answering, "No, it is three-fifteenths!" may have served in creating doubt in him that the three parts he pulled out were each one-third, the correction he made in pulling out a 5/15-stick after trashing the 3/15-stick was initiated by him. In sum, making one-third of a stick into five-fifteenths was a novel experience for him.

After Jason pulled the 5/15-stick from the 15/15-stick, Laura explained that it was one-third "Because there are three whole pieces there ... and there is one there...." She obviously knew the 5/15-stick was called "five-fifteenths," and her explanation why it could be called "one-third" indicates that she made a unit containing three composite units each of which contained five elements independently of an explanation made by Jason. Apparently, she could use one-third in making independent explanations, but not one-fourth, one-fifth, etc. She appeared to be very confident in her explanation, and her additional act of making a copy of the stick in the RULER and partitioning it into three parts as a representative of the whole cake, together with her confident attitude, supports the inference concerning making and comparing composite units. Her use of her concept of one-third in formulating an explanation independently of an explanation made by Jason was a recurrence of how she explained why four-twelfths was one-third in Protocol V.

The teacher was encouraged by Jason's production of five-fifteenths and Laura's explanation for why it was also one-third, so he turned to tasks designed to engender the production of a sequence of fractions commensurate to one-third and possibly to a plurality of such fractions.

Protocol X. (First Cont)

T: Can you now make another fraction? You already have one-third, and you already have five-fifteenths. Can you make another part so that you can pull out one-third?
J: (Shakes his head "no" and Laura sits quietly looking at the teacher.)
T: Can't do another one?
J: Uh Huh (no).
T: Last week you had a lot for halves. So I bet that you can have more number names[9] for a third.
J: (Sits in deep concentration for approximately ten seconds. He then drags a 3/3-stick from the bottom of the screen upward, activates PARTS but does not use it because the stick is already partitioned. He then fills the three parts different colors.)
L: (Sits looking straight ahead with no apparent overt indications of mental activity.)
J: (Activates PARTS and partitions each part of the 3/3-stick he colored into three parts. He then pulls the first three parts.) Three-ninths.
L: Three-ninths.
T: (To Laura.) You want to label it?
L: (Labels it "3/9" using LABEL.)
T: Can you make another one from the same family?
J: (Takes the mouse and starts.)

---

[9] The teacher introduced the phrase without considering the implications of its use. In general, the term refers to number words that serve as symbols. A number word is a symbol if it refers to a numerical scheme that can be used to generate a result of the scheme in the absence of perceptual input.

T: No, let Laura do the next one!
L: (Drags a stick from the copies Jason made to the middle of the screen and partitions each of the three parts into four parts.) Four, four, four.
T: Before you pull it out, what will be the fraction that you are pulling out?
L: Four, ah, four...
J: Twelfths (Almost simultaneously with Laura.).
T: How did you know that?
J: Because...
T: Wait, wait, wait. Let Laura. We have to take turns because we cannot all talk at the same time.
L: Four, four, and four make twelve. Four and eight and twelve. (She then labels the part "4/12" and pulls it out of the 12/12-stick she made.)
T: Now this is interesting. You have three-ninths, four-twelfths, and five-fifteenths (Pointing at the respective parts.). Can you think of something that will be ... that the numbers will be even smaller than what we have?
J: Um-hmm (yes).
T: What would you do? Which one are you going to try?
L: I know one... Maybe not. Three-seconds! Three-twos!
T: You go ahead (Gesturing toward the computer.).
J: I don't know what she is talking about.
L: (Partitions each part of a 3/3-stick into two parts. She then pulls the first two parts out of the 6/6-stick she made.)
T: Go ahead. Pull it out and label it, please.
L: Two, two... (Subvocally utters number words.) six!
T: (To Jason.) So this is what she has to label?
J: Because she had, there are six pieces and there are two. They are kind of a part.
L: (Labels the part "2/6.")

The claim that Jason's act of partitioning each part of the 3/3-stick into five parts in Protocol X was a creative act is corroborated by the fact that initially he said that he could not do another one. After the teacher reminded him that he had done a lot for halves last week, and asserted that he could have more number names for one-third, that apparently reoriented him in such a way that he considered it possible to make a partition other than the one he had already made. He had eliminated the perturbation that drove his generation of five-fifteenths, and the teacher's provocation served in his reestablishment of a goal to generate another fraction. That he sat in deep concentration for approximately 10 seconds does indicate that partitioning each part into three parts did not immediately occur to him. I suggest that his experience was more or less one of being in a state of perturbation – in a search mode – but with nothing appearing in his consciousness. Recursive partitioning operations were evoked, and to partition each of the three parts into three parts appeared to him suddenly. This is indicated by the activity in which he engaged that marked the end of the search period. All at once he knew what to do. Moreover, after the teacher asked, "Can you make another one from the same family?" he immediately initiated activity, which indicates that he was now aware of how to proceed. That is, he seemed to abstract the operations involved in his operating.

The teacher sensed that Jason now knew what to do and that Laura also could engage in mathematical activity. So, he stopped Jason from acting by saying, "No, let Laura do the next one!" It might seem as if he should have made this suggestion earlier, but his suggestion was made on the basis of his interpretations of not only Laura's language and actions, but also on the basis of his interpretation of her body language as well. Her apparent confidence when saying "three-ninths," when coupled with her confident attitude explained above, led the teacher to ask her to make the next fraction. She immediately partitioned each of the three parts of the 3/3-stick into four parts, and said that it was four-twelfths of the whole stick. My interpretation is that she also knew that it was one-third of the stick because when the teacher asked the children to think of something so that the numbers would be smaller, she generated "three twos." What she meant was that she would partition each one of the three parts into two equal parts. She counted how many parts the 6/6-stick comprised before she could say "two-sixths," and then she said "six" rather than "sixths" because she had just counted the six parts.

Both children seemed to be poised to make the generation of fractions commensurate with one-third systematic in that they could generate them sequentially, one after another. So, the teacher asked them to arrange the fractions that they had made in a systematic order.

Protocol X. (Second Cont)

T: Can you now arrange the screen so that we will have two-sixths, then we will have three-ninths, then we will have, you know ... a sequence.

J: (Arranges the sticks so that the stick corresponding to "2/6" is on the bottom, then "3/9," then "4/12," then "5/15." Laura was active during this time, making several suggestions.)

T: Now comes the question. Without making the stick, can you think and tell me what will be the next one in the sequence that will be one-third? Which one will it be?

L: The next to highest or to lowest?

T: You started with the two-sixths, then three-ninths...

J: (Points to a 3/3-stick at the very top of the screen over all the others.) That one will be the lowest.

T: That's right. That's the one first, so you want to put it down here (underneath the 2/6-stick).

J: (Moves all the sticks upward to make room for the 3/3-stick and then drags it beneath all of the others and labels it "1/3.")

T: (While Jason is arranging the sticks.) While you are working, think what will be the next one, the next one in the sequence upward.

L: (After talking to herself.) Six thirty-sixths!

T: Six thirty-sixths? We will wait for Jason. OK, she is saying the next one will be six thirty-sixths. (To Jason.) What do you say? (He didn't ask Laura to explain because he was waiting for Jason to complete the rearrangement of the sticks.)

T: OK, Laura said that the next one that you are going to put here will be six thirty-sixths.

L: (Grabs the mouse.) I know ... (Plays with copies of a 3/3-stick remaining at the top of the screen.)

T: Wait, wait, wait. Jason, what do you say?

J: Seven twenty-oneths. Three times seven is twenty-one, and twenty-one comes before thirty-six.

L: (Just as Jason is starting to say "Seven twenty-oneths.") Oh, oh, I know!! (After Jason is done explaining.) six times three is ... six-eighteenths. Six-eighteenths (With confidence.)!

T: So, it will be six-eighteenths now?

J: Yeah, six-eighteenths.

T: And what will come next? After the six-eighteenths, what will come?
J: (Puts his head down and thinks.) I don't know...
T: (To Laura.) Do you want to do the six-eighteenths?
L: Yes.
T: (To Jason.) Think of the next one while you work.
J: (After about three seconds.) I got it. I got the next one!
L: (Activates PARTS, dials to "6" and clicks on each part of the 3/3-stick at the top of the screen. She then pulls six parts and labels it "6/18.")
T: Beautiful. (To Jason.) You have another one? Tell us what it is.
J: Seven twenty...
T: Seven twenty-firsts! Before you do it, can you think of what will be the next one?
L: I think, there will be eight in there, and eight in there, and eight in there (Looking up and pointing with her finger three times as if seeing a 3/3-stick in her visualized imagination.). Eight twenty-fourths!!
T: (To Jason.) This is what you thought of? What will be the next one?
L: (Again looking upwards as she points three times.) nine, nine, nine. Nine twenty-sevenths!
T: (To Jason.) I wanted Jason to say it. OK, what will be the next one after nine twenty-sevenths?
J: Ten-thirtieths.
L: (Again points three times in the air as Jason is answering "Ten-thirtieths.") eleven, eleven, eleven. Eleven thirty-three!

Laura went on in the same way, generating "12/36," "14/42," "15/45," "16/48," "17/51," "18/54," "19/57," and "20/60." She used her multiplicative computational algorithm to calculate those products (e.g., "3×17") that she could not quickly find using mental addition. Jason could not keep pace with her fast calculations, and this is one of the first times that she appeared to be the more powerful of the two. Jason definitely was aware of the sequence of fractions being calculated because he guessed "16/48" after Laura had said "15/45." He was aware that Laura could produce fractions of the sequence faster than he could and appeared abashed that he could not keep up with her. Laura's confident way of operating demonstrates that when she enacted her current ways and means of operating, she too could operate independently and confidently. She became involved in a way that was quite similar to the way Jason became involved in that she engaged in independent mathematical activity with a playful orientation, i.e., mathematical play.

Both Jason and Laura had established a scheme for producing a sequence of fractions, each commensurate with one-third. The operations of the scheme for Jason included recursive partitioning and the activity of the scheme was to calculate the numerosity of the parts produced by using recursive partitioning operations. The activity of Laura's scheme also was to calculate the numerosity of the three composite units produced by partitioning, but in doing so, she used her standard multiplicative algorithm mentally. Whether she engaged in recursive partitioning is problematic because she abstracted a pattern in operating. When generating a given fraction of the sequence of fractions, she first generated the next number in her number sequence and used that to partition the 1/3-stick. She then used her computational algorithm to find the numerosity of the partition of the 3/3-stick if she were to partition each of the other two parts. Although she definitely distributed the operation of partitioning across each part of the 3/3-stick, the inference that this units-coordination constituted a recursive partition is not warranted because it was Jason who generated five-fifteenths and three-ninths to start the sequence. It was only then that Laura generated four-twelfths. Her production of this fraction was based on her assimilation of Jason's language and actions using her units-coordinating scheme and her part-whole fraction scheme. For example, when Jason generated "three-ninths," she also said "three-ninths" in recognition of Jason's results. This recognition, when coupled with her production of "four-twelfths" immediately afterwards, does indicate that she distributed partitioning into three parts across each one-third and produced "nine" as the total number of parts. But it does not indicate that she engaged in recursive partitioning because making that inference requires an independence in the initiation of operating.

Producing Fractions Commensurate with Two-Thirds

The teacher proceeded to test if Jason and Laura could independently find another fraction for "two-thirds." In particular, he was looking for indications that Laura would engage in recursive partitioning in doing so.

Protocol XI. Fractions commensurate with two-thirds.
T: Copy the cake and put it into thirds. Make a three-thirds stick. And fill it with different toppings, please. (For the children, this meant to color the parts different colors.)
L: (Makes a copy of the stick in the RULER, partitions it into thirds, and fills the two outer thirds with different colors.)
T: Copy another one.
L: (Makes another copy of the stick in the RULER.)
T: Make two-thirds of that cake (Points to the 3/3-stick Laura had colored.).
J: (Fills the middle third with the same color as the first third and pulls the first two-thirds from the stick.)
T: Now comes the question. You gave me, like, twenty different thirds (Referring to the children making fractions commensurate with one-third.). Can you give me now a different two-thirds than you have here? A different two-thirds of the cake?
L: I know. (The teacher nods, so she continues. She partitions the extra copy of the stick she made into three parts and pulls out the last two parts.)
T: OK. That's very good. But, can you give me another fraction. Can you give me a fraction that will be two-thirds out of the whole, but with a different partitioning, a different number of pieces? (Both children sit quietly for approximately 20 seconds.) Can you find a way to partition the cake so that you will be able to pull out two-thirds?
L: I know.
J: (Following Laura.) I know.
T: (Encourages Laura to carry on.)
L: (Makes a copy using the stick in the RULER. She partitions it into three parts and fills the outer two parts with the same color. She then partitions the middle part into two equal parts.)
T: I want you to pull out two-thirds.
J: I know.
T: (To Jason.) Just wait a little. (To Laura.) That's a very good direction to start with.

J: (Intervenes in spite of the teacher's admonition to "Just wait a little".) Put another one in here (Partitions the first one-third into two parts.), and another one in here (The last one-third.), and then that would be two, two, and two and six in all (Pointing appropriately to the screen but without marking the parts.).
T: Now, can you pull out the two-thirds?
L: Wait, wait, wait. (Partitions each of the two outer thirds into two equal parts, pulls the four parts she just made, and joins them together.)
T: How much is it of the whole thing now that you joined it together? How much is it?
J&L: (Together.) Four-sixths!
T: (To Jason.) You want to make another one? Another one that will be two-thirds? Start with a full cake and do it a different way.
J: (Partitions an unmarked copy of the stick in the Ruler into twelve parts. He then colors the first four parts.)
T: How many are you going to take?
J: Eight.
T: Why eight?
J: Because eight is two-thirds of twelve!!

After the teacher asked "Now, can you pull out the two-thirds?" Laura did partition each of the two outer thirds into two equal parts and pull out the four parts she just made. But she did so only after Jason directed her to partition the two outer thirds as well as the middle third. Why she partitioned only the middle third of the stick into two parts can be explained by her making two of one-third, which is how she interpreted making two-thirds of the stick in a different way. Jason, on the other hand, first conceptually partitioned each of the three parts and only then established a fraction commensurate with two-thirds. His partitioning of an unmarked copy of the stick in the RULER into 12 parts when it was his intention to make another one that would be two-thirds, along with his explanation, "Because eight is two-thirds of twelve," corroborates the claim that Jason could indeed engage in recursive partitioning and that he could use it to produce fractions commensurate with two-thirds as well as one-third and one-half.

An Attempt to Engage Laura in the Construction of the Unit Fraction Composition Scheme

Jason was absent for the teaching episode held on the 10th of January, and it permitted Laura to be the primary actor in solving the situations of learning posed by the teacher. Even though Laura assimilated Jason's language and actions in Protocols IX, X, and XI and thereafter acted as if she had constructed the unit fraction composition and the commensurate fraction schemes, we do not know whether Laura had constructed these schemes, primarily, because she did not solve a situation of learning independently of Jason's solutions. Jason's absence in this teaching episode forced the teacher to confront the lacuna in Laura's reasoning. His strategy was for Laura to establish one-half of one fourth, and then measure the stick using

6 The Unit Composition and the Commensurate Schemes

MEASURE so Laura would be confronted with explaining why "1/8" appeared in the Number Box. But it proved to be difficult for Laura to independently establish that one of two parts of a 1/4-stick was one-half of one-fourth when the remaining 3/4-stick was hidden from her view.

Protocol XII. Laura's explanation that one-half of one-fourth is three and a half.
T: (After Laura drew a stick and agreed that it was to be thought of as a pepperoni pizza.) Now we will start with a four-fourths stick. Do you want to prepare one, a four-fourths stick?
L: (Dials Parts to "4" and clicks on the stick she drew.) Can I change it to different colors?
T: Yeah, why not.
L: (Fills the first three parts of the 4/4-stick different colors.)
T: First, can you pull out one-fourth of the stick?
L: (Pulls out the third one-fourth of the stick, presumably because she liked its color, purple.)
T: OK, now here comes the surprise. (Using COVER, covers the last three parts of the 4/4-stick, leaving the first part visible. He then establishes that Laura knew that the visible part was one-fourth of the stick as well as the part she pulled.) Let's say that both of us have to share it. Show me your share and tell me how much it will be of the whole pizza.
L: Just that one piece (Points to the 1/4-stick she pulled out.)?
T: You can use whatever you want. You will have to show me, we will have to share one-fourth of the pizza.
L: (Repeats the purple 1/4-stick to make a 4/4-stick. She then fills the first two parts.) Two-fourths of the pizza will be one child.
T: All right. I'll repeat the question because I can see it was my mistake. We can only share the one-fourth. Take this away (The 4/4-stick she just made.) because we don't have a whole pizza (To share.). (After Laura trashes the 4/4-stick she made, he asks her to pull the visible part out of the partially covered 4/4-stick.) Here is the question. It's only you and me, but we have only the one-fourth. We have to share this one (The 1/4-stick.). Can you show me your part, and tell me how much will it be of the whole pizza?
L: (Dials PARTS to "2" and clicks on the 1/4-stick.)
T: Now, what is your share?
L: Umm, umm (Fills the first part with a color different than the second.).
T: What type of a pizza is that one?
L: One-half of a fourth.
T: So, how much is it of the whole pizza? That is very good.
L: Umm (After about ten seconds.) three and a half!!
T: Three and a half – what?
L: Well, that's one-half, and then there's the whole one (The three covered parts and the one-half of one-fourth.).

It was unexpected that Laura would repeat the purple 1/4-stick to produce a 4/4-stick. Her goal seemed to be to produce a whole pizza to share equally between two children – "Two-fourths of the Pizza will be one child." Along with repeating the purple 1/4-stick, the goal is indicative of a reversible partitive unit fraction scheme because she regarded the whole stick as a multiple of one of its parts and knew before iterating that two of the four parts would be the share of one child. Had she actually pulled two of the parts from the 4/4-stick, the indication would be stronger, so it is necessary to corroborate the inference in further teaching episodes. On the basis of her current actions, I can only hypothesize that a partitive unit fraction scheme had emerged as a result of her coordinating the operations of disembedding and of iterating fractional parts of fractional wholes in the commensurate fraction tasks with Jason [cf. Protocols IX, X, and the continuations of X].

Recursive partitioning still seemed to be beyond her because she focused on the complement of the 1/2-stick she pulled out (three and a half) when asked how much one-half of one-fourth was of the whole stick. But focusing on the complement of the 1/2-stick in the whole stick is a part of recursive partitioning, so I explore what the consequences of focusing on the complement were in the continuation of Protocol XII.

Protocol XII. (Cont)

T: I see, umm, how much is that piece (Points to the 2/8-stick Laura made by partitioning the 1/4-stick into two parts.) of the whole pizza?
L: One-fourth.
T: One-fourth. And you took one-half of the one-fourth. You said it's one-half of one-fourth. Can you think of a fraction name for that piece?
L: Four and a half (The four-part stick and the one-half of one-fourth she pulled out.)!
T: Four and a half?
L: Cause there will be four of them and then one-half of it.
T: I don't think I see what you are saying. That is why I am asking questions... Can you use that one (The partially covered 4/4-stick.) to show me what you mean?
L: (Shakes her head "no.")
T: Can you pull out your part (From the visible 2/8-stick)?
L: (Pulls out one of the two parts of the 2/8-stick she refers to as one-half of one-fourth.)
T: Well, how much is this of the whole pizza?
L: It would be half of one-fourth.
T: Can you think of a way to find out how much it is of the whole pizza?
L: I can measure it!
T: Go ahead!
L: (Measures and "1/8" appears in the NUMBER BOX.)
T: Can you explain to me why?
L: Yeah. Because if you would put half on all of them, on all of, umm, and then if you'll half them all, then they would be one-eighth because there are eight pieces!

This was the first time that Laura was observed making an explanation of the sort that Jason had been capable of in the case of commensurate fractions. Her comments, "three and a half" and "four and a half," were harbingers of her explanation in the last two lines of the continuation of the Protocol. She definitely distributed the operation of partitioning into one-half across the four parts of the 4/4-stick when explaining why "1/8" appeared in the NUMBER BOX. In that three of the four parts were not visible, it would be necessary for her to operate on a re-presented 4/4-stick. Had she independently produced "one-eighth" without first measuring the stick she purported to be one-half of one-fourth, then the inference that she made a recursive partition in doing so would be indeed strong. As it is, all that can be said at this point in the teaching episode is that she distributed the operation of partitioning into half across the parts of a re-presented 4/4-stick when it was her goal to explain why one-half of one-fourth of the 4/4-stick is also one-eighth of the

4/4-stick. Whether this operation might engender recursive partitioning was tested by the teacher in Protocol XIII.

Protocol XIII. A test of recursive partitioning with Laura.
T: How much is this out of the whole pizza (Points to the visible part of the partially covered stick, which is a 2/8-stick.)?
L: One-fourth.
T: We started with one-fourth, but now can you give it another name?
L: (After approximately ten seconds.) I don't know.
T: Can you point to your part?
L: (Points to the right most one-eighth of the visible two-eighths.)
T: How much is it?
L: One-eighth.
T: Can you point to mine?
L: (Points to the left most one eight of the visible two-eighths.)
T: How much is this one?
L: One-eighth.
T: So, how much is this whole piece all together?
L: One-fourth!

It would seem that Laura would simply combine the two one-eighth parts of the stick into two-eighths. Instead, she used her basic part-whole fraction scheme to answer the teacher's question "So, how much is this whole piece all together?" One could argue that Laura simply did not form the goal of finding a fraction other than one-fourth for the visibly partitioned 1/4-stick, and that is certainly a possibility. However, on the basis of her past behavior, I find it more likely that she was yet to use recursive partitioning as a means to reach the goal of finding a fraction other than one-fourth for the visibly partitioned 1/4-stick. The teacher continued on, exploring this possibility.

The Emergence of Recursive Partitioning for Laura

Protocol XIII. (Cont)
T: (After Laura had erased the mark on the visible 1/4-stick.) Let's say we were lucky the first time because we were only two. Now, we are not that lucky anymore. We are three people all together; it's you and me and Mr. Olive. He would also like to get some pizza.
L: (Takes the mouse and starts to make another copy of the stick in the RULER.)
T: Oh, no, no, no. We have only one-fourth of the pizza. All the rest is out. Can you show me your share, my share, and Mr. Olive's share, and tell me how much of the whole pizza is your share?
L: Out of that one piece right there (Points at the visible part of the partially covered stick.)?
T: Yeah. Only that one-fourth of the whole.
L: (Dials PARTS to "3" and clicks on the visible one-fourth of the partially covered stick. She then fills the left-most part she made purple, which is her preferred color, and the middle part green.) OK. (Looks at the teacher with confidence.)
T: Which one is yours?

L: The purple.
T: So, how much is your share out of the whole pizza?
L: OK. Umm, one-twelfth!
T: One-twelfth!! How come? Why?
L: Because there are four (Simultaneously puts up four fingers.) spots, and you put three in each one, and uh, four times three is twelve!
T: Did you say four spots? I was not sure I heard you right.
L: Yeah. (Re-explains) well, there are four pieces of pizza, and then there are three pieces in each, and then, and then three and four makes twelve.
T: I see. That's very nice, so what is my share?
L: One-twelfth.
T: And what about Mr. Olive?
L: One-twelfth.
T: All right. So, how much is the one-fourth in terms of twelfths? For all the three of us 0together?
L: Three-twelfths.

The explanation that Laura gave for why she said "one-twelfth" – "Because there are four (simultaneously puts up four fingers) spots, and you put three in each one, and uh, four times three is twelve." – Indicates that she partitioned the 4/4-stick when it was her goal to find how much of the whole stick her share was. The comment, "Because there are four spots." When coupled with simultaneously putting up four fingers, indicates that she visualized four spaces. Her comment, "you put three in each one" further indicates that she inserted a unit of three into each of these spaces. One can also think of her inserting the *operation of partitioning into three parts* into each of the spaces. In that these operations were carried out to serve the goal of finding how much of the whole stick one-third of one-fourth constituted, this is the first time that the possibility was opened that she engaged in recursive partitioning operations. Because of her vivid language, I infer that she was aware of inserting a unit of three into each of four units when making a units coordination. In fact, I infer that she had interiorized the operation of units-coordination. The awareness she exemplified is certainly what is needed to infer recursive partitioning. The inference is plausible that she had constructed a *unit* fraction composition scheme. The case would be stronger, however, if she had said how much the three parts she made were of the whole pizza without the teacher specifically asking her how much the one-fourth was in terms of twelfths. So, her saying "three-twelfths" cannot be taken as indication that she knew that the three parts she made were three-twelfths of the whole pizza independently of the teacher's question.

Laura definitely formed the goal of finding how much her part (one-twelfth) was of the whole stick. But, in doing so, the goal of finding how much one-third of one-fourth was of the whole pizza was only implicit in her activity. There was no indication that she explicitly formed the goal of finding how much of the whole stick one-third of one-fourth was, because she shared one-fourth of the stick among three people. This sharing activity would be sufficient for her to produce a share for herself and for the two other involved individuals and to then form the goal of finding how much her part was of the whole pizza without intentionally engaging in the operations of finding one-third of one-fourth and then asking herself how much *that* piece was of the whole stick. For this reason, I was not confident in imputing a unit fraction composition scheme to her at this point in the teaching experiment. In fact, in the very next task of the teaching episode, Laura conflated one-sixteenth and four-sixteenths of a whole pizza.

Protocol XIV. Laura's conflation of one-sixteenth and four-sixteenths.

T: OK, and here comes the question. What is your share, or my share, or Mr. Olive's share of the whole [four people were to share the 1/4-stick]?
L: Four-sixteenths.
T: Is that my share?
L: No, that's my share.
T: Umm, can you tell me why?
L: Because four times four is sixteen.

There was an indication of recursive partitioning operations in the task following Protocol XIV: the teacher told Laura that her share was one-thirty second, and asked her to figure out how many people would have to share the visible one-fourth of the partially covered stick in order that she could get one thirty-second. She immediately dialed PARTS to "8" and clicked on the visible part. In answering the teacher's question concerning why she knew how to do that, she said, "Because eight times four is thirty-two." In other words, she could produce the partitioning operation, eight, given a result, one thirty-second, of the partition. After she produced eight as the partitioning operation, the most significant event of the teaching episode occurred. The teacher asked her how much three-fourths is in terms of thirty-seconds, and she answered, "twenty-four thirty-seconds" because "eight times three is twenty-four." The implications of this unexpected answer remain to be explored in subsequent teaching episodes.

Laura's Apparent Construction of a Unit Fraction Composition Scheme

Of interest in the teaching episode on the 8th of February was whether Laura could engage in the productive thinking that is necessary to produce a unit fraction composition scheme. To begin the investigation, after Laura had drawn a stick the length of the screen, the teacher asked Jason to make two halves in the stick and Jason partitioned the stick into two parts using PARTS.

Protocol XV. Laura's enactment of the composition of two fractions.

T: Let's say this is Jason's part (Pointing to the left-most one-half of a stick.). You see that this one (The whole stick.) is the whole stick, and you took half of it. Now, you are going to take half of Jason's part.
L: Right now? What do I do?
T: Laura, you take one-half of this one-half. (Points to the left most one-half of the stick.)
L: Half of this half (Picks the mouse up.). Half of that half, half of that half … OK. (Clicks on PARTS and clicks on the left most half of the stick. After coloring each of the two one-fourth parts she made, pulls out one of the two parts.)

T: (Points to the part Laura pulled out.) You could label it in terms of the w-h-o-l-e stick (Runs his finger along the whole stick)?
J: (After a few seconds.) I know (Smiling.).
L: (After approximately 10 seconds.) A half of one-half.
T: That's right. That's a good one. So how much is it of the whole?
J: I know it!
L: I don't know!
T: Can you use the computer to tell you?
L: Measure it. (Clicks on MEASURE and then on the 1/4-stick, and "1/4" appears in the number box.)
T: Why is it (One-fourth.)?
L: Because if you had all, if you had "halved" this one (points to the right most one-half of the original 2/2-stick), this one would be one-fourth (Pointing to the 1/4-stick she pulled out.). Half of that half, half of that half ...OK.

After reenacting the teacher's language – "Half of that half, half of that half …," it is not surprising that Laura partitioned the left-most part of the 2/2-stick into two equal parts. Laura could obviously give meaning to "half of that half," at least enactively. But, typically, she could not say how much that was of the whole stick.

There is a crucial difference between *explaining* why one-half of one-half is one-fourth, and in *producing* one-fourth as referring to the stick created by taking one-half of one-half. Producing one-fourth, in this case, is based on the operations for making a composite unit containing two composite units each of which contains two singleton units prior to operating. Explaining why one-fourth is one-half of one-half involves using her concept of one-fourth in a way similar to how she used her concept of one-third in Protocol V. But it is only similar because, as indicated in the continuation of Protocol XIII, Laura provided indication that she was explicitly aware of the results of using her units-coordinating scheme prior to activity. For Laura, "one-fourth" meant that she was to partition the stick into four equal parts, so she could project the one-fourth part she made into the stick by projecting it into each one-half of the stick to make two parts of the stick each partitioned into two parts. This represented progress, but it still fell short of recursive partitioning because it was an explanation rather than a production. After operating, Laura seemed explicitly aware of the two units of two that she made, but she seemed unable to produce this unit structure by recursive partitioning.

Nevertheless, her operations in Protocol XV indicated an awareness of the operation of iterating in the following continuation of Protocol XV.

Protocol XV. (Cont)

J: I know another way, I know how to… (After Laura had made the explanation at the end of Protocol XV.).
T: (To Jason.) Can you show it another way? (After considerable activity by Laura during which she labeled the 1/4-stick she pulled out "1/4.") You want to show us another way how to know it is one-fourth (To Jason.)?
J: (Copies the 1/4-stick twice and drags one copy over the right most one-half of the 2/2-stick with its left endpoint at the midpoint of the 2/2-stick. Visually, the 2/2-stick is now a 4/4-stick because the left-most one-half was already marked at its midpoint.) Because there are four of them right here and only one is filled in. That's one out of four and that makes it one-fourth. (He then drags the copy of the 1/4-stick he placed over the right-most one-half of the 2/2-stick to the trash.)

L: (Drags the extra copy of the 1/4-stick Jason made immediately above the 2/2-stick with left endpoints coinciding. She then activates PARTS and clicks on the right most one-half of the 2/2-stick, partitioning it into two equal parts. The 2/2-stick now looks like a 4/4-stick. She then repeats the 1/4-stick she placed above the 2/2-stick to complete a 4/4-stick.) Here, you see, that would be one-fourth because there are four parts, if you would have halved each. (She also justified why the 1/4-stick was indeed one-fourth of the 4/4-stick by commenting that the stick she made by using REPEAT was the same length as the 4/4-stick.)

It was a surprise that Laura decided to partition the right most one-half of the 2/2-stick into two parts and then to repeat the 1/4-stick four times to justify why the 1/4-stick was one-fourth of the original stick. At this point, she seemed to be explicitly aware of the iterative operation of repeating the 1/4-stick four times and of why she did it – "Here, you see, that would be one-fourth because there are four parts. If you would have halved each." So, she seemed aware of partitive and iterative operations, which serves to indicate a partitive unit fraction scheme. That the stick produced by iterating was the same length as the original stick seemed a logical necessity for her. So, it cannot be said that Laura was incapable of abstracting the operations she used. Rather, her necessity to actually use her units-coordinating scheme to produce a result appeared to cause her difficulty in finding how much a fraction of a fractional part of a stick was of the whole stick.

Nevertheless, Laura independently produced one-half of one-fourth immediately following the continuation of Protocol XV.

Protocol XVI. Laura finding how much one-half of one-fourth is of the original stick: a contextual solution.

T: Take half of this piece. (Points to the left-most one-fourth of the original 2/2-stick. By this time, the left-most one-half of the original 2/2-stick was marked into two parts and the right-most one-half was blank.)
L: That is right. Let's see, that would be...one, two (In synchrony with moving her left hand and then her right hand.) And that one will be (Eyes upward, mumbling to herself.). (She then turns and points to the teacher.) I know what yours would be. I know what yours would be. (Sits quietly for approximately 80 seconds while the teacher and Jason interact concerning the situation.)
T: (To Laura.) Do you know? Do you want to say?
L: (Puts two fingers up and moves them in synchrony with uttering.) 2, 4, 6, 8 (Gesturing toward the stick with her two fingers.). It would be one-eighth.
T: (Again, after interacting with Jason for approximately 50 seconds.) Laura, how did you come to know it will be one-eighth?
L: (Leans toward the screen enthusiastically and refers to an 8/8-sick Jason had made during the approximately 50 seconds she sat idly.) Because if you have, if you had just that one whole piece (Points to the left-most 2/8 of the 8/8-stick, and places her right index finger on the left endpoint of the stick and her right thumb on the mark at the end of the second one-eighth of the stick.) You can just copy it. (Moves her extended right index finger and thumb along the 8/8-stick as if she is making copies of initial 2/8-stick. When moving, she places her right index finger and thumb so that they span each successive 2/8-stick.) I mean four pieces, and you halved it, so you have two in each, and two times, so two and four are...

Laura's goal was to find how much one-half of one-fourth was of the original stick. She finally produced a unit fraction, one-eighth, for one-half of one-fourth. In doing so, she distributively partitioned each one-fourth of the 4/4-stick into two parts, which constitutes an anticipatory use of her units-coordinating scheme. Her explanation, "Because if you have, if you had just that one whole piece (points to the left-most two-eighths of the 8/8-stick …) you can just copy it. … I mean four pieces, and you halved it, so you have two in each, and two times…" does corroborate that she viewed the whole stick on which she was operating as a 4/4-stick, and that she halved each fourth. It also indicates that she enacted iterating the 2/8-stick to complete an 8/8-stick. But her solution occurred in the context of *explaining* why one-half of one-half was one-fourth, so it could not be judged as an original solution.

Nevertheless, it is quite possible that her anticipatory use of her units-coordinating scheme would engender the construction of recursive partitioning in the context of finding a fraction of a fraction. So, the teacher continued on, exploring whether she could find one-half of one-eighth.

Protocol XVI. (Cont)

T: That is so nice. So, what do you think will be your next step, after he labels it. (Jason was trying to use Label to label the piece "1/8.")
L: Ah, half on that one (Points at the first part of the 8/8-stick.)?
T: Good, very good (Raises "thumbs up."). Can you tell me what will be the label, how much will it be out of the whole stick?
L: It will be, hold on, hold on. Let's see (After about 10 seconds during which she touches each part of the 8/8-stick with the cursor.) one-sixteenth!
T: (To Jason.) What do you think?
L: I know how to explain it.

Laura independently enacted partitioning into two parts across each part of the 8/8-stick by touching each part with the cursor. Afterwards, her comment, "I know how to explain it," indicated the significance of her activity. So, her goal of finding how much one-half of one-eighth was of the original stick evoked the operation of distributing partitioning into two parts across the parts of the 8/8-stick, which is what an explicit units coordination means in the context of connected numbers. Coordinating the operations of partitioning into eight parts and partitioning into two parts was at least partially achieved: given the results of the first partitioning, she could carry out the second partitioning to achieve the goal of finding one-half of one-eighth. She had made a local modification in her reversible partitive fraction scheme by calling up her units-coordinating scheme when taking one-half of one-eighth. Whether this local modification constituted an accommodation awaits further investigation.

Progress in Partitioning the Results of a Prior Partition

A primary issue in attributing a fraction composition scheme to a child is whether partitioning, say, one-fourth of a stick into three equal parts symbolizes partitioning each one of the four-fourths into three parts. That is, a key to the establishment of partitioning operations as recursive operations is to establish them as *symbolic operations*. Laura's partitioning one-fourth of a stick into three parts in Protocol I symbolized partitioning each one of the four-fourths into three parts for *Jason*, because he explained, "See, if we would have had it in that ... four, four, four, and four – sixteen. But you colored three, so it is three sixteenth." On the basis of the analysis of Laura's language in the continuation of Protocol XIII concerning "spots," it was apparent that she had constituted her units-coordinating operations as symbolic operations in the sense that she could carry them out in re-presentation. But this is not to say that taking, say, a half of a third would symbolize taking a half of each of three thirds. We have seen that Laura could anticipate partitioning, say, each fourth into two parts when there was an immediate past experience to draw from. The teaching episode held on the 1st of March provides an opportunity to engage in further analysis of the symbolic nature of Laura's partitioning operations.

In Jason's case, the goal of the teacher was to find if he could take the results of a recursive partition as input for further partitioning. For example, if he were asked to partition one-sixth of a 6/6-bar into three parts and then partition one of the three parts into three more parts, would he use eighteen as a symbol for the results of partitioning one-sixth of a 6/6-bar into three parts? Leading up to Protocol XVII, the children made a bar and a copy of the bar, partitioned the copy into thirds, and then broke the partitioned bar into three parts. The teacher asked the children how much one-third of the 1/3-bar would be of the whole bar before they began to work on the computer: "Now, before you do it, how much do you think the partition will be after you finish? When you finish breaking it and you get one piece?" Both children knew that it would be one-ninth and Jason explained, "You break them in pieces, in three, each one of those boxes, and then another box there will be six, and another one will be nine." He then broke the upper most 1/3-bar into three pieces as in Fig. 6.2.

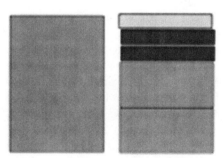

Fig. 6.2. The situation of protocol XVII.

Protocol XVII. Finding one-third of one ninth.

T: OK, Laura, now it is your turn. But before you do anything, the next turn is going to be thirding the one-ninth. How much of the whole will you have after you finish?

J: (Almost immediately.) One-eighteenth!

L: (Takes the mouse.) (Points to each of the 1/9-bars from the top down with the mouse cursor while subvocally uttering number words.) 3, 6, 9. (She then touches the cursor to the middle 1/3-bar.) twelve. (She then starts over after Jason says "one-eighteenth," counting by three until she reaches "twelve," and then starts over once again touching each 1/9-bar with the mouse cursor) 3, 6, 9 (Touches the middle 1/3-bar with the mouse cursor.) 12. (She then puts up three fingers on her partially visible left hand while subvocally uttering number words.) 15. (She continues on in this way until she has covered each 1/3-bar with three placements of the mouse cursor. She then points to the lower-most 1/9-bar, pauses and then points with the mouse cursor as described.) One twenty-seventh!

T: One twenty-seventh. (To Jason.) And you said one-eighteenth. All right, let's see... Do you want to explain first?

J: OK, ah, I counted two boxes (Points to the middle 1/3-bar.).

T: OK, that's sound. Very good. So, explain to me what you have done, and then explain what was the problem.

J: Ah, I thought, in my mind, I made three small, three pieces (Pointing at the 1/9-bars as diagrammed in Fig. 6.2.), and um, and then three here and three there (Points at the bottom two 1/9-bars.), and that's nine. And then nine plus, ah, two is eighteen, because there's nine in here (Points to the top three 1/9-bars.), so there's supposed to be nine in here (Points at the middle 1/3-bar.). So, I thought it was eighteen. One twenty-seventh.

T: Why is it one twenty-seventh?

J: Because there are three boxes. And I thought there were only two. I counted two of them.

T: So you counted only these (Points to the two unpartitioned 1/3-bars.)?

J: Um-hmm (Yes.).

T: All right, Laura. How did you come to one twenty-seventh?

L: I had three (Points with the mouse cursor to the upper-most 1/9-bar.), so it is 1, 2, 3; 4, 5, 6; 7, 8, 9 (Pointing to each 1/9-bar.), and to each box like that (Points to the middle 1/3-bar.), till I come up with the answer.

Even though Jason said "one eighteenth" rather than "one twenty seventh," it is indicated by his explanation that he took the result of partitioning each of the three 1/9-bars into three parts each, which is nine, as a given – "Ah, I thought, in my mind, I made three small, three pieces (pointing at the 1/9-bars as diagrammed in Fig. 6.2), and um, and then three here and three there (points at the bottom two 1/9-bars), and that's nine." It is also clear that he projected nine into the two unpartitioned 1/3-bars – "And then nine plus, ah, two is eighteen, because there's nine in here (points to the top three 1/9-bars), so there's supposed to be nine in here (points at the middle 1/3-bar). So, I thought it was eighteen." Changing his answer to "one twenty-seventh" was done, not simply to agree with Laura, but rather as a logical necessity once he realized that he was to find what part one of the small pieces was of the whole of the three 1/3-bars. It was in this sense that he took the results of recursive partitioning, nine, as a given in further partitioning.

Jason, on the one hand, operated in a manner that is compatible with what one would expect of a child who has constructed recursive partitioning. Laura, on the other hand, did not take nine, the results of counting the parts produced by partitioning each 1/3-bar into three parts, as a given and project it into each of the other 1/3-bars. Rather, she mentally partitioned the last two 1/3-bars into three parts and then each one of these imagined parts into three parts as she counted over each one of them. Presumably, because she did not know the number word sequence "3, 6, 9, 12, 15, 18, 21, 24, 27," she resorted to counting by one to generate the next number word of the sequence past "nine."

It could be argued, however, that she took the results of partitioning a 1/3-bar into three parts as a given when she counted the 27 parts. When counting, she placed the mouse cursor at three specific places on the 1/3-bars, and then mentally partitioned the indicated part of the bar into three parts. What this means is that she projected three equal units into the 1/3-bar and made a connected number, three. She then used each of these three units as material for further partitioning into three parts. So, the parts of the connected number, three, that she made were not simply perceptual unit items. Rather, they were implementations of her arithmetical unit items contained in her numerical concept, three. So, her counting activity may have served as an occasion for her to reinteriorize the operation of making a partition using the results of a preceding partition. It is a distinct possibility, because she first partitioned the whole bar into thirds, then each third into ninths, and then each ninth into twenty-sevenths. That she was aware of the thirds as thirds, the ninths as ninths, and the twenty-sevenths as twenty-sevenths indicates that she was aware of the cooccurrence of the three levels of units she produced. If she was aware of the level of units on which she was operating *while* operating, this would mean that she took the results of prior partitioning of a partition as input for partitioning further. In Protocol XVII, Laura seemed to finally construct recursive partitioning and the unit fraction composition scheme.

Protocol XVII. (Cont)
T: (To Laura.) All right. Do you want to make them?
L: (Makes a copy of the lower-most 1/9-bar and drags it to the bottom of the original bar. She then repeats it upward nine times so that the resulting 9/9-bar exactly covers the original bar [cf. the horizontal partitions in Fig. 6.3]. She then tries to partition each 1/9-part of the 9/9-bar into three horizontal parts using PARTS, but PARTS was not designed to partition the individual elements of a prior partition.) (After making three attempts.) Why isn't it doing it again?
T: Do you have any suggestions, Jason?
L: (Immediately after the teacher's comment.) Because all of them are going this way (Moves her hand horizontally back and forth across the 9/9-bar.).
T: So?
L: I can do them up and down!
T: All right!
L: All I have to do is to... (Activates vertical PARTS which is already dialed to "3." She then clicks on the 9/9-bar. The results were as pictured in Fig. 6.3.)
T: So, how many pieces do you have here altogether (The 27/27-bar.).
J: Twenty-seven.
L: (Nods.)
T: (To Laura.) Pull out.
L: (Tries to pull a 1/27-bar from the 27/27-bar, but she is unsuccessful.)

Fig. 6.3. Laura's introduction of vertical PARTS.

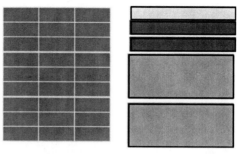

Fig. 6.4. The result of partitioning one-ninth into thirds.

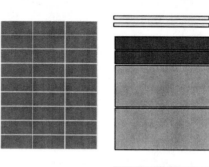

T: (Eventually.) Why don't you go and use this one (Points to the uppermost 1/9-bar of the copy.)? Do the thirding on that one, on the yellow one.
L: (Using horizontal PARTS, clicks on the uppermost yellow 1/9-bar of the copy, breaks the result into three parts and drags the lower part to the bottom of the screen as shown in Fig. 6.4.)
T: So, this (The lowest 1/27-bar.) is how much of the whole?.
L: One twenty-seventh.
J: One twenty-seventh.
T: OK (To Jason.) now it's going to be your turn. You have to third, to do the thirding of the one twenty-seventh. Don't do that first. Think how much are you going to get after you finish.
J: (Puts his head in his hands in deep concentration.)
L: (Looks intently at the screen and points to it in synchrony with subvocally uttering number words.) 1, 2, 3; 4, 5, 6. (Looking away from the screen, whispers.) three, twenty-seven. (Moves a finger in the air as if she is using her paper and pencil algorithm. She abruptly looks downward and ceases to move her finger, but now whispers.) Three times seven is twenty-one; two, three times two is six. Add the two, so it's twenty-one and sixty. I know.
T: (To Jason.) You don't know?
J: I don't know.
T: (Turns to Laura.) what did you say?
L: One eighty-one.
T: How did you know that?
L: Because all I had to do is twenty-seven divided by, I mean times three.
T: How did you know to do that? That is very nice.
L: Because if you put these three together (Points ambiguously to the three long, narrow horizontal 1/27-bars made by breaking the uppermost 1/9-bar of the copy into three parts.) They are equal one of those (Pointing at the bottom most vertically partitioned 1/9-bar.) and there are twenty-seven of those!

At the beginning of Protocol XVII (Cont), up until Fig. 6.3, Laura enacted making a 9/9-bar and attempted to partition each of the nine parts of the 9/9-bar she made into three parts each. Her confident attitude solidly indicates that she was aware of the results of partitioning each 1/9-bar into three parts in Protocol XVII. Introducing vertical PARTS and partitioning the 9/9-bar into a 27/27-bar provided an occasion for Laura to demonstrate that she understood that a 1/27-bar was indeed a 1/27-bar regardless of whether it was one of the 27 parts produced by horizontally partitioning each of the nine horizontal parts, or by vertically partitioning each of these nine parts.

When Laura explained how she arrived at "one eighty-one," she combined the three 1/27-bars she had produced by horizontally partitioning the yellow 1/9-bar of the copy and equated these three parts with the bottom-most vertically partitioned 1/9-bar. She recognized that a vertical partition of a 1/9-bar into three parts and a horizontal partition of the yellow 1/9-bar into three parts constituted identical partitionings operationally, and that there would be a total of 27 such units of three. She then proceeded to use these 27 units of three in organizing counting by ones (1, 2, 3; 4, 5, 6). The anticipation of performing counting in triples twenty-seven times led to the activation of her classroom algorithm for multiplication, which she executed mentally while looking away from the computer screen.

Laura's anticipation of counting 27 modules of three was made possible by her activated units-coordination scheme. She interpreted this counting as three times twenty-seven, which allowed her to use her algorithm meaningfully. At the point in the continuation of Protocol XVII where she changed from counting by threes to calculating using her paper and pencil multiplying algorithm, a case can be made that she abstracted coordinating the elements of the two composite units twenty-seven and three. That is, she abstracted projecting each unit of three into the units of 27, and "ran through" the coordinating process in thought, producing twenty-seven threes. This mental activity is indicated by her comment, "Because all I had to do is twenty-seven divided by, I mean times three." Laura seemed to finally construct recursive partitioning and the unit fraction composition scheme.

Discussion of the Case Study

At the outset of the teaching experiment at the beginning of the children's fifth Grade, I hypothesized that the differences in Jason's partitive fraction scheme and Laura's part-whole fraction scheme would be manifest in the schemes the two children constructed to produce fractions commensurate with a given fraction. I can now use my conceptual construct of the splitting operation as a rationale for this hypothesis. The splitting operation involves a composition of partitioning and iterating. At the end of their fourth grade year, both operations were present, but separated, in Jason's partitive fraction scheme, whereas Laura's part-whole fraction scheme was based only on partitioning operations. So I surmised that the splitting operation could be more easily brought forth in Jason's case.

A corroborating surprise occurred in the teaching episode held on the 25th of October in that Jason constructed the unit fraction composition scheme. This is the first fraction multiplying scheme that I observed in the two children, and it occurred as if Jason had constructed it over the summer vacation between his fourth and fifth grades in school.

The Unit Fraction Composition Scheme and the Splitting Operation

I explained the fraction composition scheme in an account of how Jason found that three-fourths of one-fourth was three-sixteenths without mentioning the splitting operation. Rather, I focused on recursive partitioning as the crucial operation that enabled Jason's behavior, and explained that it was embedded in his reversible partitive fraction scheme as an operation of the first part of his scheme. Jason used recursive partitioning throughout the five teaching episodes that were held with the children during the month of November, and it was this operation that served as the basic source of his productive and independent mathematical activity. In fact, I consider the splitting operation as implicit in recursive partitioning.

When it was Jason's goal to find how much three small parts of a stick were of the whole stick in the teaching episode held on the 25th of October, I assumed that he regenerated an image of the 4/4-stick and considered the three small parts as hypothetical parts of it. This re-presentation is only partial, because to find how much the three small parts were of the whole stick, he would need to know how many of these small parts constituted the length of the whole stick. Because his units of one were iterable units, it would not be necessary for him to actually iterate one of the three parts and complete the re-presentation. He was aware that he could engage in such iterative activity, so he did not need to run through the activity to realize its potential results. The potential results were available prior to operating, which is to say that he engaged in the operation of splitting.

What I have said up to this point concerning the splitting operation could be also realized, if Jason considered the partitioned part that contained the three parts being iterated enough times to partition the whole of the 4/4-stick. These operations also constitute splitting operations with the important proviso that a composite unit of four is involved as well as the unit of one. What this means is that Jason could establish three levels of units in his construction of the unit fraction composition scheme. In either case, the splitting operation is implicit in the unit fraction composition scheme, and it is this operation along with three levels of units that makes recursive partitioning possible. Therefore, Jason and Laura were working at distinctly different learning levels with respect to the splitting operation.

Independent Mathematical Activity and the Splitting Operation

On the one hand, Laura did not construct recursive partitioning in November in spite of the best attempts of the teacher to bring it forth. Jason, on the other hand, had already constructed the splitting operation at the outset, and it was manifest in his construction of recursive partitioning and the unit fraction composition scheme. This primary difference in the two children was striking, and it occurred throughout the five teaching episodes conducted during November. For example, in the teaching episode held on November 8, Laura, on the one hand, attempted to partition a stick into 12 parts without using 12 by partitioning the stick into 11 parts and then pulling one part from the stick and joining it to the 11-part stick. Jason, on the other hand, first partitioned the stick into three parts and then each of these three parts into four parts. In this case, Laura did engage in independent mathematical activity, but her way of operating contraindicated the possibility that she had constructed recursive partitioning operations. It might be conjectured that Laura simply did not think to operate in the way Jason operated. In fact, once her units-coordinating scheme was activated in the situation by means of observing Jason operate, she did know what to do and proceeded quite smoothly. However, it was characteristic that Laura needed to reenact an explanation or a recursive partitioning made by Jason, or for there to be visual cues in her perceptual field before she could engage in the actions that were needed to be successful in explaining why a fraction such as one-third was commensurate to, say, four-twelfths after she measured the 4/12-stick. Jason could independently engage in the operations that were necessary to produce such explanations or actions, and, beyond that, he could independently produce a unit fraction that was commensurate with, say, three-fifteenths. I consider such independent productions as necessary in order to judge that a child has constructed either a commensurate fraction or a unit fraction composition scheme.

Independent Mathematical Activity and the Commensurate Fraction Scheme

Throughout the November teaching episodes, the teacher attempted to bring forth the commensurate fraction scheme within Laura. However, as I have just indicated, Laura could only use her units-coordinating scheme in recognition or in the reenactment of prior recursive partitioning actions or explanations made by Jason; she could not use it in the absence of such elements. For example, in Protocol VII, after Laura had made a 15/15-stick by first partitioning a unit stick into three parts then each part into five parts, the teacher pulled out three parts of the 15/15-stick in a test to find if Laura could explain why three-fifteenths is also one-fifth after the teacher said that it was one-fifth. Laura said that she did not agree that it could be one-fifth. It was not until Jason explained that it took five of the 3/15-sticks to make the 15/15-stick by pulling a 3/15-stick from the 15/15-stick and making copies and aligning five

3/15-sticks directly beneath the 15/15-stick that she said, "I get it!" Her saying that she got it was made possible by her units-coordinating scheme and that she knew that three times five is fifteen. However, using her units-coordinating scheme to establish a posthoc explanation is quite different than using the scheme in a creative production of the explanation. This difference is especially apparent when Laura, in the very next task, looked disconcerted when "1/5" appeared in the number box upon measuring a 3/15-stick she had pulled out from a 15/15-stick. The difference seems to reside in Jason's ability to take the structure of a unit of units of units as a given template for partitioning and Laura's necessity to produce a unit structure involving a sequence of composite units as a result of operating.

In response to this constraint in Laura's operating, the teacher started asking the children to produce fraction sticks commensurate to a 1/2-stick (Protocols VIII and IX). The expectation of the teacher was that Laura would use her units-coordinating scheme productively to produce a sequence of fraction sticks each commensurate with the 1/2-stick. Similar to other situations, Laura, on the one hand, initially asked, "What am I supposed to do?" Jason, on the other hand, said, "I think I know how," and proceeded to make a 2/4-stick by using recursive partitioning. The experiment failed in the case of Laura in that she partitioned the unit stick into ten parts and took out five because she knew that five is one-half of ten. However, this way of operating did constitute productive activity on Laura's part because she then posited 48 as a possibility. So, when Laura independently generated a situation that was made possible by her part-whole fraction scheme, she became involved in a way that was quite similar to the way Jason became involved. For example, when Jason made the 200/200-stick by partitioning a unit stick into 50 parts and then each part into four parts, one could legitimately say that he engaged in independent mathematical activity with a playful orientation, or mathematical play. Laura could and did engage in mathematical play, but her possibilities were more restricted than Jason's.

Laura did not provide any indication of the splitting operation, recursive partitioning, a fraction composition scheme, or a commensurate fraction scheme during the month of November. However, she could operate as if she had constructed these schemes in the context of mathematical interaction with Jason when her units-coordination scheme was called forth. But this was not enough for her to engage in independent mathematical activity and she remained dependent on what Jason said or did, what the teacher said or did, or the context of the situation, in order to know what to do to be successful. However, in that she abstracted how she operated in the second continuation of Protocol X and produced a sequence of fractions through twenty-sixtieths, the teacher decided to pursue the possibility that she could construct a fraction composition scheme after she returned from the Christmas Holidays.

An Analysis of Laura's Construction of the Unit Fraction Composition Scheme

The second half of the teaching experiment in the children's fifth grade year was geared primarily toward investigating Laura's construction of the unit fraction composition scheme. Laura seemed to construct this scheme during the spring semester of her fifth Grade while we worked with her. However, I could not infer that she constructed the fraction composition scheme until the 1st of March (Protocol XVII). I now turn to an analysis of her constructive itinerary.

As a result of the teaching episode held on the 10th of January (Protocols XII and XIII), I inferred that Laura had become explicitly aware of partitioning each of the four parts of a 4/4-stick into three parts each (Protocol XIII, Second Cont.). She specifically said, "Because there are four (simultaneously puts up four fingers) spots, and you put three in each one, and uh, four times three is twelve!" The significance of this comment is that it followed on from the first time Laura was able to explain why one-half of one-fourth was one-eighth (Protocol XII, Cont) without relying on an explanation or action made by Jason. This was a major change and it occurred over the Christmas holidays. She could now use her units-coordinating scheme productively in formulating an explanation. I explained the progress that she made by saying that she had interiorized the operation of units coordinating and that this process of interiorization was not observed in her use of her scheme in either fraction composition or commensurate fraction situations that she engaged in prior to the Christmas holidays. Nevertheless, I look to her mathematical activity in these preholiday situations for indication of any local progress that might be a harbinger of her progress.

Laura's engendering accommodation. The first situation that seems important in Laura's progress is that she explained why 1/3 came up in the NUMBER BOX when she measured a 4/12-stick, where the explanation was made independently of Jason's explanation (8th of November, Protocol V). Even though her explanation was based on Jason's partitioning a stick into 12 parts by first partitioning the stick into three parts and then each part into four parts, she did not wait until Jason made an explanation before she made an explanation. Her ability to make an explanation independently of Jason's explanation in the case of one-third reemerged in the teaching episode held on the 29th of November (Protocol X). There, she explained why a 5/15-stick Jason made by partitioning each of the three parts of a 3/3-stick into five parts was also a 1/3-stick. This situation was very similar to the situation in the 8th of November teaching episode, and it confirms that she could use her concept of one-third in the way that I explained following Protocol V. That is, given the results of partitioning each part of a three-part stick into four (or five) parts, and given that she then had reason to believe that the 5/15-stick was one-third of the whole stick, she could regenerate the partitioning acts that Jason carried out and review their results. She used her concept of one-third in reviewing the results in that she constituted the stick now marked into 12 (or 15) uniform parts as three composite units containing four (or five) parts. There was nothing in her visual field that would suggest that the marked stick could be constituted as three composite units containing four (or five) parts. There was only the history of how Jason made the partition of a partition, so it was these operations that she would need to regenerate for herself in visualized imagination.

The crucial aspect of her operating was that she took the results of her re-presentation of Jason's partitioning actions as input for further operating. What this means is that she took the three composite units she made when reenacting Jason's partitioning actions as material that could be used with her concept of one-third. After Protocol V, in an account of this use of one-third, I commented that Laura used her concept, one-third, to monitor making the three composite units into three units that she could consider as three component parts of a stick, one of which could be pulled out. The way I explain monitoring includes both *feed-forward* and *feed-backward*.

Feed-forward was involved in that one-third activated a regeneration of Jason's partitioning of a partition. It also involved reprocessing the three composite units produced by her regeneration of Jason's partitioning actions. Reprocessing amounts to reinteriorizing the three composite units of four or five elements as component units of her concept of three. What I mean by reinteriorizing the three composite units of four or five elements as component units of her concept of three follows. Her concept of three was, of course, already a numerical concept that was comprised by a composite unit containing three units, which in turn contained records of experience that were interiorized at what I consider as the third level of interiorization. That is, the records were made by unitizing images produced by using records that were made by unitizing images produced by using records that were already interiorized the first time. I have explained how this process of interiorization works in Chap. 3.

Using her concept of three in regenerating the experience of partitioning a stick into three parts at the level of re-presentation, and then in partitioning each part into four (or five) parts, entails making records of making three composite units of four (or five) elements. Once these composite units were made, she then reprocessed the three composite units in making three component units from which she could pull one. In reprocessing, she would focus her attention on each composite unit of three, which requires unitizing the elements of each one of the composite units and then uniting the result together into a unit using the uniting operation. This process is an act of abstraction in that it leaves behind the particular sensory material implied by the figurative image, and creates a slot that can be filled with sensory or figurative material comprising units of three.

This feed-forward system generates a feedback system. When reprocessing the three composite units generated by reenacting Jason's partitioning operations in re-presentation and focusing on a particular composite unit, Laura would "see" four (or five) individual composite units. So, a check would need to be made concerning how many composite units are being made. Moreover, a check would need to be made at each point of making a composite unit of four (or five) units concerning whether three such units had been made. That is, Laura would need to set herself apart from operating and make a distinction between the levels of units she was producing. If a check was not made, then a conflation of units would emerge. Laura did discriminate between the four units she was taking as a unit while making three composite units and the three composite units. This check might seem to be trivial, but it is essential in generalizing the process to fractions other than one-third.

At the time I conducted the analysis of the 9th of November teaching episode, my question was whether the modification in Laura's way of operating with one-third was an accommodation. At that point, I looked to whether she could make similar explanations in the case of other unit fractions. No such explanations were forthcoming prior to the Christmas holidays. In fact, she relied on Jason's explanations for other such fractions (cf. Protocol VI for "1/9" and Protocol VII for "1/5"). However, she made another such explanation in the case of one-third in the teaching episode held on the 29th of November. So, in retrospect, I now consider her explanation as indicating an accommodation in her part-whole fraction scheme, even though she did not make similar explanations in the case of other unit fractions. That she did indeed abstract the system of operations involving one-third is solidly indicated by the way in which she generated the sequence of fractions, "8/24, 9/27, 10/30, 11/33, 12/36, 13/39, 14/42, 15/45, 16/48, 17/51, 18/54, 19/57, and 20/60" in the second continuation of Protocol X on the 29th of November teaching episode. Given this abstraction of operating in which she held three constant, and systematically increased by one the number of parts into which each of the three parts was to be partitioned, it is, in retrospect, surprising that she did not also make such an abstraction in the case of one-half. This surprising result only serves to confirm the special role that three played in her progress toward constructing recursive partitioning operations. The limits of her accommodation are also revealed in her attempts to produce fractions commensurate with two-thirds (cf. Protocol XI). In contrast to Jason's independent way of operating in the case of two-thirds by initiating the partitioning of each third into two parts and then taking two of these three parts as four-sixths, Laura attempted to make two in one-third for another way to partition the stick so that she could pull out two-thirds. She could use Jason's explanation in finishing her partition and pulling four of the six parts out of the stick as another way of making two-thirds, but this was the result of interlocked interactions with Jason. So, her accommodation apparently did not generalize to two-thirds any more than it generalized to unit fractions other than one-third.

Laura's metamorphic accommodation. It was a surprise that Laura was explicitly aware of partitioning each of the four parts of a 4/4-stick into three parts in the Teaching Episode held on the 10th of January (Protocol XIII, Second Cont). I have given an account of how she interiorized Jason's partitioning of a partition in the case of one-third, and it is these partitioning actions that constitute units coordinating in the case of connected numbers. To explain the progress she made over the Christmas holidays, it is necessary to attribute a metamorphic accommodation to her in the case of her part-whole fraction scheme. Not only could she make an explanation for why one-half of one-fourth was one-eighth in the continuation of Protocol XII, she also tried to produce the fractional part that one-half of one-fourth was of the whole pizza when she said that it was three and a half of the whole pizza and then four and a half when the teacher indicated to her that three and a half was not correct. When a part of a stick (a 3/4-stick) was hidden from view, she re-presented that stick and referred to it in an indication that she was aware of there being three hidden parts – "Well, that's one-half, and then there's the whole one (the three covered parts and the one-half of one-fourth)." What she did not do was partition her re-presentation of the three hidden parts into two parts each and then operate on these parts as if they were visible to her in establishing that one-half of one-fourth was one-eighth. This would have constituted recursive partitioning operations, which she was yet to construct. What she could do was to use her concept of one-eighth to generate, but not to regenerate, partitioning each of the four parts, the one visible and the three hidden, into two parts each and pull one of these parts from the eight and call it "one-eighth." There was also indication of a reversible partitive fraction scheme in the Protocol XII for the first time we had worked with her during the whole of her fourth Grade and up to the 10th of January of her fifth Grade.

Laura's construction of the partitive fraction scheme. Laura's part-whole fraction scheme contained operations she could use to make what goes by the conventional names "proper fractions" and "unit fractions." In saying this, I do not imply that her operations led her to make fractions she could operate with as though they were proper fractions or unit fractions. Rather, these terms are used for identification purposes only. So, the accommodation that she made in her part-whole fraction scheme was local in that it pertained to only one-third. My assumption is that the interiorization of these operations as well as the feed-forward and -backward systems that she constructed for one-third produced a systemic perturbation in her part-whole fraction scheme: the operations for making one-third were interiorized operations at a level above the operations for making other fractions, and they included material the other operations did not include (composite units). The perturbation induced by this rupture in the scheme sustained the activation of those operations that produced the rupture. The feed-forward and -backward system contained the activated operations and regulated the functioning of the operations that originally produced the perturbation until the perturbation was eliminated. I do not make any assumptions concerning the scope of the unit fractions that were reinteriorized, but apparently one-eighth was among these fractions.

The question arises why the operation of iteration, which is essential in reconstituting her part-whole scheme into a partitive fraction scheme, emerged. When she reprocessed, or reviewed, the elements of the composite unit she produced as a result of her enactment of a partition of a partition at the level of re-presentation, she would sequentially focus her attention on each of the composite units she produced, thus constructing at least the ability to repeat the composite units involved in reprocessing. When she used her numerical concepts in reprocessing the material she produced in the reenactment of partitioning a partition, and considering that her numerical concepts were essentially uniting operations, she would unite the unit items she created in her review into composite units. So, given that she used one-eighth to produce four composite units each containing two parts of the imagined stick, she could take these four units as a composite unit containing the four composite units, each of which contained two parts. This unit of units of units that she produced would then contain units of two that were iterable. The iterability of the unit of two would be inherited from the units of the numerical concepts she used in generating the explanation. In consideration of the iterability of the composite unit of two, Laura's ability to produce an image of a unit of units of two prior to operating is essential because it is this structure from which the iterability of the unit of two was abstracted. The unit of two then refers to this structure.

Laura's Apparent Construction of Recursive Partitioning and the Unit Fraction Composition Scheme

In spite of her substantial progress, Laura was yet to construct recursive partitioning operations and the fraction composition scheme as of January of her fifth-grade year. She could now make independent explanations for why, say, a half of a half was a fourth (Protocol XV), and, in context, she could even *produce* one-eighth as one-half of one-fourth and one-sixteenth as one-half of one-eighth in the teaching episode held on February 8. The basic difference in recursive partitioning and Laura's explanations can be seen in Jason's ability to take a unit of units of units as a given prior to operating, whereas, in Laura's case, this structure had to be called forth by the particular situation of explanation. However, the apparent progress she made on the 8th of February apparently had its consequences, because in the teaching episode held on the 15th of February, I finally inferred that she had constructed recursive partitioning and the fraction composition scheme. The inference was made in retrospective analysis as a result of Laura's operating rather powerfully when finding one-third of one-twenty-seventh. She made an abstraction in that, rather than count by three twenty-seven times, she recognized that she could use her computational algorithm for multiplying – "three times seven is twenty-one; two, three times two is six. Add the two, so it's twenty-one and sixty. I know." This is reminiscent for how she abstracted her multiplying algorithm for finding the sequence of fractions commensurate with one-third in the teaching episode held on the 29th of November. So, even though she behaved like a child who had constructed the unit fraction composition scheme, it is entirely possible that the operations that generated a unit of units of units was called forth by the particular situation in which Laura found herself.

Acknowledgments I would like to thank the editors of the Journals of Mathematical Behavior and Mathematics Thinking and Learning for granting permission to publish parts of earlier versions of this chapter in this book.

Chapter 7
The Partitive, the Iterative, and the Unit Composition Schemes

John Olive and Leslie P. Steffe

The fourth-grade teaching experiment with Joe and Patricia constituted a "replication" of the teaching experiment with Jason and Laura while they were in their fourth and fifth grades. We use scare quotes to indicate that the intent is not to repeat the experiment with Jason and Laura under the exact same conditions. Rather, the intent is to generate observations that can be used not only to corroborate the previous observations, but to refine, extend, and modify them as well. In a teaching experiment, the teacher does not establish a hypothetical learning trajectory at the outset of the experiment for the entire experiment. Rather, the teacher/researchers hypothesize what the children might learn in the next, or even in the next few, teaching episodes based on their current interactions with the children and their interpretations of it, and it is the testing of these hypotheses in the teaching episodes that, in part, constitutes the experimental aspect of the teaching experiment. Both the possibilities that are opened by the particular children and the constraints that the researchers' experience that emanate from within the children provide new observations that can be retrospectively analyzed to generate a "replicate" case study.

So, the fourth-grade case study of Joe and Patricia differs in intent from the fourth- and fifth-grade case studies of Jason and Laura. The primary concern in the case studies of Jason and Laura was to explain the construction and architecture of the children's basic fraction schemes: Jason's construction of the unit fraction composition scheme and the unit commensurate fraction scheme and Laura's construction of recursive partitioning. In the current case study, the primary goal is to compare and contrast the fraction schemes that Joe and Patricia constructed during their fourth grade with the fraction schemes that Jason and Laura constructed during their fourth grade and fifth grade, which includes accounting for observed differences in their constructive itineraries.

In what follows, we analyze 22 protocols that were selected from the teaching episodes that started on 2nd of November and proceeded on through the 3rd of May of the children's fourth-grade year. The protocols were selected to demonstrate the children's construction of recursive partitioning and the splitting operation, the partitive and iterative fraction schemes, and the fraction composition scheme, none of which were constructed by Jason and Laura during their fourth-grade year. Joe worked alone with his teacher up to February when Patricia joined him. Patricia had been working with another child, Ricardo, who moved from the school district, so she was paired with Joe at that time. In the first part of the analysis, we focus on the time period when Joe worked alone with his teacher.

Joe's Attempts to Construct Composite Unit Fractions

In the 2nd of November teaching episode, we worked with Joe on the construction of composite unit fractions. At this point in the teaching experiment with Jason and Laura, neither child could use the number of iterations of a 3-stick to produce a 24-stick$_8$ to establish the 3-stick as one-eighth of the 24-stick$_8$ [cf. Protocol VIII. (Cont.) of Chap. 5].[1] Rather than use the 3-stick as a unity, Jason said that a 3-stick was three-eighths of a 24-stick$_8$. This conflation of the unity of the 3-stick and the three unities it contained was interpreted as a necessary error. For comparison purposes, we decided to analyze Joe's ways and means of operating with the same type of tasks in an attempt to explore if he could use his units-coordinating scheme in the construction of composite unit fractions. Our overarching goal in this initial analysis was to explore the levels of units Joe had constructed and how he could use those levels in the construction of composite unit fractions. Initially, Joe made a similar conflation as the two other children when he thought that a 10-stick was one-tenth of a 40-stick$_4$.

Protocol I. Joe's initial conflation.
J: (Makes four copies of an unmarked 10-stick, joins them together and then erases the three marks that indicated the original four 10-sticks. He measures the unmarked stick, and "40" appears in the number box.)
T: What fraction of the 40-stick did you use to make the 40-stick?
J: (Shrugging his shoulders.) I don't know.
T: I bet I know what you did. I think you used the 10-stick. (Copies the 10-stick above the unmarked 40-stick.)
T: And what you did was…you repeated it…let's see (repeats the 10-stick to make a 4-part stick equal to the 40-stick). You repeated it four times!
J: (Smiles and nods his head.)
T: So what fraction of the 40-stick is the 10-stick?
J: One-tenth.
T: That's the 10-stick. (Pointing to the first part of her 4-part stick.) How many of it do we have? (Counts each part of the stick she just made.) We have one, two, three, and four. So the 10-stick is?
J: One-fourth.

---

[1]The notation, "n-stick$_m$," introduced in Chap. 5, is used to denote an *m*-part *n*-stick. In this case, an eight-part 24-stick.

7 The Partitive, the Iterative, and the Unit Composition Schemes

T: One-fourth of it. Very good! Do you want to make up another problem for me? I like that.
J: (Clears the 40-stick and the 4-part stick from the screen. He then makes nine copies of the 5-stick and erases all marks. The teacher then measures Joe's stick. It measures "45.")
T: What fraction of the 45-stick did you use to make the stick?
J: One-ninth.

When the teacher repeated the 10-stick to make a 40-stick$_4$, she reenacted what Joe had actually done. The reason she made this reenactment is that Joe frequently came from his classroom to the teaching episodes seemingly depressed. His depression seemed functional and specific to his classroom because, once his teacher/researcher established rapport with him, he became alert and engaging. After the teacher reenacted Joe's actions, he initially said that the 10-stick was one-tenth of the 40-stick$_4$. The teacher's demonstration in counting the number of parts in the 40-stick$_4$ as the number of 10-sticks used helped Joe connect the number of parts with the appropriate fraction word. Joe's realization that he used one-ninth of the 45-stick$_9$ in making it indicates that he now interpreted the teacher's question appropriately and realized that the fraction was given by the number of sticks he used to make the 45-stick$_9$ and not by the numerical size of the stick. The continuation of this teaching episode indicates that Joe's new interpretation of a unit fraction (in terms of the number of repetitions needed to construct the target stick) was very explicit and not only a matter of learning a way of using language.

Protocol I. (Cont)
T: (Sets a new problem. She makes an 18-stick.) The stick I used to make the 18-stick is one-sixth of the 18-stick. Which stick did I use and how many times did I use it?
J: (Copies the unmarked 3-stick above the unmarked 18-stick and repeats it to make a 6-part stick the same length as the 18-stick.)
T: That's exactly what I was thinking! So what fraction of the 18-stick is the...(Pointing to the first part of Joe's stick.)?
J: Three.
T: Three-stick?
J: (Subvocally counts the parts.) Six.
T: It's one-sixth of it. Very good!
J: I know how many times you're gonna use it.
T: How do you know that?
J: Because you said it was one-sixth and all you've got to do is take off the one and you'll have six.
T: That's very good! You knew that you had to use what six times?
J: (Looks up to the ceiling, thinking.) Umm...I don't know. I used the three six times. I guessed the three. I just knew that I had to use something six times.
T: So, if I told you that the stick that I used to make the 18-stick was one-half of the 18-stick, what would you tell me? What stick would I have to use to make it?
J: The 9-stick.
T: How many times would I need to use it?
J: (Copies the 9-stick below the unmarked 18-stick.) Two times! I don't have to do that one.

Protocol I. (continued)

T: That's excellent. You're so good. I have to ask you this one. What if I told you that the stick I used was one-third, one-third of the 18-stick, which stick would I use?
J: The 6-stick.
T: The 6-stick. And how many times would I need to use it?
J: Three.
T: That's excellent! That's very good.
J: All you have to do is turn it, do it over…do it backwards.
T: What do you mean, do it backwards?
J: You said "one-third" instead of "one-sixth." You turned it backwards. Put one-third and it will be the 6-stick instead of the 3-stick.
T: That is excellent!
J: You will use it three times instead of six times.

Joe's voluntary explanation for how he came up with the reverse situation, "All you have to do is…do it backwards," is an indication of his involvement in the activity and the heightened level of his mental activity. He was aware of his thinking and could explain it explicitly. In the context of connected numbers, Joe used his units-coordinating scheme in generalizing assimilation to figure out which sticks had been used and how many times they had been used to create a connected number. Through effective modeling of both the language of fractions and the iterative operations that Joe had used to verify his choice of stick, the teacher was able to provoke a modification in Joe's use of unit fraction words. The modification was a realization by Joe that the number of times he used a stick to make a longer stick provided him with the fraction name of that stick. But this was more than the use of language that the following statement by Joe might imply: "Because you said it was one-sixth and all you've got to do is take off the one and you'll have six."

In fact, the continuation of Protocol I led us to hypothesize that he modified his units-coordinating scheme to produce a reversible composite unit fraction scheme. Before that Protocol, we had already hypothesized that he could operate with three levels of units and that operating with three levels of units opened the way for him to modify his units-coordinating scheme into a reversible composite unit fraction scheme. But before imputing such a scheme to him, we used the tasks in the continuation of Protocol I to explore whether Joe used his units-coordinating operations to establish a connected number as a composite unit containing six composite units, each containing three units as a result of iterating a 3-stick six times.

Attempts to Construct a Unit Fraction of a Connected Number

In the 9th of November teaching episode, Joe corroborated that he had constructed the operations necessary to use the number of iterations of a 3-stick to produce an 18-stick$_6$ to name the unit fraction the 3-stick is of the 18-stick$_6$. Joe verbalized this way of operating explicitly as well as acted it out using the actions of TIMA: Sticks.

7 The Partitive, the Iterative, and the Unit Composition Schemes   175

A set of number-sticks had been created and placed at the top of the computer screen which was separated by a long thin segment that stretched across the full width of the screen that separated the screen into an upper and a lower part (see Fig. 7.1). The length of each stick was a multiple of the shortest stick. The shortest stick was designated as the unit stick (the 1-stick) and each of the other sticks were named as an *n*-stick where *n* could be any number from two to ten. Sticks created by repeating or joining copies of sticks from this ordered collection of number-sticks were also named in the same manner (e.g., a stick created from four iterations of the 6-stick was a 24-stick$_4$).

Fig. 7.1. A set of number-sticks in TIMA: Sticks.

Protocol II. Establishing sticks as fractions of a 24-stick.

T: (While Joe has his eyes closed, the teacher makes a 24-stick below the separator and erases the marks from her 24-stick.)
T: The stick that I used was one-third of the length of the stick I have right here. (Pointing to the unmarked 24-stick.)
J: (Measures the stick and "24" appears in the number box. He then smiles to himself and counts down the set of number sticks ending on the 8-stick. He copies this stick and repeats it three times to make a stick the same length as the 24-stick.)
T: That is right!
J: You said one-third, so what will be…three times eight is twenty-four.
T: Think of a stick you could use to make the 24-stick and tell me what fractional part of the 24-stick it would be, and I will try to tell you what size stick it is and how many times I should use it.
J: Close your eyes. (Trashes the 3-part 24-stick and looks at the set of number sticks.) Ok. I didn't have to do nothing… It's umm… It's one-sixth.
T: The stick that you used is one-sixth of the 24-stick?
J: (Nods his head.)
T: So, I want something, I want a stick that when I repeat it six times would give me…
J: No!
T: Would give me the twenty-four.
J: (At the same time as the teacher is speaking.) One-fourth!
T: Oh! You used the one-fourth stick?
J: (Nods his head.)
T: You used one-fourth, so I want a stick that when I repeat it four times will give me the twenty-four, and I think that is the 6-stick! What do you think?
J: (Nods his head.)

Joe knew to use the 8-stick for one-third of the 24-stick because "three times eight is twenty four." We regard Joe's interpretation of one-third as something that when multiplied by three gave the total number as a modification of his units-coordinating scheme because of the way he was able to pose the problem for the teacher and the self-correction he made in the process.

Joe hesitated in naming the fraction – "It's umm...It's one-sixth" – when posing his problem for the teacher. We suggest that he was trying to figure out both the numerosity of an imagined stick and the number of times he would have to use it to produce twenty-four. That Joe used the numerosity of an imagined stick to generate the fraction word rather than the number of times he would need to iterate the stick indicates that he was aware of the two numerosities. This awareness was confirmed when Joe realized his mistake as soon as the teacher voiced her interpretation of one-sixth. At this point he made a self-correction rather than accept the teacher's actions. This self-correction is interpreted to mean that Joe was aware of the operation of iterating as well as what was being iterated prior to action. It is in this sense that Joe was constructing meaning for unit fraction language through iteration of his connected numbers. He was aware of the 6-stick as one-fourth of the 24-stick because he knew that the 6-stick iterated four times would produce a 24-stick. But whether "one-fourth" referred to the 24-stick as a fractional whole and to a 6-stick as one out of the four parts of the 24-stick is equivocal.

For example, in the continuation of the teaching episode following Protocol II, Joe successfully identified the 4-stick as one-seventh of the 28-stick, but when asked by the teacher what fraction of the 28-stick two 4-sticks joined together would be, Joe responded with "One-fourteenth...because you add one-seventh and another seventh it makes fourteen." Joe seemed yet to interpret the 4-stick as one out of the seven equal parts of the 28-stick and use the seven equal parts as material for further operating, where the operations are partitioning and disembedding.

Partitioning and Disembedding Operations

Our goal of bringing forth Joe's use of his units-coordinating scheme in constructing meaning for composite unit fractions had been successful in the sense that the meaning he attributed to, say, one-fourth was his anticipation of iterating, say, a 6-stick four times to make a 24-stick – a 6-stick was one-fourth of the 24-stick because it could be iterated four times to make the 24-stick. What seemed to be lacking in this concept of one-fourth was the ability to partition into four equal parts the results of iterating the 6-stick four times and then disembed one of these four parts from the four parts. It might seem that iterating the 6-stick four times would produce a composite unit containing four composite units each of which contains six units. However, there is a distinction to be made between producing such a unit structure and in using that unit structure as material in further operating. We have already suggested that Joe could produce the unit structure by means of iterating, but he was yet to make an explicit correlation between the number of iterations and the number of 6-sticks in

7 The Partitive, the Iterative, and the Unit Composition Schemes

the 24-stick, an explicit one-to-four comparison of the 6-stick and the 24-stick, and an explicit six-to-twenty-four comparison. To make such a correlation and comparison, he would need to partition the unit structure into four parts, disembed one of these parts from the four parts, and then make part-to-whole comparisons at three levels of units. So, we decided to introduce intervening tasks, designed as transitional tasks between Joe's use of his units-coordinating scheme and the use of his number concepts in partitioning the results of using that scheme.

In the 16th of November teaching episode, we attempted to provoke Joe's partitioning operations by asking him to make fractional parts of sticks where the fractional parts were not multiples of a given unit stick. The task was to draw a stick that would be one-fourth of the 27-stick that Joe had created by repeating a 9-stick three times. The teacher had affirmed that there was no stick on the screen that was one-fourth of the 27-stick.

Protocol III. Making a stick that is one-fourth of a 27-stick.

J: (Draws an estimate that is just a little more than one-half of a 9-stick and iterates this estimate four times to make a stick just longer than two 9-sticks.) About...a stick longer than that.
T: You need a stick longer than what?
J: That one. (Pointing to the 4-stick he just created.) I know what I can do.
J: (Draws another estimate stick, located on the screen between the 3-part 27-stick and the 4-stick he had just produced. He is not sure how much longer to make this estimate than his previous one. He ends up drawing a stick that is a tiny bit longer than one-third of the 27-stick. This estimate is directly above the 3-part 27-stick [cf. Fig. 7.2].)
J: (Iterates his estimate three times to make a 3-part stick that is longer than his target stick – the 27-stick$_3$. He does not make the fourth iteration but trashes the new stick. Joe makes a third estimate about two-thirds of a 9-stick. He iterates it four times to get a stick about two unit sticks shorter than the target stick. Joe makes his fourth estimate longer than his third estimate and iterates it only three times and realizes that a fourth iteration would take him past his target stick [cf. the next-to-the-bottom stick with the arrow pointing to it in Fig. 7.3].)

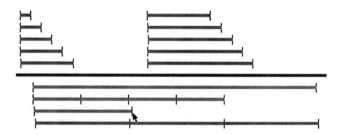

Fig. 7.2. Making estimates for one-fourth of a 27-stick.

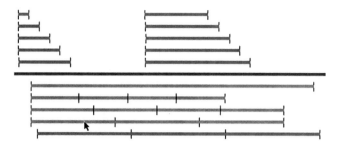

Fig. 7.3. Joe's estimates for one-fourth of a 27-stick.

Joe's estimates and adjustments in the above problem for one-fourth of the 27-stick were not very accurate. He did not seem to produce the results of iterating an estimate four times in visualized imagination and compare the imagined stick with the target 3-part 27-stick. Each trial was essentially independent of the preceding trials and Joe relied on actually iterating an estimated stick to make a stick that he could compare with the target stick. The protocol continues with the teacher making an estimate for one-fourth of the 27-stick.

Protocol III. (Cont)

T: This one (Pointing to the first part of the repeated 9-stick.) is one-third of the 27-stick, right? (Joe is not looking at the teacher or the screen.) And the one you made is too little. (Trashes the 3-part stick that is among the 4-part sticks. She points to the longer of the 4-part sticks that Joe made in Fig. 7.3.) This one is too short, so what I have to do now is draw a stick that is longer than this guy. Do you agree?

J: (Nods agreement. He is now intently looking at the screen. The teacher draws an estimate that is about one unit longer than Joe's. She repeats it four times. It is only slightly longer than the target stick. Joe takes the mouse.) A little bit longer.

T: You think we need one longer? But mine was longer than the green stick.

J: (Does not listen to the teacher. He draws an estimate that is just a tiny bit shorter than the teacher's and repeats it four times. It is the same length as the target 27-stick.)

T: That is it! (Joe is smiling broadly.) That is fantastic!

J: Let's see how big that is. (Joe cuts off the first part from his estimated 4-part stick and drags it up to the 5-stick. He then moves it over between the 6-stick and 7-stick.)

J: A little bit shorter than the 8-stick. No. A little bit shorter that the 7-stick.

T: So its somewhere between the 6-stick and the 7-stick. That's fantastic.

In the continuation of Protocol III, Joe seemed to become aware of the partitioned nature of the resulting stick as he iterated because, after recognizing that the teacher's resulting stick was a little bit longer than the target, he made an appropriate adjustment to the teacher's estimate. Joe also wanted to know what *size* stick he had made because he made visual comparisons with the known sticks to figure out that his one-fourth of twenty-seven was just a bit shorter than the 7-stick. He did not use numerical relations to come to this conclusion, only visual comparisons.

7 The Partitive, the Iterative, and the Unit Composition Schemes

As the teaching episode progressed, the teacher attempted to further provoke Joe's partitioning operations by asking him to make fractions of a stick that was drawn freehand, e.g., make a stick that is one-seventh of an unmarked stick. Without a known multiplication fact to solve the problem, we hypothesized that Joe would need to mentally partition the unmarked stick into seven equal parts in order to make a reasonable estimate for one-seventh of the stick. The screen display consisted of an ordered collection of number sticks from one to ten units long as in Fig. 7.1. A mystery stick was drawn freehand below the separator and Joe was to choose an estimate that would be one-seventh of this mystery stick. He chose a 2-stick from the set of number sticks and repeated it seven times directly below the mystery stick. His resulting stick was approximately two-thirds of the mystery stick. He made a second estimate, without acting, that the 3-stick would be one-seventh of the mystery stick because "it's about twenty-seven – twenty-one, I mean."

Rather than asking Joe to justify this second, verbal estimate, the teacher continued using the 2-stick Joe had originally chosen, making ten repetitions to make a stick that was about one unit stick short of the mystery stick. Joe confirmed that he was thinking of the mystery stick as a 21-stick when he then exclaimed "Twenty-one (Joe combined the ten iterations of the 2-stick with the remaining part of the mystery stick)! That's what I said!" and that the 3-stick would be one-seventh of the 21-stick because "If you use three seven times you might get twenty-one." Joe then copied the 3-stick below the 20-stick and repeated it seven times, making a stick a tiny bit longer than the mystery stick. With the teacher's encouragement, Joe then drew a stick below the first part of his 7-part 21-stick, making it a very tiny bit shorter than a 3-stick. He repeated this estimate seven times to make a stick almost exactly the same length as the mystery stick (cf. the bottom marked stick in Fig. 7.4).

The preceding task may have provoked Joe's use of his numerical concepts in partitioning as well as in iterating because he could imagine a stick iterated seven times as being equal to the mystery stick. Joe had, in fact, produced a way of operating that would enable him to use the results of iterating a stick in partitioning. His iterative operations provided him with a way of positing the length of a stick in terms of a given unit stick as a result of the imagined iteration. Based on his making a

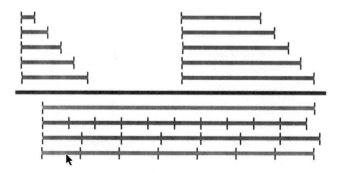

Fig. 7.4. Finding one-seventh of a mystery stick.

second estimate (without acting) that the 3-stick would be one-seventh of the mystery stick, we infer that Joe was starting to construct an equipartitioning scheme as a result of his iterating operations. In retrospect, because we cannot infer that Joe had constructed an equipartitioning scheme up to this point in the teaching experiment, we cannot yet confirm the earlier hypothesis that he had constructed a composite unit fraction scheme[2] as an accommodation of his units-coordinating scheme and had constructed three levels of units that he could take as given in further operating. Still, his estimating activity had major consequences for his construction of a partitive fraction scheme.

Joe's Construction of a Partitive Fraction Scheme

The first appearance of Jason and Patricia's equipartitioning schemes entailed both children using their numerical concepts in partitioning a stick (cf. Chap. 5, "The Equipartitioning Scheme") without any concerted attempts in the teaching experiment to provoke equipartitioning in these two children. The situation was quite different in Joe's case in that his construction of equipartitioning followed on from his use of his anticipatory iterative operations in the teaching episode held on December 7. In this important episode, Joe also constructed a partitive unit fraction scheme and used it in the production of fraction language for nonunit proper fractions.[3]

The screen in TIMA: Sticks was set up in a new arrangement for this teaching episode. Only four unmarked sticks were available at the top of the screen. These were the unit stick, 3-stick, 5-stick, and 7-stick. All were unmarked and colored differently and Joe did not know how many times the unit would need to be iterated to produce them. A long thin segment separated this set of sticks in the top part of the screen from the rest of the screen. The teacher had created a blue stick below this separator. Joe was to find the stick that was one-fifth of this blue stick (cf. the bottom stick in Fig. 7.5). Joe looked at the four possible sticks and chose the green 3-stick (second from the left). He copied the green stick below the bottom stick and repeated it five times to make a 5-part stick the same length as the bottom stick.

Fig. 7.5. Find a stick that is one-fifth of the long blue bottom stick.

---

[2] A composite unit fraction scheme entails transforming three parts out of twelve parts, for example, into one composite part out of four composite parts.
[3] Proper fractions are commonly thought of as fractions that do not exceed the fractional whole.

7 The Partitive, the Iterative, and the Unit Composition Schemes

Joe made an accurate choice for one-fifth of the long blue stick. We hypothesize that in order for him to have made this choice he would have had to anticipate iterating the green stick. Checking his choice by repeating his stick five times to make a partitioned stick with the same length as the target stick corroborates the inference that iterating was an anticipatory operation. This spontaneous way of checking his estimate for one-fifth of a stick implies an iterative fraction concept: one-fifth of a given quantity is the amount that, repeated five times, will regenerate the given quantity. Joe had indeed begun to construct a partitive unit fraction scheme.

Following this first task, Joe set a problem for the teacher to solve. He copied the yellow 5-stick (third from left) and iterated it four times to make a 4-part stick below the separator. He erased all the marks and filled this stick with a new color, pink. He then asked the teacher to find one-fourth of his pink stick. The teacher chose the yellow stick and repeated it four times to check that it was one-fourth of the pink stick, thus emulating Joe's iterative strategy.

The teacher then asked Joe to find one-third of his pink stick. Joe at first chose the light blue 7-stick (the right most stick) which was the longest of the given set of sticks. He repeated a copy of this stick three times to check if it was one-third. The resulting 3-part stick was just a bit longer than Joe's pink stick, so Joe threw away this 3-part stick and then made a series of estimates for a one-third stick, beginning with a stick that was just slightly longer than the one-fourth stick that was still on the screen. He increased the length of each subsequent estimate, repeating each estimate three times to check against the length of his pink stick. His fourth estimate produced a 3-part dark blue stick almost exactly the same length as his pink stick.

Joe's series of estimates was a remarkable advancement beyond his primitive attempts to draw a stick that was one-fourth of a 27-stick in Protocol III. Had the teacher asked Joe to mark the blue stick only once to estimate one-fifth of the stick and to prove that the estimate was right and had Joe done so, that would constitute prima facie indication that he had constructed an equipartitioning scheme as well as a partitive unit fraction scheme. Still, his repeated estimates, although they did not entail Joe marking off an estimate for one-third of the pink stick using one mark and then disembedding the estimated part from the pink stick, do indicate a projection of the estimates into the pink stick. For this reason, we infer that Joe had, indeed, begun to construct a partitive unit fraction scheme. The inference is corroborated in Protocol IV.

The teacher asked Joe to pull out one of the three parts from his dark blue 3-part stick. Protocol IV begins at this point in the teaching episode.

Protocol IV. Joe's production of "two-thirds" and "one whole one."

T: Can you make a stick that is twice as long as this one? (Pointing to the pulled out one-third.) (Joe repeats the 1/3-stick. The teacher laughs.) That's good! What do you think that is? Do you want to give it a name?

J: (Inadvertently clicks REPEAT one more time, adding another one-third to the two 1/3-sticks he had created.)

T: Do you remember what fraction of the blue stick was that one? (Pointing to the first part of the 3-part stick that Joe had made.)

J: One-fourth. No, one-third.

T: That was one-third. That's very good. Now, when we have one that is twice as long (She starts to cut the extra piece off the 3-part stick.) what do you think we should call it?
J: Two-thirds?
T: Two-thirds! That's good. Could you make one that is three times as long?
J: I can draw it this time. (Joe draws a stick below the 2/3-stick, stopping first at the 1/3-mark, then at the end of the stick, and then continuing past the end about the same as one-third more. He thus has a stick approximately the same length as his original stick. He then pulls the first part out of the 2/3-stick and adds it to the end of the 2/3-stick to check if his freehand stick is three times as long as the 1/3-stick, as shown in the bottom of Fig. 7.6. His freehand stick is a tiny bit short.)
T: What fraction of it do you think this one is now? (Pointing to the 3/3-stick.)
J: (Thinks for 5 seconds.) One whole one.
T: It's one whole one, right! But it's three of, three times as long as....(Pointing to the first one-third.)
J: This.
T: One-third, right.

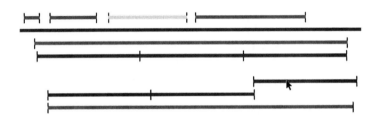

Fig. 7.6. Estimating a stick that is three times as long as a 1/3-stick.

We had used the language "twice as long as" with respect to fraction sticks for the first time in the above protocol. Joe interpreted the phrase as meaning he needed to make a 2-part stick that included the original stick that he was using to make a stick twice as long as the original stick. He had indeed generalized his operation of iteration in the case of discrete units in the continuous case, and his unit fractions inherited their iterability from his iterable unit of one.[4] Furthermore, Joe drawing a stick that was three times as long as the 1/3-stick by stopping at the 1/3-mark, at the end of the 2/3-stick, and then again after he believed he was done, corroborates the inference that Joe indeed projected units into the sticks he was making fractional parts of by drawing estimates. It also indicates that he could mark a fractional part of a stick off by using one mark, disembed that part, and iterate it in a test to find if the iterated part was a fair share had he been presented a task like the one in Protocol II of Chap. 5.

---

[4]"Twice as long" means "two more" for children who are yet to construct the operation of iteration. Including the stick being iterated in the iterations is indication that each stick produced in the iterations is an instantiation of the abstracted unit used in the iterations. There is only one iterable unit and the sticks produced in the iterations can be regarded as identical one to the other.

7 The Partitive, the Iterative, and the Unit Composition Schemes

His use of the term "two-thirds" to name the stick he made that was twice as long as the 1/3-stick and naming the three repetitions of the 1/3-stick as "one whole one" suggest that he could establish meaning for nonunit proper fraction words by means of iterating a partitive unit fraction. Following Protocol IV, the teacher made a stick from three repetitions of the 7-stick and then joined a 1-stick to this to make a stick that was the same length as a 22-stick. She erased all the marks and asked Joe to find one-eleventh of her stick, which was colored red. Joe first tried the 3-stick and quickly realized that it was too long. He drew a succession of six estimates, each one getting a little bit shorter before arriving at an estimate for one-eleventh that the teacher was willing to accept. He iterated each estimate 11 times to check each estimate by comparing them with the red stick. Protocol V begins when the teacher asks Joe to pull one part out of his 11-part stick that was almost the same length as her original 22-stick, an *unmarked* red stick.

Protocol V. An important element in Joe's construction of a partitive fraction scheme.

T: (Asks Joe to pull out one of the parts of the 11/11-part stick and Joe does so.) What fraction of the whole is that? One piece?
J: (Thinks for 5 seconds.) Umm. One-eleven?
T: One-eleventh. Can you make me a stick that is five times as long as that (Pointing to the pulled-out part.) one-eleventh?
J: (Draws a stick below the 1/11-stick, extending his stick beyond the one part. He counts five parts along the 11-part stick while continuing to draw his stick and tries to line the end of the drawn stick up with the end of the fifth part. Unfortunately, the beginning of the stick he is drawing is not lined up with the beginning of the 11-part stick, only with the one pulled-out part. Joe repeats the 1/11-stick five times to make a 5-part stick above his estimate. His estimate is just a tiny bit short.)
T: So which one is five times as long as that? (Pointing to the first part of the new 5-part stick.)
J: (Points to the 5-part stick then to the estimate he drew.) The bottom. (Meaning his estimate.)
T: That is five times as long as what?
J: One of those. (Pointing to a part of the 5-part stick.)
T: What was the name of it?
J: One-eleventh.
T: (Questions Joe's decision concerning the two sticks, pointing out that his estimate was a little shorter than the 5-part stick. Joe acquiesces and agrees that the 5-part stick is the one that is really five times the one-eleventh.)
T: What do we want to call it? What fraction of the red stick is that stick that you just made?
J: Let's see. Eleven. (Pointing to the 11-part stick that is almost the same length as the red stick.) This is five. (Pointing to the 5-part stick.)
J: (Thinks for 13 seconds.) Six. One-sixth.
T: How do you say that?
J: No, one-fifth. 'Cause if you use umm...Wait! (Thinks for another 13 seconds.) I don't know.
T: Do you want to tell me what you were thinking?
J: If you use it six times you will get eleven (Meaning to iterate a 1/11-stick six times and join the result to the 5-part stick. The teacher pulls out the first part of the 5-part stick.)
T: This part right here was what?

J: One.
T: One-eleventh, right. Do you agree with me?
J: No! Five-elevenths!
T: What is five-elevenths? (Joe thinks.) Which one? Show me. (Joe points to the 5-part stick.) Why is it five-elevenths?
J: Because it's five and it's part of eleven.
T: That's very good! That's five parts of the eleven. That is fantastic!
T: If I asked you to make a stick that is eight times as long as one-eleventh, what fraction of the red stick do you think it would be? (Joe thinks for 5 seconds.) You can make it and look at it if you want.
J: Umm. I think eight-elevenths.
T: Eight-elevenths! You're so smart!
T: What if I say, "make me a stick that is twelve times as long as one-eleventh"? (Joe thinks for 5 seconds.)
T: Can you make me a stick that is twelve times as long as one-eleventh? (Joe shakes his head.) No? Why not?
J: 'Cause it goes over eleven.
T: It goes over eleven. (Laughing.) How much?
J: One.
T: One what?
J: One...stick over eleven.
T: One stick over eleven. (The observer whispers for the teacher to accept that answer.) That is fantastic! It's your turn to make up a problem for me.

In his response to the teacher's request to make a stick that was five times as long as the 1/11-stick, Joe interpreted the "five times as long" as meaning he needed five of those sticks, end-to-end. Joe used his concept of iterable length units in assimilating the teacher's multiplicative language. Moreover, choosing an unmarked estimate when the teacher asked him to point to the stick that was five times as long as the 1/11-stick indicated that a stick did not need to be marked into five parts for it to be five times as long as the unit stick that was used to produce it. This was an important event because it indicated that the length of the stick was relevant as well as the numerosity of the parts. Had he chosen the 5-part stick it might have indicated that he relied on the numerosity of the parts to make the desired stick. So, both numerosity and length were involved in his production of a stick that was five times as long as the 1/11-stick.

When the teacher questioned his choice of which stick was five times as long as his 1/11-stick, it may have oriented Joe to believe that he had chosen the wrong stick: "Let's see. Eleven (pointing to the unit stick), this is five (pointing to the 5-part stick)" in response to the teacher's question to name the 5-part stick as a fraction of the whole. Joe was now in a state of perturbation as indicated by the 13 seconds he spent in deep thought. The numerical comparison left him with a difference of six, so he chose one-sixth as a possible answer. He then rejected that answer because he had five parts in the stick he was trying to name, so he chose one-fifth. So, "one-fifth" referred to the numerosity, five, of the 5-part stick at least momentarily. Likewise for "one-sixth."

Once the 1/11-stick was pulled out of the five-part stick, Joe was able to make the quantitative comparison again and realized that he had five of those elevenths, so taking them as one thing gave him five of one-eleventh. In saying, "Because its five and its part of eleven," Joe made a part-to-whole quantitative comparison, which is indicative of a partitive fraction scheme. It was at this point in the teaching episode that Joe made the connection between the results of iterating a unit fraction and its relation to the whole stick because he was able to re-embed the 5-part stick within the 11-part stick. Joe had now formed a scheme that he could use to generate proper fractions as iterations of a unit fraction. Naming a stick that was eight times as long as one-eleventh as an 8/11-stick without making it is solid indication that he could now generate meaning for nonunit proper fraction words. Joe's fraction scheme, however, was limited to producing fractions that could be embedded within the fraction whole. His comment that "it is five and it's part of eleven" relates back to this limitation. The necessity of the part-whole comparison is solidly indicated by Joe's belief that the resulting stick could not be more than the whole as indicated by his claiming that he could not make twelve-elevenths even though he knew it would be one more than eleven. This constraint also corroborates the claim that length was a constitutive part of his meaning of proper fraction number words.

Joe's Production of an Improper Fraction

An important event occurred in the teaching episode held on 10th of February: Joe began to reorganize his partitive fraction scheme into a scheme to generate fractions greater than the fractional whole. In contrast, neither Jason nor Laura constructed the iterative fraction scheme throughout their fourth grade. In fact, in Laura's case, we did not judge that her partitive unit fraction scheme had emerged until the 10th of January of her fifth grade year (cf. Chap. 6, "An Attempt to Engage Laura in the Construction of the Unit Fraction Composition Scheme"). The reasons for this crucial difference between Joe's constructive timeline and Jason and Laura's timelines are explored following Protocol VI.

Protocol VI begins after Joe had created first three-fifths and then four-fifths of an unmarked stick by making a mark at the appropriate point on the stick by using a copy of a 5-part stick as a guide. The 5-part stick and a stick with one mark four-fifths of the ways along the stick were visible on the screen (cf. Fig. 7.7).

Fig. 7.7. Joe's mark for four-fifths of a stick.

Protocol VI. Joe producing six-fifths of the 5/5-stick.

T: That's really neat! Now I'm really hungry. I want you to make me another one. I want you to make me six-fifths of that candy.
J: Can't!
T: Why not?
J: You only got five of them.
T: Five what?
J: Fifths.
T: You only got five-fifths. So is there any way of making one, do you think?
J: Make a bigger stick.
T: Make a bigger stick. How much bigger do you think it should be?
J: One more fifth.
T: Ok. Do you want to show me?
J: (Pulls the end part out of the stick that has a mark at the four-fifths position only, and joins this one piece to the end of his stick to make a stick one-fifth longer than the original see Fig. 7.8.)

Fig. 7.8. Making six-fifths of a stick.

Significantly, Joe was able to interpret the teacher's request for six-fifths of the candy as being one more fifth than the whole bar. The novelty for Joe was to envision a *longer stick* that would *include* the whole stick. We regard this as a modification of his partitive fraction scheme, for which a fractional part had to be included in the whole of which it was part. One reason this modification was possible for Joe was that his partitive fraction scheme included unit fractions whose iterability had been inherited from his iterable unit of one. Joe knew that six is one more than five, so six-fifths could be one more fifth than five-fifths. But iteration by itself would be insufficient to produce six-fifths of the unit stick unless there was a reversal of the inclusion relation between the part and the whole.

Joe's production of the 6/5-stick was qualitatively different than Jason's production of improper fraction language (cf. Chap. 5, Protocols XVIII and XIX) while Jason was in his fourth grade. In Jason and Laura's case, the teacher initiated the production of fraction language by asking the children to make a stick that was twice as long as a 6/11-stick. After Laura produced it, the children were then asked how much that stick was of the unit stick. The children were not asked to produce a stick that was twelve-elevenths of the unit stick. In contrast, Joe was asked to produce a stick that was six-fifths of the unit stick without any indication about how he should make the stick. The teacher was involved in Joe's production and asked Joe timely questions, but she did not indicate to Joe how he was to proceed. It was Joe who pulled the end part out of the stick that was marked at the four-fifths position and joined that part to the end of the stick to make a stick one-fifth longer than the original.

Because the bottom stick was marked into five parts in Protocol VI, we do not know what would have happened if the stick had been a blank stick like the one presented to Jason in Protocol XX of Chap. 5. If Joe would have first partitioned the stick into five parts and then pulled out one of the five parts and joined it to the end of the stick, it would have indicated that he could partition a stick with the goal of iterating a resulting part to form a fractional number. Recall that a fractional number is a connected number that takes its fractional meaning from the part of which it is a multiple. The relation to the whole of which it is a potential part is inferential in that it is established by means of reasoning. The continuation of Protocol VI provides an opportunity to further investigate Joe's modification and to investigate the question of whether Joe's partitive fractions had emerged into fractional numbers.

Immediately prior to the continuation of Protocol VI, Joe had successfully estimated one-seventh of a stick and had used his estimate to mark off all seven-sevenths on the original stick. He had then pulled out four-sevenths of the stick to give to an observer. The teacher asked him to make a stick that was nine times as long as the 1/7-stick. Joe cut off the first part of the 4/7-stick. He then repeated this 1/7-stick nine times to make a 9-part stick. The following is an excerpt from the ensuing conversation.

Protocol VI. (Cont)
T: How long is that stick?
J: (Joe thinks for 3 seconds) Nine-sevenths.
T: Why? (Joe thinks for 15 seconds) You are right. It is nine-sevenths. But why do you think it is nine-sevenths?
J: Because it was...you were making these, the sevenths (pointing to the parts of the 9/7-stick). So each of these would be one-seventh.

In the excerpt, Joe was able to work with a unit fraction as both a part of a whole and a unit part freed from the whole. After repeating a 1/7-stick nine times to make a stick nine times as long as one-seventh of the original whole stick, Joe was able to name the resulting stick as nine-sevenths "because...you were making these, the sevenths, so each of these would be one-seventh." This comment indicates that one-seventh was freed from the whole because he did not name the 9-part stick he made as nine-ninths and because he explicitly said that each of the nine parts would be one-seventh. So, the 1/7-stick for Joe was an iterable unit[5] that he used to generate a composite unit of nine, where each one of the nine units was one-seventh of his 7/7-stick. This suggests that Joe had constructed nine-sevenths as a *fractional number*. That is, it suggests that nine-sevenths was a multiple of one-seventh and that its relation to the whole was inferential in that it was established by means of reasoning. So, Joe seemed to have constructed the operations that are necessary to transform his partitive fraction scheme into an iterative fraction scheme for generating improper fractions.

We return to a discussion of Protocol VI to explain what these operations might be. In the discussion, we assume that Joe had constructed the operations that produce an improper fraction for illustrative purposes. Because Joe had constructed the explicitly nested number sequence, he knew that a unit of six units of one contains one more unit than a unit containing five units of one. In this case, there would be no necessity to consider the extra unit as also belonging to the original five

---

[5]To judge whether a unit fraction is an iterable unit requires the observation that the child uses it to produce an improper fraction. In the case of the partitive fraction scheme, a unit fraction inherits its iterability from the iterable unit of one.

units. In contrast, when joining a 1/5-stick to the five units of the 5/5-stick, to know that the 6-part stick produced was six-fifths he would need to know that the 1/5-stick belonged to both the new 6-part stick and to the original 5/5-stick. This entails splitting the 5/5-stick because he would need to conceive of the 5/5-stick as a unit whole at once partitioned into five parts and as realized as a multiple of any one of these parts. This also entails the operations that produce a unit of units of units (the 6/5-stick conceived of as a composite unit containing the 5/5-stick and the 1/5-stick), which is the unit structure that underpins a true conception of multiplication.

Patricia's Recursive Partitioning Operations

The 15th of February teaching episode was the first time the teacher taught Patricia during her fourth grade. The purpose of the teaching episode was to explore Patricia's operations to find out if they were compatible with Joe's operations. What follows indicates that Patricia had constructed recursive partitioning operations. After Patricia partitioned a stick into nine equal parts using PARTS, she was asked to make one-half of a 9/9-stick. She independently pulled the middle 1/9-stick from the partitioned stick using PULL PARTS and marked it into two equal parts again using PARTS. She put this marked 1/9-stick back over the middle ninth of the 9/9-stick and cut the stick into two equal parts using CUT. The teacher then asked what fractional part of the whole sticks one-half of the ninth would make. After an interval of time during which the teacher moved one-half of the 1/9-stick to the beginning of the original stick, Patricia counted along the nine-ninths by twos and said "one-eighteenth."

She was then asked how to make one-twenty-seventh from the 9/9-stick. After thinking for 30 seconds, she pulled out a 1/9-stick and used PARTS to partition it into three equal parts. She then repeated this partitioned 1/9-stick nine times using REPEAT to make a stick with twenty-seven parts. Using PULL PARTS, she then pulled out one of these parts and called it one-twenty-seventh! To know to pull a 1/9-stick out from the 9/9-stick, to use PARTS to partition it into three parts, and then use REPEAT to make a stick with twenty-seven parts, prior to operating, Patricia would have needed to mentally insert three equal parts into each ninth. So, rather than the teacher imposing the recursive nature of PARTS, Patricia produced recursive partitioning as a means of reaching her goal to make one-twenty-seventh from the 9/9-stick.

The Splitting Operation: Corroboration in Joe and Contraindication in Patricia

Corroboration of the splitting operation in Joe occurred in the next teaching episode on the 22nd of February. In a task similar to the one posed to Jason and Laura in Protocol XII of Chap. 5, Joe and his new partner, Patricia, were asked to produce a stick such that a given 9-stick was nine times longer than their stick.

7 The Partitive, the Iterative, and the Unit Composition Schemes

Protocol VII. Joe's splitting operation.

T: Patricia can you find a stick so that this one here (Pointing to the 9-stick just marked by Joe.) is nine times as long as it?

P: (Repeats the original stick, intending to make a stick nine times as long as the original. She ends up with a stick with 72 parts and knows that she only repeated eight times. Patricia pulls out nine parts and goes to join these to her 72-part stick but is interrupted by the teacher.)

T: The question was, Patricia, can you find a stick or make a stick so that this one right here (Pointing to the original marked stick.) is nine times as long as it?

P: You mean nine times as long?

T: This one (Pointing to the original marked stick.) is nine times as long as the stick.

J: Oh, I can do that! That's easy! (Smiling.)

T: How?

J: Just do one of these. (Points to the first share of the original, 9-part stick.)

P: (Patricia doesn't understand.)

T: (To Joe.) Tell her what you mean.

J: You said that this (Pointing to the original stick.) is nine times as long as it, so just put one of these. (Pointing to the first share.) That's (Pointing to the whole stick.) nine times as long as it.

P: (Still doesn't understand so the teacher asks Joe to show her. Joe pulls out the first share from the 9-part stick.)

J: This long stick (Pointing to the 9-part original.) is nine times as long as this little one. (Pointing to the pulled out first share.)

J: (Moves the little stick underneath the 9-part stick, counting each part as he does so.) See. One, two, three,...,eight, nine.

P: Is that what you meant? That little piece? (The teacher nods.)

P: Oh! I thought you meant the big stick.

For Joe to realize that the 9-stick was nine times as long as any one of its parts indicates a composition of his partitioning and iterating operations. Furthermore, when posing a problem for the teacher in Protocol II above, Joe was aware that he did not have to do anything because an unmarked 24-stick was present on the screen. He was aware that four 6-sticks together constituted a 24-stick without having to act.

Patricia, on the other hand, attempted to make a stick by iterating a copy of the 9-stick nine times. Even though her interpretation of the teacher's request could be explained by the ambiguity of the teacher's phrase, she still maintained her initial interpretation after the teacher's attempt to clarify her statement and after Joe's solution and explanation. It would appear that the splitting operation was not available to Patricia at this point in the teaching episode. This seemed to be an anomaly because we thought that recursive partitioning implied the splitting operation. So, we continued to ask Patricia to make fractions of fractions to explore whether the splitting operation would emerge.

Protocol VII. (First Cont)

T: (Asks Patricia to make ten-eighteenths of the original nine-part stick.)

P: (Repeats the original stick, making an 18-part stick and fills ten of those parts.)

T: (To Joe.) Can you make a stick so that this one here (Pointing to the original 9-part stick.) is three times as long as it?

Protocol VII. (continued)

J: (Thinks for 10 seconds then smiles. He erases the first two marks from Patricia's 18-part stick and pulls out the stick so formed. He moves this stick under the original stick and repeats it three times to make a stick the same length as the original 9-part stick.)
T: That's really good. So what is that stick? How long is it?
J: One-third.
T: That's good. Is there another name for it, Patricia?
P: (Focusing on the 9/9-stick Joe made.) The whole?
J: (Joe is smiling and thinking to himself.)
T: The piece that he worked with. (Pulling out the piece that Joe used.) This one right here.
P: (After 3 seconds) Three-ninths.
T: Three-ninths. (To Joe.) Do you agree with that?
J: (Does not respond at first. He thinks seriously, then nods his head in agreement.)
T: Why?
J: (Shrugs his shoulders.)
T: Don't just agree with what she says! Why?
P: (Drags Joe's piece underneath the parts of her stick. It lines up with three of them.) Yes, three-ninths.
J: It can be three-ninths!

For Joe to have solved the task of making a stick such that the 9-stick would be three times as long as his stick in the way that he did, he must have conceived of the 9-stick as three times a 3-stick which was embedded in the 9-stick. That he said that it was one-third of the 9-stick indicates that he considered the stick that he was looking for was one-third of the 9-stick prior to acting. The fact that he established this stick by erasing two marks from Patricia's 18-stick indicates that he regarded each part of the 18-stick as being the same as any one of the parts of the 9-stick,[6] and that three of these parts, any of which repeated three times would produce the 9-stick. Joe was able to free the parts of Patricia's stick from the whole stick that they constituted, and use them as disembedded parts of the 9-stick. In agreeing with Patricia's name for his unmarked 1/3-stick as three-ninths of the original stick, Joe probably projected the three parts back into his stick to reconstitute the three one-ninths. Corroboration for this conjecture came toward the end of the same episode. In the second continuation of Protocol VII, the teacher asked the children to find a stick that was one-eighteenth as long as the 9-stick. A 3-stick that was the same length as the 9-stick was still available on the screen.

---

[6]Patricia made the 18-stick using two 9-sticks, so for Joe, each of its parts was of length equal to each part of the 9-stick.

7 The Partitive, the Iterative, and the Unit Composition Schemes

Protocol VII. (Second cont)

T: The original stick was a 9-stick, right? The question is, can you find a stick that is one-eighteenths as long as the 9-stick?
P: Oh yeah! I see.
J: (Places marks on the first third of the 3-stick that would make six pieces in that first third, not all equal, though.)
T: What are you doing, Joe?
J: (Shrugs his shoulders.) I tried.
T: What were you thinking of, Patricia when you said, "I see"? Can you show us?
P: (Erases all the marks from the stick Joe was working on and dials "18" in PARTS to make eighteen parts in the stick.)
J: I should have thought of that!
T: (To Joe.) So, show me the piece that is one-eighteenths of that. What is the stick that is one-eighteenths of the 9-stick?
J: The purple one?
P: (Pulls the last piece out of her 18-part stick.) This is one-eighteenth of the whole stick, right there.
T: (To Joe.) Do you agree with that?
J: (Joe nods.)
T: Why?
J: Because when you said make a stick that this is nine times bigger than it, that's what I did.

Joe attempted to solve the task of making a stick that would be one-eighteenths of the 9-stick by marking one part of the 3-stick into six parts. Using his multiplicative knowledge that three times six is eighteen, he projected six parts into one of the three parts of the 3-stick using recursive partitioning. Patricia solved the task by simply erasing all marks from Joe's 3-stick and putting eighteen parts in the blank stick, using PARTS. She then pulled one of these eighteen parts out of the stick. Joe was surprised by the simplicity of her solution because he had constructed the solution using recursive partitioning. He did agree that the stick Patricia pulled out of her 18-part stick was one-eighteenths of the 9-stick: "Because when you said make a stick that this is nine times bigger than it, that's what I did." This statement, when coupled with his previous behavior in the continuations of Protocol VI indicates that he had, indeed, established a *splitting operation* – a composition of his partitioning and iterating operations. Patricia's solution to the one-eighteenths of the 9-stick task only indicates that she used a partitive fraction scheme to find one-eighteenths of a given stick, partition the stick into eighteen equal parts, and disembed one of those parts. The above protocols cast doubt on the present availability of the splitting operation in Patricia's case.

A Lack of Distributive Reasoning

Joe had constructed the operations that produce three levels of units and operated as if he was aware of all three levels of units prior to operating further using the three levels. Furthermore, he routinely engaged in recursive partitioning and operated in such a way that we attributed the splitting operation to him as well. In the 1st of March teaching episode, we were interested in whether these operations were sufficient to produce distributive partitioning (cf. Chap. 4, "Levels of Fragmenting"). Initially, Joe was asked to share a pizza, that had been cut into four parts, among six friends, but he was unable to make the partitioning. So, the teacher asked Joe to share four slices of pizza among eight friends. The following protocol begins at the point where Joe realizes that he can put two parts into each part of the 4-part stick that represents the four slices of pizza. The teacher was interested in whether Joe could name the share of two of the eight people as both "two-eighths" and "one-fourth."

Protocol VIII. A problem with finding another name for two-eighths of a pizza.
J: Oh! I know how to do it. (Dials "2" in PARTS and makes two equal parts in each of the four parts of his stick.)
T: That's really good! So, how much does each person get? Show me the share of one person.
J: (Points to the first part of the 8-part stick.)
T: Show me. Show me that that is the share of one person.
J: (Pulls out that first part and repeats it eight times to make a stick the same length as the 8-part stick.)
T: How much is that share? The share of one person?
J: One-eighth.
T: How much would the two of us get together?
J: (Sits quietly, thinking.)
T: How much of the pizza would you and I get?
J: Two-eighths.
T: Two-eighths. How much of the whole pizza...is there another name for it?
J: (Thinks for 7 seconds)
T: Can you think of another name?
J: Two-fourths.
T: You think so?
J: No, umm. Eight-fourths?
T: Let's see it. Let's see if it is eight-fourths. Pull out the share of the two of us together.
J: (Pulls out the first two parts of his 8-stick, one at a time. The teacher asks him to join them together. He does so.)
T: So how much of the whole pizza is that? How much did you say it was?
J: Two-eighths.
T: Two-eighths. (Asks Joe to erase the mark in their share.) Now how much of the whole pizza is that?
J: Eight-fourths.
T: Let's see it. Show me that it is eight-fourths.
J: (Joe repeats the share four times to make a 4-stick the same length as the 8-stick.)
T: How much is it?
J: Eight-fourths.
T: Is it?!
J: Let me think about it. (Joe thinks for 20 seconds) I don't know.

Joe was not able to share four slices of pizza among six friends, which contraindicates that he had constructed distributive reasoning, and his behavior is consistent with the findings of Lamon (cf. Chap. 4). He was yet to realize apparently that taking one-sixth of each of four parts is equivalent to taking one-sixth of all four parts together. He did not engage in this kind of reasoning when sharing the four parts among eight people even though he iterated the one-eighth he pulled out eight times to verify that it was indeed one-eighth. After partitioning each of the four parts of his stick into two parts, he knew he had made eight parts, so he could answer "two-eighths" as how much the share of two people was of the whole pizza. It was a surprise that he did not also say "one-fourth" because, based on the operations that he had already demonstrated, he could restructure the bar into a unit containing four composite units each of which contained two individual units and operate on this unit structure. The sharing task was novel with respect to the operations that he used up to this point in the teaching experiment, and the operations of which he was capable seemed suppressed.

Patricia, in the first part of the 3rd of March teaching episode, like Joe in the 1st of March teaching episode, when given sharing situations involving a sliced pizza, she was only able to share three slices among six people. In that situation, like Joe, she halved each slice. She had no way of approaching the task of sharing four slices among six people. She tried to overcome the perturbation by changing the constraints of the task: either she relaxed the requirement that all people get the same amount or she ate the extra two slices after she had halved each of the four slices and given six of these halves to her guests. She also could not think of a way to share five slices among four people. It appears from these last two teaching episodes that neither Joe nor Patricia's assimilation of the tasks evoked distributive reasoning in the two children. So, we are forced to infer that the operations of distributing one partition over the elements of another partition to make a distributive partitioning are more involved than recursive partitioning. They entail partitioning the whole of, say, four bars into six parts by partitioning each bar into six subparts and assembling one of the six parts by taking one of the subparts from each bar. It would seem as if the operations that produce three levels of units, recursive partitioning operations, and splitting operations would be sufficient to engage in distributive reasoning.

Emergence of the Splitting Operation in Patricia

In the 3rd of March teaching episode, Patricia was able to make a stick such that a given stick was forty-eight times as long as the one to be made. This was the first indication of the splitting operation. She also said that the stick she made was one-forty-eighth of the given stick.

Protocol IX. Patricia's splitting operation.
T: (Draws a long stick across the top of the screen.) I'm thinking of a stick and the stick that I am thinking of, when I compare it with this stick (Pointing to the long stick on the screen.), this one is forty-eight times as long as the stick I am thinking. Find me that stick.
P: (Thinks for 6 seconds and then dials up "48" in the PARTS button. She makes forty-eight parts in the given stick and then pulls one part out.) Is that the stick?

Protocol IX. (Cont)
T: That's right! How did you know that was the stick I was thinking of?
P: Because you said this (Pointing to the long stick.) was forty eight times as long as the stick, so one of these must be one-forty-eighth!
P: (Sets a problem for the teacher. She draws a very short stick and says that this stick is ninety-nine times as long as the stick she is thinking of. It becomes clear that she is thinking of a very, very tiny stick. She wants the teacher to put ninety-nine parts in her small stick, and she says it will be all black. The teacher does so and tries to pull one part out. Patricia also tries to pull just one part out of the black stick.)

Patricia interpreted the teacher's language appropriately when finding a stick so that the one the teacher had drawn would be forty-eight times as long as the stick Patricia was to find. This was a change from the 17th of February teaching episode where Patricia was unable to understand Joe's explanation for taking one part out of a 9-part stick in response to a similar request. The above episode does indicate that Patricia had constructed the splitting operation because she must have split a stick into forty-eight parts in order to produce a stick such that the given stick was forty-eight times as long as the stick to be produced. She also knew that the stick to be produced was one-forty-eighth of the given stick. So, not only did Patricia split the given stick into forty-eight parts, but also she engaged in reciprocal reasoning.

For Patricia, one-forty-eighth was a *unit fraction* as distinct from a *partitive unit fraction*. The stick she started with was forty-eight times as long as the stick she made by means of splitting the stick into forty-eight parts and the stick she made was one-forty-eighth of the stick she started with because it could be iterated forty-eight times to produce the stick she started with. Further, the reciprocal relation between the stick she started with and the stick she made was established prior to action as indicated by her comment, "Because you said this (pointing to the long stick) was forty-eight times as long as the stick so one of these must be one-forty-eighth!" This claim is corroborated after Patricia playfully selected "99" when setting a problem for the teacher. Although it is not indicated in Protocol IX, she was convinced that when one part was pulled out of the 99-part stick, that the part was too big and must be more than one part.[7] Her judgment corroborates the claim that the reciprocal relation between the whole and the part was anticipatory and that she established it prior to action.

There was indication in the 15th of February teaching episode that Patricia had constructed recursive partitioning, so it was a surprise to the researchers that there were no indicators of the splitting operation in the 22nd of February teaching episode. However, given Patricia's performance in Protocol IX in the 3rd of March teaching episode, we do infer that the operations that produce splitting were available

---

[7]Note that when pulling one part out of the 99-part stick, the computer creates a beginning and an ending tick mark next to each other so the thickness appeared to be a lot more than one-ninety-ninth of the small stick.

7 The Partitive, the Iterative, and the Unit Composition Schemes

to her in Protocol VII even though they were not provoked. The splitting operation does not suddenly emerge in children in the context of solving a task without there being operations available that permit the emergence.

Emergence of Joe's Unit Fraction Composition Scheme

The emergence of the fraction composition scheme was observed early on in Jason's fifth grade without previous behavioral indication of recursive partitioning or of splitting (cf. Chap. 6, "The Unit Fraction Composition Scheme"). In fact, we inferred that recursive partitioning and splitting were embedded in Jason's reversible partitive fraction scheme[8] and that they were the operations that transformed that scheme into a unit fraction composition scheme. Now, we can test the hypothesis that children who are yet to construct a unit fraction composition scheme embed recursive partitioning and splitting operations in their reversible partitive fraction scheme in the construction of a fraction composition scheme.[9] In Protocol X, which was extracted from the 8th of March teaching episode, Joe initially did not know how to show the teacher one-half of one-fifth of a 5-part stick.

Protocol X. Making one-half of one-fifth and of one-tenth.

T: (A stick is at the top of the screen partitioned into five equal parts.) Show me one-half of a fifth of that stick.
J: (Thinks for 40 seconds) I don't know.
T: Just draw what you think one-half of a fifth of that stick is going to be and then it will be your turn to make up a problem for me.
J: (Joe goes to draw a stick directly underneath the first part of the 5-part stick but stops after making the mark at the beginning of the stick. He then uses PULLPARTS to pull out the first part of the 5-part stick. He dials "2" in the PARTS button and marks this 1/5-stick into two equal parts.)
T: So what's a half of a fifth? Show me.
J: (Joe points to one of the two parts in the 1/5-stick.)
T: Pull it out and let me see it.
J: (Joe pulls out the second part from the 1/5-stick.)
T: You are so smart! How much of the whole stick is that?
J: (After 2 seconds thought.) One-tenth. (Smiles.)
T: Why is it one-tenth?
J: (Joe thinks but does not respond.)
T: You are right: that is one-tenth. But why is it one-tenth?
J: I don't know.

---

[8]For a scheme to be a reversible scheme means that any result of the scheme can be taken as a situation of the scheme and that the activity of the scheme can be reversed to produce a result of the scheme, which is a possible situation.
[9]For these operations to be embedded in Joe's reversible partitive fraction scheme, they must be operations used in assimilation.

Protocol X. (continued)

T: (Following the above, the teacher uses PARTS to partition a stick that Joe had made into ten parts and pulls out one of the parts as her response to Joe's challenge to her to make one-tenth of his unmarked stick.)

J: Make one-twentieth of that. (Pointing to his stick.) No, that's too easy. All you've got to do... is make a half...

T: Make one-twentieth? Can you make one-twentieth of this (pointing to the 10-part stick) without erasing the marks?

J: (Joe dials "2" in the PARTS button and partitions the teacher's pulled out 1/10-stick into two equal parts. He pulls out the second part and smiles as if he had made a 1/20-stick.)

T: (Orients Joe to make one-twentieth using the 10-part stick rather than the 1/10-stick.) Let's try to make that (Pointing to the 10-part stick.) one-twentieth without making a new stick. Show me one-twentieth of that (Pointing to the 10-part stick.) without erasing any of the marks.

J: (Joe moves his 1/20-stick up above the first part of the 10-part stick and uses it to mark the middle of that first part. He then moves his 1/20-stick above the second part of the 10-part stick and uses it as a guide to mark the middle of that part. The witness wants Joe to use PARTS.)

T: I know you don't like using PARTS, but how would you use it? How would you use it if you had to? I know you don't like using PARTS.

J: (Joe thinks for 10 seconds.)

T: Just show me.

J: I can't use it because I can't erase the lines. (Meaning the hash marks.)

T: (Miss-hears Joe.) What do you mean, you are going to erase the lines? I don't understand.

J: (Joe goes to the PARTS button and dials "2." He clicks on the third, unmarked part of the original 10-part stick. It is partitioned into two equal parts as a result of his click. He continues clicking the cursor in each of the remaining tenths of the stick, ending up with a repartitioned stick with twenty parts!)

In response to the teacher's request to make one-half of one-fifth, Joe sat for 40 seconds thinking and then said that he did not know what to do. Given his later comment that "I can't use it because I can't erase the lines (meaning the hash marks)," Joe initially did not know what computer actions he could take to make one-half of one-fifth. This constraint provided an occasion to infer that Joe did not mentally engage in recursive partitioning when he sat thinking for 40 seconds. Once he realized he could use PULLOUT, his reversible partitive fraction scheme was evoked and he pulled out a 1/5-stick, partitioned it into two parts, and pulled out one of the parts and said that it was one-tenth of the original stick. His inability to explain why indicates that the visual graphics were essential in the evocation of visually based recursive partitioning and that he was unaware of how he operated. One might say that there was a reorganization of his perceptual field in that he visually projected two parts into each part of the 5-part stick in one fell swoop as a result of experiencing an insight.

When the teacher asked, "can you make one-twentieth of this (pointing to the 10-part stick) without erasing the marks?" she essentially replicated the test of her hypothesis that Joe could mentally engage in recursive partitioning without using the computer actions. Rather than comply with the teacher's directives, Joe partitioned the teacher's pulled-out 1/10-stick into two equal parts, pulled out the second of the two parts and smiled as if he had made a 1/20-stick. Apparently, Joe had used recursive partitioning when posing his task to the teacher – "Make one-twentieth of that (pointing to his stick). No, that's too easy. All you've got to do… is make a half…." Consequently, it was not necessary for Joe to actually partition each part of the 10-part stick into two parts, producing a 20-part stick, in order to engage in recursive partitioning. For this reason, we hypothesize that recursive partitioning was embedded in the first part of Joe's reversible partitive fraction scheme and that recursive partitioning operations were interiorized, numerical operations for Joe. However, whether recursive partitioning was restricted to one-half awaits further investigation. In any event, Joe's actions in Protocol X opened the possibility for him to take a result of his reversible partitive fraction scheme, e.g., one-tenth, as material for input in the case of other fractions. In this sense he was beginning to construct the scheme as a unit fraction composition scheme.

Joe's Reversible Partitive Fraction Scheme

We inferred in Protocol X that Joe's partitive fraction scheme was a reversible scheme based on how he operated when finding what fraction one-half of one-tenth of a stick was of the whole stick. Corroboration of this inference occurred at the beginning of the 13th of March teaching episode, when Joe independently generated a reversible fraction situation. He presented Patricia with part of an imagined stick and she was to create the whole from it.

Protocol XI. Reversibility in Joe's partitive fraction scheme.
J: (Joe draws a stick about 2 in. long.) That's half my stick.
P: (Patricia dials "2" in PARTS and makes two parts in Joe's stick. She pulls out half.)
T: Is that right?
J: (Shaking his head.) No. I said that (Pointing to his original stick.) was half my stick.
P: (Erases her mark from Joe's original stick and repeats it to make a stick twice as long as Joe's. She then pulls one part out of this 2-part stick.)
T: Which is the stick that Joe was thinking of?
P: The bottom. (The one she just pulled out.)
J: No, the top. (The 2-stick.)
P: Oh, yes.

Joe's rejection of Patricia making one-half of the stick that he had drawn as the stick she had to make corroborates our interpretation that Joe's partitive fraction scheme was reversible. Patricia's partitive unit fraction scheme, on the other hand, did not appear to be a reversible scheme that she could use to produce the whole from one-half. However, based on the fact that she had constructed reciprocal reasoning as well as recursive partitioning, our hypothesis, at this point in the case study, is that she interpreted Joe's language as requesting that she make one-half of the stick he drew without other available operations being evoked. This hypothesis is confirmed in Protocol XII.

Protocol XII. Reversibility in Patricia's partitive fraction scheme.
P: (Starts by making a stick about 6 in. long.) This is a fourth of my stick.
J: That's one-fourth of the stick?
P: (Patricia nods her head.)
J: Can't make that; it'll be off the screen. (At the teacher's request, Joe attempts to make Patricia's stick. He uses REPEAT and keeps track of the number of iterations even though he cannot see them. He checks that he has four parts in this repeated stick by dragging the stick across the screen so as to bring each part into view as he counts them.)
J: (Joe takes his turn to give Patricia a problem. He draws a stick about 1 in. long.) That's a fourth of my stick.
T: No, no. You cannot use her clue. You have to make up your own clue.
J: That's a fifth of my stick.
P: (Patricia repeats Joe's stick to make a 5-part stick.)
T: Is that right?
J: (Joe nods his head.)
T: How long is that stick that you just made?
P: Five of those little pieces. It could be 5 in. because an inch is kinda like one of those…
T: Oh! I see. No, in terms of what he did, what he made. Now, how long is that stick?
P: Five-fifths.

In the 5th of April teaching episode, Patricia presented a task to Joe that compelled us to impute reversibility to her partitive fraction scheme.

Patricia had spontaneously posed a reversible fraction task to Joe for the first time in Protocol XII. When coupled with her response to Joe's problem to make a stick that is one-fifth of a given stick indicates that she constructed reversibility in her partitive unit fraction scheme that she only appeared to lack in the previous episode. In retrospect, it is possible that her attempts in Protocol XI to make a stick such that a given stick is one-half of the one to be made, along with Joe's correctives, evoked a change in Patricia's partitive fraction scheme. She did make a stick that was twice the length of the one Joe said was one-half of the stick to be made. Moreover, after she pulled out one of the two equal parts of this stick, she thought that stick was the desired stick. Joe, however, told her that the desired stick was the other one and she agreed. Apparently, this interaction with Joe was sufficient to bring forth an accommodation in her partitive fraction scheme in that it now included reversibility. In Protocol XII, she knew that the stick she made by repeating Joe's stick five times was a 5/5-stick. So, she was explicitly aware of the reciprocal relationship between the 1/5-stick and the 5/5-stick, a relationship that is based on the splitting operation, which she had constructed by Protocol IX.

Fractions Beyond the Fractional Whole: Joe's Dilemma and Patricia's Construction

Given that Joe had constructed a stick that was six-fifths of a marked 5/5-stick in Protocol VI, we inferred that he used the operations that produce three levels of units, at least in the context of producing that stick. Up to this point in the case study, there have been no behavioral indications that Patricia had constructed those operations. But given the compatibility of the other operations of the two children, such as recursive partitioning and splitting, we hypothesize that Patricia had also constructed the operations that produce three levels of units and that she could produce a fraction greater than the fractional whole. The hypothesis is especially compelling because Patricia could engage in reciprocal reasoning, which is an indicator that she could use three levels of units as input for operating as well as produce them as a result of operating. Protocol XII provides an opportunity to test that hypothesis. The three continuations of Protocol XII also provide interesting data on Joe. Even though Joe had already constructed a stick that was six-fifths of a marked 5/5-stick, he experienced a dilemma – he could produce fractions greater than the fractional whole, but he could not understand why these novelties were fractions.

Protocol XII. (First Cont.)

T: (To Joe.) I would like you to make me a stick that is two times as long as the four-sevenths. (Pointing to the 4/7-stick.)

J: (Joe appears to count to himself. He appears to be saying "eight." He then repeats the 4/7-stick to make an 8-part stick.)

T: How long is that stick, Joe?

J: (Joe erases the marks from the 8-part stick.)

T: Good! How long is that stick in terms of the red stick, the stick we started out with?

J: (After 5 seconds.) Eight.

T: Eight what?

J: Sevenths. No. (Joe seems perturbed. He is moving his stick around.) I don't know. How can it be *eight* sevenths?

T: How can it be eight sevenths? Good question! (To Patricia.) How can it be eight sevenths?

P: You want me to tell you?

T: Yes.

P: Because there's seven in there (Pointing to the partitioned 7-stick.) and you used the same little pieces as in that 7-stick except that that stick is bigger and you used the same little pieces and there's eight in there. (Pointing to the unmarked 8/7-stick.)

T: (The teacher asks Patricia to show Joe what she means.)

P: (Patricia pulls out four parts from the 7-part stick and explains to Joe that he repeated that stick to make a stick with eight parts because two times four is eight, and then he erased the marks. The eight parts were the same as the little pieces in the 7-stick, so it is eight sevenths.)

Joe interpreted the "two times as long as" as the teacher intended and doubled the 4/7-stick. He counted to eight so he knew that he would have eight parts in his resulting stick. However, he said "eight" when asked by the teacher how long the stick was that he made. His immediate response of "sevenths" to the teacher's question of "eight what?" indicates that he did regard each part as a seventh, but then he experienced conflict and said that he did not know the answer. His question, "How can it be *eight* sevenths?" when coupled with Protocol VI, implies that Joe had constructed operations that were sufficient to produce a fraction stick longer than a stick taken as the fractional whole, but that the results of his operations were novel and conflicted with the expected results of his partitive fraction scheme.

It was quite surprising that Patricia not only experienced no conflict, but also that she could explain why the stick Joe made was eight-sevenths. Joe's stick was unmarked, so Patricia's explanation was not merely a perceptual reading of the eight one-sevenths. Rather, it was based on mental operations she carried out. Because she said, "The eight parts were the same as the little pieces in the 7-stick, so it is eight-sevenths," we infer that she *split* the 7-stick into seven equal parts with the understanding that any part could be used to regenerate the whole of the 7-stick. We also infer that she established the 7-stick as a composite unit containing seven parts each of which was one-seventh of the 7-stick, and, further, that she disembedded a part from the 7/7-stick and affixed it to the 7/7-stick to make an 8/7-stick. All of these operations are operations of an iterative fraction scheme and together indicate that Patricia established a unit of units of units, at least in operating. In the second continuation of Protocol XII, Joe is still trying to work out why a fraction can be greater than the fractional whole.

Protocol XII. (Second Cont)

T: Using one-seventh, can you make eight-sevenths?
J: (Joe shakes his head. The teacher waits for 25 seconds.)
T: Using one-seventh can you make me three-sevenths?
J: (Joe repeats the one-seventh three times.)
T: Very good. You are absolutely right. Can you make me seven-sevenths using one-seventh?
J: (Joe goes to repeat the 3/7-stick, but the teacher stops him and asks him to use the 1/7-stick. Joe does so, repeating a 1/7-stick seven times.)
T: Very good. Using one-seventh, can you make me ten-sevenths?
J: (Joe shakes his head "no." The teacher asks Patricia to try. She goes to repeat the 1/7-stick but REPEAT is still active from Joe's seven-sevenths so one more part is inadvertently added to Joe's 7/7-stick. The teacher asks Joe what he would call his stick now that one piece has been added. Joe checks the number of pieces in his stick using a menu item. The numeral "8" appears in the number box.)
J: Eight.
T: Eight what?
J: Eight of those. (Pointing to the 1/7-stick.)
T: And how long is that stick?
J: One-seventh.
T: (The teacher asks if the 8-part stick is longer or shorter than the 7/7-stick.)
J: Longer.

7 The Partitive, the Iterative, and the Unit Composition Schemes

T: How much longer?
J: One.
T: One what?
J: One-seventh.
T: It's one-seventh longer. Ok. How many whole sticks do we have in that one right there? (Pointing to the 8/7-stick)
J: (Counting the parts in the stick.) Eight.
T: Eight what?
J: Eight sticks! (Still refusing to name the stick eight-sevenths.)
T: Ok.

Throughout the protocol, the teacher did her best to activate Joe's operations of splitting the 7/7-stick, disembedding one part of the stick and affixing it on the end of the stick, and then uniting the 7/7-stick and the affixed 1/7-stick into a composite unit that could be produced by iterating the disembedded 1/7-stick eight times. When Joe replied "eight of those" after the teacher asked Joe "Eight what," Joe knew that the 8-stick was eight of the 1/7-sticks, but the operations that Joe enacted were essentially those of his number sequence where the 1/7-stick played the role of a unit of one, because he said "Eight sticks!" even after he said that the 8/7-stick was one-seventh longer than the 7/7-stick. The perturbation that Joe was experiencing apparently suppressed the more advanced operations that the teacher attempted to activate, and the suppression continued on throughout the third continuation of Protocol XII.

Protocol XII. (Third Cont)

T: (The teacher trashes the 3/7- and 4/7-sticks and places the 1/7-stick underneath the 7/7-stick. She then asks Joe how much the 1/7-stick is of the 7/7-stick.)
J: One-seventh.
T: Now, what I would like you to do is make me ten-sevenths. (After waiting 10 seconds for Joe to start.) Is it going to be larger than seven-sevenths or smaller than seven-sevenths?
J&P: Longer.
T: How much larger, or longer, is it going to be?
J: (After 5 seconds.) Three-sevenths.
T: That's absolutely right! Would you like to make it?
J: (Joe very slowly copies the 1/7-stick and then repeats it very deliberately ten times.)
T: That's very good. Now what was that? How long was that stick?
P: Are you asking me? (The teacher nods.) Ten-tenths! That stick right there. (Pointing to the stick Joe just made.)
T: How about you, Joe? Do you think she is right or not?
J: (Joe shrugs his shoulders.)
T: What did you just make?
J: Ten of these.
T: Ten of what?

Protocol XII. (continued)

J: Sevenths.
T: Ten of sevenths. So if we have to give it a name, what would we call it?
J: (Joe mumbles something.)
T: What did you say?
P: He called it "John!"
T: Can you give me a fraction name?
J: A seventh.
T: Eh?
J: A seventh, one-seventh.
T: (The teacher asks Patricia.)
P: For the big stick?
J: Oh! For the big stick?
T: Yes, for the big stick.
P: Ten-sevenths.
J: Ten sevenths.
T: Why is it ten-sevenths, Joe?
J: I don't know.
T: Why is it ten-sevenths, Patricia?
P: Because that little stick that we started out with is one-seventh and we made ten of those little sticks, so its ten-sevenths.

Joe's responses to the teacher's questions throughout Protocol XII and its continuations were in stark contrast to his response to a task presented to him almost 2 months prior to this episode on the 10th of February. On that occasion (cf. the first continuation of Protocol VI), Joe was asked to make a stick nine times as long as a 1/7-stick. Joe repeated the 1/7-stick nine times and named the result nine-sevenths! In that case, he appeared to experience no conflict and enacted his most advanced conceptual operations. In the interim, Joe apparently began to doubt that fractions could be greater than the fractional whole. In Protocol XII, Joe was being asked to make something that he doubted could be made. The operations that he used to make fractions greater than the fractional whole in Protocol VI in the 10th of February teaching episode apparently did not reorganize Joe's partitive fraction scheme. These operations were still available to Joe, as indicated throughout Protocol XII, but they were not a part of the operations he used to produce fractions. Fractions, for Joe, did not exceed the fractional whole.

When the teacher asked the children to make ten-sevenths in the third continuation of Protocol XII, she added a question that may have been critical for Joe to begin to resolve his perturbations. She asked if the 10/7-stick would be longer or shorter than the 7/7-stick. Both children responded that it would be longer. Joe could then state that the ten-sevenths was going to be three-sevenths longer than the seven-sevenths, but he still could not conceive of ten-sevenths as a fraction, even though he could establish a relation between ten-sevenths and seven-sevenths!

When the teacher asked, "How long is that stick (referring to the stick Joe had made by repeating the 1/7-stick ten times)?" she did not indicate a unit stick that could be used as a term of comparison. Had either of the children come close to completing their construction of fractional numbers, it would not have been necessary to indicate such a unit stick, simply because a fractional number primarily takes its meaning from the iterations of a unit fraction that produced the fractional number. Patricia's response of "ten-tenths" indicates the nascent state of her construction of fractional numbers because she did subsequently say, "*ten-sevenths*," for the length of the 10-part stick Joe made, and explained why. Joe eventually concurred with Patricia that it would be a 10/7-stick, but he still was not able to provide an explanation for calling it ten-sevenths.

Later in this same episode, Patricia made a very small stick and said it was one-ninety-ninths of the stick she was thinking of. Her posing of this task is indicative of two things. First, it is indicative of the power of her splitting operation and reciprocal reasoning in that she could apply them to what to her were quite small numbers and, reciprocally, quite large numbers. Second, it is indicative of her fascination with creating very small fragments of a stick and operating using those fragments. Joe, using his reversible partitive fraction scheme, started to make her stick by repeating a copy of the tiny stick. The teacher stopped him after making 14 iterations. Both children correctly named this stick fourteen-ninety-ninths of the stick Patricia was thinking of. The teacher then posed the problem of making a stick that was ten times as long as the 14/99-stick. Protocol XIII provides the children's solution to this problem and raises again Joe's doubt that fractions can be greater than the fractional whole.

Protocol XIII. How can a fraction be bigger than itself?
T: The stick that I am thinking of is ten times as long as this one, without using the calculator.
J: As this one (pointing to the 14-part stick)? (Makes several guesses including seventy ninety-ninths, ninety-nine, and ninety-nine ninety-ninths.)
P: (Uses the calculator to find out what is ten times fourteen.)
T: (With "140" showing in the calculator, to Patricia.) You stop now. (To Joe.) you tell me the measure of the stick, tell me how long the stick is, that I'm thinking of.
J: Ninety-nine? No! One hundred-forty ninety-ninths?
T: Very good.
J: I still don't understand how you could do it. *How can a fraction be bigger than itself?*
T: That's a really good question. Think about that for next time.
P: I know how. Because you always have the same little stick you started off with!

Joe's initial response of "ninety-nine" to the teacher's question for the measure of the stick that was ten times as long as the 14/99-stick indicates that ninety-nine ninety-ninths was the largest possible entity that he considered as a fraction involving ninety-ninths. He acquiesced after Patricia used the calculator to find the product of ten and fourteen and said "one hundred-forty ninety-ninths," but he still was not convinced. Joe's critical question – How can a fraction be bigger than itself? – indicated his uncertainty and doubt that was introduced by the functioning of his recently constructed operations throughout Protocols VI and XII. He could conceive of the possibility of a fraction like one hundred-forty ninety-ninths because he knew he could iterate a 1/99-stick one hundred-forty times. However, he was still unable to explain to himself why this result should be considered as a fraction. He was yet to modify his partitive fraction scheme in such a way that it would allow entities like one hundred-forty ninety-ninths to be conceived of as a fraction.[10]

Patricia's explanation, "Because you always have the same little stick you started off with!" indicated that her unit fraction one-ninety-ninth was freed from ninety-nine ninety-ninths in the sense that she could use it as a unit in constructing other fractional numbers in the same way that she used her unit of one in constructing other whole numbers.[11] Her explanation also indicates that the units she produced by iteration were identical to each other, which is a critical realization in the production of fractions greater than the unit whole. These considerations warrant the inference that she had constructed the operations that produce an *iterative unit fraction scheme*. Whether or not she had actually produced this fraction scheme is an open question at this point in the teaching experiment. Joe also had constructed such operations, but he was yet to reorganize his partitive fraction scheme into an iterative fraction scheme.

Joe's Construction of the Iterative Fraction Scheme

In the 19th of April teaching episode, a pizza-baking situation was introduced with the goals of activating, in the two children, the operations that produce the iterative fraction scheme and, in so doing, leading them to establish meaning for improper fractions. A rectangular region had been created on the screen to represent an oven for baking pizza, and a stick inside the oven represented a pizza. *A pizza could be cut into only eight equal slices.* One child decides how many people come to the Pizza Restaurant and how many slices each person wants. The other child, the baker, has to bake enough pizzas to feed the group and then must show one person's share. Both children have to say how much of one pizza each person gets and how much of one pizza was eaten by all of the people. Protocol XIV begins with the first problem posed by Joe, where six friends each order two slices of the 8-slice pizza.

---

[10] One possible explanation for the stark differences in Joe's willingness to conceive of a fraction greater than the whole could be classroom experiences in the intervening months of which we have no knowledge. Traditional approaches to teaching fractions in the elementary grades emphasize that fractions are always *parts* of a whole.

[11] Nevertheless, it was still one out of the ninety-nine equal parts.

7 The Partitive, the Iterative, and the Unit Composition Schemes

Protocol XIV. Twelve slices make 12/8 of a pizza.
J: Six people each have two slices.
P: (Copies one stick out of the oven and uses PARTS to partition it into eight parts. She then wants to use PULLPARTS, but the teacher stops her.)
T: How many pizzas does she need to bake?
J&P: Two
T: Why?
J: Because there's only eight and she'll need four of those (Of the second pizza.) and there'll be four left over.
P: (Patricia copies a second stick out of the oven and partitions it into eight parts. She breaks this stick up into its eight parts.)
T: So how much of the whole pizza does each person get?
P: Two-sixteenths.
T: How much of *one* pizza does each person get?
P: What?
T: Joe, what do you think? How much of one pizza does each person get?
J: Two slices.
T: Two slices, but how many slices did we have in the pizza to begin with?
J: Eight. And then you had to get another pizza and you'll need four out of that pizza.
T: So, if you had to give me a fraction to tell me how much all the people, how much of the pizza all the people had together...
J: Twelve-eighths.

Patricia's answer "two-sixteenths" followed on from the teacher redirecting her to use two pizzas rather than one to make the pizza needed for the six people. Originally, her intention seemed to be to iterate one-eighth of a stick twelve times to establish the pizza needed for the six people. Using two pizzas to make the share of one pizza for one person introduced an element into the situation she was yet to resolve – she had two fractional wholes in her visual field rather than one and she was yet to modify the situations of her iterative fraction scheme to include more than one fractional whole. It would appear that the presence of two fractional wholes permitted Joe to establish a fraction beyond one fractional whole. He seemed to be in the process of resolving his question posed in Protocol XIII.

Protocol XIV. (Cont.) Joe has an insight into improper fractions.
J: Nine people and each person gets two slices. (A new problem.)
T: How many pizzas does she need to cook?
J: (Thinks for 5 seconds.) Two (Holds up two fingers.)
T: Two?!
J: No, wait...three.
T: Why?
J: Because eight plus eight is thirteen, I mean sixteen, and you need three (sic) more, so another pizza.
P: (Copies three sticks, puts eight parts in each and breaks one stick up into its eight pieces. She pulls two of those eight pieces down below the other sticks.)

Protocol XIV. (continued)
T: So what is that, Patricia?
P: The share of one person.
T: (The teacher is confused so Joe points to the two pieces Patricia moved down.)
J: She means those two.
T: Oh! Those two? So how much of the whole pizza is the share of one person?
P: Of just one whole pizza?
T: Uh-huh.
P: Two eighths
T: How much pizza do all people get, put together?
P: (Counting to herself.) Eighteen.
J: Eighteen.
T: Eighteen what?
J: (Inaudibly.) Two-eighteenths. (Loudly.) Eighteen-eighths!
T: Eighteen-eighths? What does that mean?
J: No, ahh…
T: You are right, but what does that mean?
J: You've got eighteen of the… (Points to the broken stick.) eight pieces.
T: Ok.
J: (Joe mumbles something to himself. It could be "wait a minute." He then looks up with wide eyes!)
W: How many eight-eighths do we have in eighteen-eighths?
J: (Puts up two fingers.) Two.
P: (Agrees.)

Joe's engagement in the tasks in this teaching episode became pronounced when he was able to clarify the teacher's confusion concerning Patricia's actions. Joe's first response of two-eighteenths for the amount of pizza eaten by all could have been the result of his interpretation of the teacher's question using his partitive fraction scheme (one person's part of all eighteen pieces). The fact that he corrected himself and offered eighteen-eighths as the amount of one pizza eaten by all nine people indicates self-regulation in his way of operating. This self-regulation results from evoking the operations that produce an iterative fraction scheme[12] and using them to reconstitute the current situation as a situation of such a scheme. Joe was in the process of reorganizing his partitive fraction scheme into a new scheme. Note that Joe's new scheme was not being constructed as a reorganization of his partitive fraction scheme using operations *within* his partitive fraction scheme. Rather, new operations – splitting and the operations that produce three levels of units – that emerged from outside of his partitive fraction scheme were being used.

After Joe answered "eighteen-eighths" and the teacher asked him what eighteen-eighths meant, he again experienced a perturbation as indicated by his comment, "No, ahh…." Unfortunately, the teacher confirmed his answer, but he was able to

---

[12] Split one pizza into eight equal parts and iterate one part eighteen times.

explain, "You've got eighteen of the...eight pieces." His reaction to his own explanation indicates that he had a resolving insight at this point in the teaching episode – he could actually have eighteen of eight pieces by baking more than one pizza. Joe's correct response of "two" to the witness's follow-up question concerning how many eight-eighths we had in eighteen-eighths serves to corroborate that insight.

Before this insight, Joe could full well *split* a pizza into eight-eighths (cf. Protocols VI and VII). "Eight-eighths" referred to a composite unit of eight elements each of which was one-eighth of the composite unit and each of which could be iterated eight times to produce the composite unit. Joe could also mentally iterate one-eighth as many times as he wished to produce a multiple of the unit fraction. But the result, say, eighteen-eighths, was realized only as a result of operating mentally and could not be included within a fractional whole. This was the basis of Joe's question concerning how a fraction could be bigger than itself. He could mentally produce such a fraction, but he could not include it into a fractional whole. We explain Joe's insight as his restructuring the eighteen one-eighths that he made using the results of potentially iterating one-eighth, eighteen times. But the restructuring occurred reciprocally in that the unit of three units, two containing eight units of one-eighth and one containing two units of one-eighth, constituted the eighteen units of one-eighth. His ability to view the unit containing two component units of eight-eighths and one more unit of two-eighths as eighteen units of one-eighth was crucial in his insight and, indeed, it could be said to constitute the insight.

Patricia could at least follow Joe's explanations, indicating that she may also have established quantitative operations similar to those of Joe. In Protocol XV, which follows, Joe's way of operating corroborates the way he operated in Protocol XIV, indicating that a change had indeed occurred, and Patricia's operations seem compatible with Joe's. The number of slices per pizza was changed from eight to twelve. The teacher intended to ask the children, "If we have thirteen people and each person wants three slices of pizza, how many pizzas does she (the baker) need to put in the oven?" But Joe interpreted the task as asking how much of one pizza one person would get if that person got three slices of pizza and the teacher confirmed Joe's interpretation instead of clarifying her comments. From that point on, the teacher tailored her questions to what the children did.

Protocol XV. Operating with improper fractions.

T: Another person walks in, so we have thirteen. And say each person wants three slices of pizza.
P: You said another...
J: One person? He wants three slices of pizza?
T: Yeah. (The teacher accepts Joe's interpretation.) How many pizzas does she need to put in the oven?
J&P: One.
J: Joe copies the stick from the oven and partitions it into 12 parts because each pizza now was to be cut into twelve rather than eight slices. He then cuts off the first three parts.)
T: What's that, Joe?
J: One person's, uh, person's pizza.

Protocol XV. (continued)
T: One person's share?
J&P: Yeah.
J: You said that another person comes in and he wants three.
T: So how much of the whole pizza would that person get?
P: How much of the whole pizza?
J&P: Three, uh, twelfths.
T: Very good. Now let me ask you one question, and you can choose not to answer. (To Joe.) If we put the share of that last person together with the share of the other people, together, how much would that be?
J: Twenty-seven twelfths. (Joe correctly interpreted "the share of the other people" to mean that the other twelve people got two pieces each and the thirteenth person got three pieces because this task was related to an immediately preceding task.)
T: Why?
J: Because you added three to twenty-four and that's twenty-seven, and these (Pointing to the parts of one stick.) are twelfths, and that (Pointing to the partitioned extra stick.) is twelve, so it'll be twenty-four, twenty-seven twelfths.
W: In terms of whole pizzas, can you tell me how much you sold?
P: The whole thing?
W: Yes, the whole thing.
J: How many pizzas I sold?
W: Yes, how much pizza did you sell?
J: Twenty-seven pieces.
W: Can you tell me in terms of pizzas?
J&P: Three.
W: Did you sell all of…
P: Two and three pieces!
J: (At the same time as Patricia.) Two and three-twelfths!

Joe's explanation for why it was twenty-seven twelfths along with his answer, "two and three-twelfths!" corroborates the inference we made in Protocol XIV that he had reorganized his partitive fraction scheme into an iterative fraction scheme with the proviso that he still needed multiple fractional wholes into which he could embed his improper fractions. Patricia's answer, "two and three pieces!" indicates that she structured the result into a unit of units of units, but beyond that there was no corroboration of her iterative fraction scheme that emerged in Protocols XII and XIII.

A Constraint in the Children's Unit Fraction Composition Scheme

The intent of the 25th of April teaching episode was to explore whether Patricia could construct the unit fraction composition scheme and, if so, whether she, as well as Joe, could make the necessary adaptations to their unit fraction composition

7 The Partitive, the Iterative, and the Unit Composition Schemes

schemes in order to find the fraction of a fractional whole that a unit fraction of a proper fraction of the fractional whole is. The teaching episode began with finding the fraction of a pizza constituted by one-half of one-third of the pizza. The teacher used the pizza-baking situation that was used in Protocols XIV and XV.

Protocol XVI. Finding one-half of one-third of a pizza.

T: (Copies a stick out of the oven.) Show me a third of that.
P: It's got no marks.
T: How would you show me a third of that?
P: I would go over to PARTS and put in three pieces and pull one out. (Joe is dialing "3" in the PARTS button. He partitions the unmarked stick into three parts. The teacher asks Patricia to pull out one-third and she does so.)
T: Very good! Now let me ask you a question, and then we'll move onto another task. Do you think a half of that (Pointing to the 1/3-stick Patricia just pulled out.) is bigger than it or smaller than it?
P: Than what?
T: (Fills the 1/3-stick in orange.) A half of the orange piece. Is it bigger than the orange piece or smaller than the orange piece?
J&P: Smaller.
T: (Asks them to show her half of the orange stick.)
P: (Dials "2" in PARTS and pulls out one of the two parts she made in the orange stick.)
T: That's very good. Can you tell me what that is, Joe?
J: (Eventually names it a half of the third.)
T: Can you tell me how much of the whole pizza that little piece is?
J: One-sixth.
T: Why is it one-sixth?
J: Because it's like this: It was part of one-third and because it's three times two because you've got, umm, it's three times two because it's only half and you've got to make the other half. That's how you do it.
T: That's really good. Do you want to do another problem?
J: It'll be two, um, two of those (Pointing to the half of one-third.) in each one (Pointing to the three parts of the pizza stick.), and just count them up and it'll be six.
T: (Does not think Patricia has followed Joe's explanation, but Patricia says she understands.)
P: I got it. Each of those little thingy-ma-jigs, they each have two of those, um, orange lines, and so two times, umm, wait...
T: Take your time.
P: Two times three is six, so it'll be one-sixth.

Joe's response of one-sixth to the teacher's question of how much the one-half of one-third was of the whole pizza was very quick. Joe may have said "one-sixth" because he knew that two times three is six, but his explanation indicated that he engaged in recursive partitioning and mentally inserted a partitioning into two parts into each of the three-thirds of the whole pizza when it was his goal to find how much of the whole pizza one-half of one-third made. He had indeed constructed a unit fraction composition scheme. Patricia seemingly assimilated Joe's explanation of how he partitioned the 3-part stick using the two as a partitioning template. She also said, "two times three is six, so it'll be one-sixth," which constitutes corroboration that not only could she assimilate Joe's explanation, but also that the assimilation was sufficient for her to construct a unit fraction composition scheme. In the next protocol that took place during the last 10 min of this teaching episode, the teacher posed a problem that required making one-half of three-sevenths.

Protocol XVII. What is one-half of three-sevenths of a pizza?

T: The first person gets three-sevenths of a pizza and the second person gets half as much as the first person. So, Patricia, how much of the whole pizza is the second person ordering?
P: A half of three-sevenths. (Makes three-sevenths of a pizza and cuts it into halves.)
T: And how much of the whole pizza is that?
J: (Joins Patricia's two halves back together.) There's three-sevenths right there. That's not half of three-sevenths. That's three-sevenths.
P: A half of three sevenths. No! One of those little pieces, that's what I meant.
J: Oh!!
T: How much is the second person ordering, guys?
J: (Has cut the 3/7-stick in half again and trashes one of the halves.)
P: A half of three-sevenths.
T: How much of the whole pizza is that?
P: A half of three-sevenths!
J: No!
T: What do you think, Joe?
J: (Starts thinking hard, letting go of the mouse and looking down at the table.)
J: Fourteen something.
T: Let's see. Why did you say that?
J: The last one we did you had to double it, uh double, because we did half of...one-third. Just like the last one, so I did that one times two, too.
T: So you think it works this time too? Why don't you do it? See if it works.
J: (Looks for the whole stick.) Where's the whole stick? (He moves the 7-part stick underneath the unmarked one-half of 3/7-stick. He repeats this unmarked stick four times and pauses. The right endpoint of his repeated stick is now aligned with the end of the sixth part of the 7/7-stick. He makes one more repeat, making a stick that goes beyond the 7/7-stick by a small amount, (one-fourteenth). He smiles.)

Patricia made one-half of three sevenths, but she could not find what fractional part of the whole stick it was. She kept repeating that it was one-half of three-sevenths of a pizza. Joe conjectured that he had to double the seven because he was taking half of the three-sevenths and he related this process to taking half of the one-third (he doubled the three-thirds to get six-sixths when he put two parts in each third). However, the only way he had of finding the fractional quantity was to iterate the one-half of three-sevenths piece in an attempt to recreate the whole 7/7-stick. This was a natural use of his reversible iterative fraction scheme. The resulting stick, however, was longer than the 7/7-stick.

A discussion about why the repeated stick did not match the 7/7-stick followed on from Protocol XVII. Joe thought it was because the half of the 3/7-stick was not

7 The Partitive, the Iterative, and the Unit Composition Schemes   211

exactly a half. Because of his reversible iterative fraction scheme, given an unknown part of the whole, he expected to be able to reproduce the whole from the part to find the fraction the part was of the whole, which constituted creative but speculative mathematical behavior. Both children needed to make an accommodation in their iterative fraction schemes[13] in order to construct a fraction composition scheme more general than their unit fraction composition scheme. The accommodation would involve the use of distributive reasoning. For example, in order to partition a 3-part stick into two equal parts in a way that would be useful in finding the fraction that one-half of three-sevenths of a stick is of the whole stick, they would need to partition each of the three parts into two subparts and take a set of three subparts as one-half of three-sevenths. At that point, they would need to continue mentally partitioning each one of the remaining four-sevenths into two subparts and accumulate the subparts to find that there are fourteen subparts in all, thereby establishing that one-half of three-sevenths is three-fourteenths. We demonstrated in Protocol VIII that neither of the children had constructed distributive reasoning at that time.

Fractional Connected Number Sequences

The opening minute of the 28th of April teaching episode provides solid corroboration that both children had constructed an iterative fraction scheme. Joe spontaneously named all the fractions of a 12-part pizza stick and Patricia realized that this could go on indefinitely and chimed in with "infinity twelfths!"

Protocol XVIII. Fractional connected number sequences for twelfths.
J: (Joe makes a 12-slice pizza (stick) and the teacher puts it at the top of the screen above a thin cover that acts as a separator. The teacher asks them to find all fractions of that pizza.)
P: One-twelfth, two-twelfths,..., nine-twelfths, ten-twelfths. (Pause.)
J: Eleven-twelfths, twelve-twelfths, the whole. (With Patricia.) (Pause.)
J: (Smiling.) Thirteen-twelfths, fourteen-twelfths, fifteen-twelfths, sixteen-twelfths, seventeen-twelfths...
P: Infinity twelfths!

Both children seemed quite aware that they could keep on sequentially producing the fraction immediately following a current fraction. For this reason, we refer to the sequence of fractions for twelfths that the children produced as a fractional connected number sequence. Producing it involves the children's number sequence for one. In fact, Joe seemed to abstract his number sequence for one in the process of producing the fraction immediately following twelve-twelfths and Patricia seemed to abstract her number sequence by listening to Joe producing the fractional number

---

[13] In Protocol XVIII, Patricia demonstrates that she has constructed an iterative fraction scheme.

sequence to seventeen-twelfths. The fact that Patricia said "Infinity twelfths" is an indicator that her number sequence was indeed in a state of activation and use. Once abstracted, the children's number sequence for one was generalized in the sense of a generalizing assimilation. The following protocol corroborates this inference.

Protocol XIX. Finding one-half and one-third of twelve-twelfths.

T: (Asks the children to take turns making each of the fractions from the 12-part stick. Patricia starts by pulling out one part, Joe follows pulling two parts. They continue taking turns until they have one-twelfth through eleven-twelfths on the screen. Patricia then copies the whole stick for twelve-twelfths. They explain that they would make thirteen-twelfths by taking thirteen of the 1/12-stick or by adding on one-twelfth to the 12/12-stick. They would make fourteen-twelfths by adding two-twelfths to the twelve-twelfths.)
T: I'm thinking of a fraction, but I'm not sure we have it there. Do we have a half there?
P: Yeah, we have six.
T: What do you mean?
P: A half of the 12-stick?
T: What do you mean, six?
P: (Locates the 6-stick in the staircase of sticks.) This stick, right here.
J&P: (Both children explain how six is half of twelve (six plus six is twelve). Patricia demonstrates by repeating a copy of the 6-stick to make a 12-stick.)
T: You guys did really well. Do we have a 1/3-stick? Do we have one-third of that pizza up there?
J: Yes. Four.
T: Which one is it?
J: Four.
T: What do you mean "four"?
J: The four...four-twelfths.
T: The four-twelfths. You wanna show me that? You want to prove to me that that's true?
J: (Measures the 4/12-stick and "1/3" appears in the number box.)

The children reasoned interchangeably with fractions and with whole numbers. Joe found the stick that was one-third of the original stick by finding a 4-stick saying that it was four. But he was able to name the stick as a 4/12-stick when asked what he meant by "four." Being able to switch back and forth between reasoning with whole numbers and fractional numbers is further indication that their fractional connected number sequence included their whole number sequences as an integrated and constitutive aspect.

It is interesting that Joe interpreted the teacher's request to "prove" that his chosen stick was one-third of the 12-stick by using the MEASURE function. Both children indicated that they were able to produce commensurate fractions for unit fractions of a 12-part stick. Protocol XX picks up where Patricia offers the 3-stick as another possible fraction of the original 12-stick.

7 The Partitive, the Iterative, and the Unit Composition Schemes   213

Protocol XX. Transforming three-twelfths into one-fourth.
P: You can also use three...
J: Three four times.
P: To make a 12-stick.
T: Let's now work with the fractions. Can you think of a different way...a different name for three-twelfths?
J: (After 3 seconds.) One-fourth.
T: Why?
J: Because three times four is twelve.
P: If you uh, if you uh, what's it called? We did this in math. What's it called? You reduce. If you reduce three-twelfths. Like, how many.... You divide by three. Three will go into three and three will go into twelve. So three divided by three is one and three divided by twelve is four. So it's reduced to one-fourth. (Patricia drew the numerals on the table with her finger as she talked.)
T: (Asks Patricia to show her what she means. Joe wants to use a calculator to show the teacher what Patricia meant. Patricia decides to demonstrate with Sticks. She copies the 3/12-stick and fills it in a different color. She copies the 12-stick (Her name for the whole stick.) and also fills all the parts in a new color. She pulls the 12-stick under the 3-stick.)
P: Three-twelfths, Ok. And so. So if you like had... You have to see what number can go into three and go into twelve equally. Three will go into twelve four times; three will go into three, one time. So three divided by three is one. (She copies the 1-stick above and to the right of her 3-stick.). If I was writing on my paper I would put one, and three divided by twelve is four, so I'd put four (Copying the 4-stick below the 1-stick.), so it would be three-fourths. Do you get it? That's the easiest way I can tell you without having paper!
T: (The screen at this point has a 1-stick above a 4-stick and a 3-stick above a 12-stick, as well as the staircase of sticks from 1- to an 11-stick, and two 12-sticks immediately below the separator.) Do you have any ideas, Joe? You said that three-twelfths was a fourth of the whole stick, of that stick. (Pointing to the stick above the separator.) You want to prove to me that that is true?
J: (Takes the mouse.)
T: That is true, by the way. Both of you are right, but I would like to see how you got that answer.
J: (Moves Patricia's 12-stick above her 3-stick, lines up the left ends, and repeats the 3-stick. The right end of the 12-stick is off the screen as Joe does one more repeat than is needed. When he drags the repeated stick fully into view and lines it up with a fully viewed 12-stick, he realizes that he clicked one too many times. He says he must have clicked five times.)
T: (Before Joe cuts off the extra, the teacher asks Joe to measure the resulting stick. He does so and five-fourths appears in the number box.) That's five...
P: Fourths.
T: So how many times did Joe click?
P: Five times.
J: I only clicked four times because I already had one up there (this is true as the first click of REPEAT creates the second stick).
P: (Checks to see how many 3-sticks are in the resulting stick. She counts along the stick, recording triplets as she goes. She gets to twelve and has four fingers raised.) That would be four, and that (Pointing to the three parts.) would be...
J: Five.
P: Five. You clicked one too many times. Do you see it? (She points to the end of the twelfth part of Joe's stick.)
J: There was already one up there, and I repeated it.

In the opening interchange in Protocol XX, Joe demonstrated how he used his whole number multiplication knowledge to produce a fraction commensurate with three-twelfths (one-fourth because three times four is twelve). Patricia, on the other hand, attempted to explain how she was taught to reduce fractions in their mathematics class: find a number that will divide the numerator and denominator and divide each by this number. When asked by the teacher to show her what she meant, Patricia lined a 3-stick above a 12-stick and a 1-stick above a 4-stick to represent the results of dividing three by three and twelve by three. That she said three-fourths instead of one-fourth indicates that her activities were not based on her operations that produce commensurate fractions. Further, there is no indication whatsoever that Patricia saw the 1-stick and 4-stick as representing the same *ratio* as the 3-stick and 12-stick. Rather, her use of the sticks can be regarded as the meaning for the numerals she would write on paper to carry out her procedure. Moreover, her dividing the numerator and the denominator by three did not seem to be related to transforming a composite of three into a composite unit of one and a composite unit containing four units of three into a composite unit containing four units of one, where "one" refers to a 3-stick with the marks erased. Rather, it was a procedure[14] that she had learned to execute and it adequately demonstrates classroom practices that are used in teaching fractions that we judge as inappropriate.

In contrast to the algorithm that Patricia used, Joe's way of demonstrating that the 3-stick was one-fourth of the 12-stick was to iterate the 3-stick four times to make a stick the same length as the 12-stick (a use of his iterative unit fraction scheme). This indicated that he transformed a composite unit containing twelve units of one into a composite unit containing four units of three. So, for him, one-fourth referred to one unit of three out of the four such units. Repeating the 3-stick one time too many was opportune for us as analysts as it provided an occasion to observe Patricia reasoning in a manner similar to the way Joe reasoned to produce one-fourth when she counted in triplets to establish the 15-part stick as five-fourths of the original 12-stick. Such reasoning stands in contrast to her procedure for reducing fractions.

Establishing Commensurate Fractions

The 3rd of May teaching episode serves to illustrate how the children used their iterative fraction scheme to transform unit fractional parts of a 24-part stick into fractions commensurate with the unit fraction, e.g., one-fourth of a 24-part stick as six-twenty-fourths of the stick. Protocol XXI begins with the task of naming as many fractions as possible of a 24-part stick.

---

[14] A procedure is a scheme in which the activity is only connected to rather than contained in the first part of the scheme. The first part of some procedures can contain operations that can be judged as mathematical concepts. In Patricia's case, however, the first part of her procedure was constituted by the words "you reduce." Her activity of dividing can be regarded as her meaning for the words.

7 The Partitive, the Iterative, and the Unit Composition Schemes

Protocol XXI. Renaming unit fractions of a 24-part stick.

T: (Draws a stick and partitions it into twenty-four parts. She places this 24-part stick at the top of the screen and draws a thin cover underneath it as a screen separator.)
T: I want you to name as many fractions as you can of that stick.
P: One-twenty-fourth, two-twenty-fourths, three-twenty-fourths…nine-twenty-fourths…
J: A half of twenty-four.
T: (To Patricia.) That's very good. (To Joe.) What's half of twenty…a whole?
J: Twelve. I was going to do that.
T: Ok. Go ahead and do that one.
J: (Copies the 24-part stick and pulls out a 12-part stick from the copy.)
T: That's very good. What is that? The one you just made?
J: Half of twenty-four.
T: A half of the 24-part whole. Is there another name for it, Joe?
J: (Does not respond.)
T: What about you, Patricia? Is there another name for it?
P: (Counts the number of parts in Joe's stick.) I know.
T: What is it?
P: Twelve-twenty-fourths. (Joe agrees.)
T: Why is that twelve-twenty-fourths?
P: Because it's twelve out of twenty-four.
T: (Asks if there are other fractions they could make.)
J: You could make a fourth of it.
T: Show me how you make a fourth of it.
J: (Pulls out six parts from the 12-part stick.)
T: That's a fourth of what, Joe?
J: The umm (Points to the top 24-part stick.) twenty-four pieces.
T: Very good.
J: (Copies the 24-stick above his 6-stick.) That's six-twenty-fourths.
P: Six-twenty-fourths. Six times four is twenty-four.
J: (Repeats his 6-stick to make a stick the same length as the 24-stick.)

Both children could quickly find unit fractional parts of the 24-stick and drop down one level of unit to units of one, and produce fraction terms using the units of one. The encouraging surprise came when Joe established one-fourth of the 24-stick by pulling six parts out of his 12/24-stick. This indicates that he regarded each part of the 12-part stick as still being one-twenty-fourth of the original stick. He was also able to spontaneously rename the 1/4-stick as six-twenty-fourths, again indicating that he was relating the six parts he pulled out of the 12-part stick back to the original 24-part stick. Patricia agreed with Joe that one-fourth was six-twenty-fourths because "Six times four is twenty four." Joe demonstrated his meaning of one-fourth as a part that can be iterated four times to reproduce the whole by iterating the 6/24-stick four times to make a stick the same length as the original 24-part stick. There was some question as to whether the children actually established one-fourth and six-twenty-fourths as commensurate, so after continuing to work with unit fractional parts of the 24-stick, the teacher turned to finding three-fourths of the 24-stick.

Protocol XXII. Establishing fractions commensurate to nonunit fractions.

T: Can you make me three-fourths of the whole?
J&P: Three...fourths.
P: Oh! I know.
J: I've got it! I've got it!
P: (Moves the 6/24-stick to the middle of the screen.)
T: Patricia, do you know the answer?
J: Yes, I know. Tell me the answer first! Tell me the answer. (To Patricia.)
P: Three of these, three of these. (Waiving the 6/24-stick around.) Six, six, six.
J: Well, what is that? Tell us how long it will be.
P: Oh! Three times six – eighteen. Eighteen-twenty-fourths.
J: (Claps his hands.) Yeah!
T: How did you know that? How did you know that was eighteen-twenty-fourths?
J&P: Because six times three is eighteen.
P: And another six would be four.
T: (To Joe.) Mister smarty-pants, can you find me five-eighths, can you find me five-eighths of that? What's five-eights of the whole?
J: (Takes the mouse from Patricia.) Oh! I think I know.
T: What is it?
J: Fifteen!
T: Fifteen what?
J: Let me see...
P: Five-eighths of the whole?
T: Ah, ah.
J: (Copies the 3/24-stick he pulled out to the middle of the screen and repeats it five times. He then drags a copy of the 24-part whole underneath his repeated stick and counts the parts in his stick.)
J: Fifteen! Clapping his hands.
T: Fifteen what?
J: (Pointing to the 3/24-stick.) Of those threes.
T: So what fraction of the whole is five-eighths of it?
J: Fifteen...thirds of a twenty-four.
T: (Repeats Joe's phrase while trashing the 24-part whole underneath the 15-part stick so Joe would focus on the 15-part stick.) How long is this stick that you just made?
J: Five of those. (Pointing to the 3/24-stick.)
T: And how long was each one of those guys?
J: Three...three-twenty-fourths.
T: Three-twenty-fourths. That's very good, and...
J: Oh! It's fifteen-twenty-fourths!
T: That's great.

Both children displayed an immediate strategy for finding three-fourths of the 24-part stick. Patricia realized that it would be three of the 6/24-stick because that stick was one-fourth of the 24-part stick. Her explanation that it will be "Six, six, six" and her response to Joe's request for how long that would be – "three times six – eighteen. Eighteen-twenty-fourths." – indicate that Patricia could use her iterative fraction scheme using composite units to generate three units of six as meaning for three-fourths. In that she knew that three-fourths in this case was also eighteen-twenty-fourths, she was able to transform three units of six out of the four such units into eighteen units of one out of the twenty-four such units. This transformation is essential in establishing the two fractions as commensurate. Joe's exclamation, "I've got it!" and his affirmation of Patricia's result indicates that he could similarly use his iterative fraction scheme to establish fractions commensurate to proper fractions.

Joe's immediate response of "fifteen" for five-eighths of the 24-part stick, followed by his explanation and test for the number of parts in the stick formed by iterating the 3/24-stick five times, is further indication that he could use his iterative fraction scheme to establish fractions commensurate to proper fractions by decomposing the proper fraction five-eighths into five of one-eighth of twenty-four. It is also evident that he was working with the three-twenty-fourths as a composite unit of three units. His exclamation "Oh! It's fifteen-twenty-fourths" before the teacher had finished asking her follow-up question indicates that he connected the commensurate fraction for one-eighth (three-twenty-fourths) to the result of iterating the 3-stick (one-eighth) five times.

The episode does indicate that both children could make unit fractional parts of a composite unit in the form of a connected number and transform these unit fractional parts into commensurate fractions. Moreover, they could iterate a composite unit fraction (one-fourth as six-twenty-fourths) three times in the production of a fraction commensurate with a proper fraction (three-fourths as eighteen-twenty-fourths). The children had constructed the operations necessary to *transform* a fraction such as three-twenty-fourths into one-eighth and use the transformed fraction to create five-eighths of twenty-four twenty-fourths by iterating the three-twenty-fourths five times. These are constitutive operations of a *unit commensurate fraction scheme*.

Discussion of the Case Study

Composite Unit Fractions: Joe

Similarly to the case of Jason and Laura in their fourth grade, we made the decision to bring forth Joe's units-coordinating scheme in an attempt to activate Joe's production of composite unit fractions. The first step was to encourage generalizing assimilation of the scheme in the context of connected numbers. In Protocol I and its continuation, after Joe learned that the number of iterations of a connected number (e.g., a 4-stick) could be interpreted in terms of fraction language (e.g., six iterations of a 4-stick produced a 24-stick, so the 4-stick was one-sixth of the 24-stick), he operated smoothly in the production of unit fraction language in this context. After only one example involving iterating a 10-stick four times, Joe iterated a 5-stick nine times in a situation that he generated for his teacher, and said that the 5-stick was one-ninth of the 45-stick upon being asked. Moreover, in the continuation of Protocol

I, not only did he know that a 6-stick was one-third of an 18-stick, but also he knew that "one-third" and "six" and "one-sixth" and "three" were interchangeable.

However, a constraint emerged that disallowed inferring that Joe had constructed a composite unit fraction scheme using his units-coordinating scheme in the context of connected numbers. What seemed to be lacking was that iterating a 6-stick three times, say, did not induce a connected number (a composite unit) containing three composite units (three 6-sticks) each of which contained six units (the parts of the 6-stick), and "one-third" did not symbolize one disembedded connected number (a 6-stick) from the connected number containing the three 6-sticks and a comparison between the two connected numbers. This is different than the reason that neither Jason nor Laura constructed a composite unit fraction scheme at the same point in the teaching experiment (cf. Chap. 5). The difficulties they experienced were explained as *necessary errors* arising from their reasoning with two, rather than three, levels of units. They could produce three levels of units as a result of operating, but seemed unaware of the structural relations among the three levels that they had produced. At this point in the teaching experiment, we were not certain if Joe was aware of the structure of three levels of units *prior* to operating, but the difficulties he had in applying partitioning and disembedding to the results of iterating made this a moot point. However, in Protocols XXI and XXII, both Joe and Patricia used their iterative fraction scheme in the production of composite unit fractions (six-twenty-fourths as commensurate to one-fourth) as well as composite proper fractions (fifteen-twenty-fourths as commensurate to five-eighths). Our goal at the outset of the children's fourth grade was for the children to use their whole number multiplying schemes (units-coordinating schemes) in the production of commensurate fractions, but our goal proved to be unviable (for Jason and Laura). At that time we were not aware of the necessity of children to take three levels of units as a given before they could construct commensurate fractions.

Joe's Partitive Fraction Scheme

The tasks we designed for the 16th of November teaching episode (Protocol III) were intended to provoke a transition in Joe from producing what proved to be only pseudocomposite unit fractions to his construction of an equipartitioning scheme that he could use to make an estimate of a unit fraction of an unmeasured stick. Joe became aware that segmenting a stick using a segmenting unit produced a partitioning of the stick. In fact, Joe was able to choose which one of the ten sticks was one-seventh of a drawn stick and verify his choice by iterating the chosen stick seven times and comparing the result with the drawn stick. Joe had, in fact, produced a way of operating that would enable him to use the results of iterating a stick in partitioning.

The consequences of Joe's progress were manifest in the very next teaching episode held on the 7th of December (Protocol IV). In that teaching episode, Joe produced a partitive fraction scheme and fraction language for nonunit proper fractions. He made a choice of which of the four given sticks was one-fifth of a drawn stick and tested his estimate by iterating the chosen stick five times. Based on his choice and on his iterations of the choice, we inferred that to make the choice, Joe would have needed to gauge the results of mentally iterating the chosen stick and to compare the image of the results with the drawn stick. Our inference that Joe engaged in equipartitioning operations was rather dramatically corroborated by his creative production of "five-elevenths" as the fraction name of a stick that he produced by iterating a 1/11-stick five times. However, the fraction name was not based on Joe's iterating the 1/11-stick. Rather, it was based on his production of the 5-part stick as a part of the 11/11-stick – "Because it's five, and its part of eleven." This necessity for the 5-part stick to be construed as a part of the 11/11-stick for Joe to call it "five-elevenths" served as a basis for imputing a partitive fraction scheme to him rather than an iterative fraction scheme. Nevertheless, establishing a partitive unit fraction scheme represented basic progress that went quite beyond the progress that Jason made during his fourth grade.

Emergence of the Splitting Operation and the Iterative Fraction Scheme: Joe

We decided to engage Joe in the task of Protocol VI in the 10th of February teaching episode to gauge his progress in the construction of the operations that produce improper fractions. Up to this point in time, we had not presented tasks to Joe that would permit an analysis of the emergence of the splitting operation, so we do not know whether Joe had constructed this operation prior to the 10th of February. However, in the 22nd of February teaching episode, the teacher asked Patricia, who had joined the teaching episodes, to find a stick such that a marked 9-stick was nine times as long as the one to be found (Protocol VII). After Patricia made an 81-stick by iterating the 9-stick, Joe pointed to the first part of the 9-stick and said, "Just do one of these." He then pulled the part out of the 9-stick and moved the pulled part along the 9-stick while counting each part. The fact that the 9-stick was marked into nine equal parts may have alleviated the necessity for Joe to posit a hypothetical stick that was a substick of the 9-stick and such that the 9-stick was nine times longer than the hypothetical part. It may have also alleviated the necessity of Joe partitioning the 9-stick to produce the stick. Nevertheless, Joe produced a 6/5-stick and a 9/7-stick in the 10th of February teaching episode and, in the first continuation of Protocol VII, given a 9-part stick, he creatively produced a stick such that the 9-part stick was three times as long as the stick he produced. Hence we infer that the splitting operation was available to Joe. Although he did have a marked 9-part stick to use, the fact that he understood that the stick he produced was one-third of the 9-part stick corroborates the inference that he mentally posited a stick that could be iterated three times to produce a 9-part stick. So, the inference that the splitting operation was available to Joe in his creative production of improper fractions is warranted.

Emergence of Recursive Partitioning and Splitting Operations: Patricia

An indication that Patricia had constructed recursive partitioning operations is contained in the 15th of February teaching episode. In that episode, she produced a partitioning of a 1/9-stick to produce a 1/27-stick and a 1/36-stick by partitioning each of the nine parts into three and four parts, respectively. These partitioning acts followed upon her making one-half of one-ninth of a 9/9-stick and producing "one-eighteenth" for how much this stick was of the 9/9-stick. Her production was based on her counting by two nine times after the teacher pulled the one-half of one-ninth part to the beginning of the 9/9-stick. This was a suggestive act, and it provoked Patricia's partitioning after she sat for about 30 seconds in deep concentration. The teacher's provocation did not lead to an immediate solution by Patricia, which indicates that her assembling the recursive partitioning operations as a means of achieving her goal was a creative act. Whether Patricia had embedded recursive partitioning within her partitive fraction scheme as a subscheme is problematic, but her creative problem solving activity in the 15th of February teaching episode does open that possibility as a result of further problem solving.

The splitting operation emerged in the 22nd of February teaching episode as Patricia interacted with Joe. This was corroborated when she independently used the splitting operation to produce a stick such that a given stick was forty-eight times as long as the stick to be made in the 3rd of March teaching episode (Protocol IX). She partitioned the stick into forty-eight parts and pulled one of them out and said that the given stick was forty-eight times as long as the stick, so one of these (one of the forty-eight parts) must be one-forty-eighth. She also posited a similar task for the teacher when she drew a very small stick and said that stick was ninety-nine times as long as the stick she was thinking of. These were powerful uses of the splitting operation, and they stand in stark contrast to her repeating a given stick nine times in the 22nd of February teaching episode to make a stick so the given stick is nine times as long as the stick to be made.

The function of the interaction with Joe seemed to be one of establishing a consensual interpretation of the phrase, "Make a stick such that a given stick is (so many times) as long as the stick to be made." The fact that she could so easily establish a consensual interpretation of the phrase and then operate so powerfully in the 3rd of March teaching episode indicates that the operations that were available to her were sufficient to produce the splitting operation if that operation had not already been constructed. Our hypothesis is that the operations that produce a split of a stick are those operations that produce a unit of units of units. In fact, the hypothesis is that to produce a split of a stick, this unit structure must be available in such a way that it can be used to assemble an image of a unit of units. The additional level of unit permits the child to "look down" or to be aware of a sequence of units of units. Patricia had already engaged extensively in the partitioning of sticks, so she could take a partitioned stick as a unit of units. Iteration of her units of one was also available to her, so once she learned that "nine times as long" meant

7 The Partitive, the Iterative, and the Unit Composition Schemes

not to iterate a particular stick nine times, but rather to find a part of the stick so that it could be iterated nine times to constitute the stick, she performed as if her splitting operation had been constructed and used for a long time.

Patricia's way of learning to engage in splitting operations was replicated in the case of the construction of reversibility of her partitive fraction scheme in the 13th of March teaching episode and in the next teaching episode held on the 5th of April (Protocols XI and XII). In the 3rd of March episode, she interpreted Joe's statement that the stick he had drawn was one-half his stick as meaning to make one-half of the given stick (indicating a lack of reversibility). However, not only did she start the 5th of April teaching episode by posing a problem to Joe of making a stick such that her stick was a fourth of the stick to be made, but also she solved Joe's posed problem (this stick is one-fifth of a stick) by repeating his stick five times, and in reply to the teacher's question of how long the stick was made, she said, "five-fifths." Her reply and her operating together are indicative of an awareness of the stick that she made as split into five parts each of which could be iterated five times to produce the stick. That she learned to operate reversibly with such minimal interaction is also indicative of her reflective awareness of how she operated, an awareness that is made possible by the production of a unit of units of units.[15] Therefore, the progresses made on 13th of March and the 5th of April corroborate the hypothesis that Patricia had constructed the splitting operation and was operating at the level of the operations that produce a unit of units of units.

The Construction of the Iterative Fraction Scheme

We have already indicated above that the iterative fraction scheme emerged on Joe's part in the 10th of February teaching episode before Patricia joined the teaching experiment. From the time when Patricia joined the teaching experiment on the 15th of February, up until the teaching episode held on the 5th of April, neither Patricia nor Joe was asked to make an improper fraction. So, it was a surprise that Patricia interacted with Joe as she did in the first continuation of Protocol XII in the 5th of April teaching episode. There, upon his teacher's request, Joe made a stick twice as long as a 4/7-stick, and then asked himself how the stick could be eight-sevenths. Patricia apparently assimilated Joe's language and actions using whatever fraction schemes that were available to her, because she explained why the stick Joe made was eight sevenths – "Because there's seven in there (pointing to the partitioned 7-stick) and you used the same little pieces as in that 7-stick except that that stick is bigger and you used the same little pieces and there's eight in there (pointing to the unmarked 8/7-stick)." Her assimilation and rather clear explanation along with her

---

[15]The hypothesis behind this statement is that one can be aware of the material on which one operates as well as of the operations that produced that material, but one is not aware of the highest level of operation until one can take the products of that operation as input for further operation.

making a 10/7-stick in the second continuation of Protocol XII and again explaining why the stick she made was a 10/7-stick, all indicate that she assembled the operations that are necessary to construct an iterative fraction scheme on-the-spot, so to speak. There were no long, drawn-out constructive itineraries in Patricia's construction of the iterative fraction scheme, which illustrates her adaptive power.

We also documented similar adaptive power in Joe's case. However, in the teaching episode held on the 5th of April, and even before, Joe experienced a conflict that was induced by the results of the rather novel operations that he used to produce improper fractions when he asked himself, "How can a fraction be bigger than itself?" in Protocol XIII. This was an internal conflict that we assumed that Joe could not work out on his own in his attempts to fit improper fractions within the conceptual network comprised by his fraction schemes. So, we introduced the pizza-baking situation in the 19th of April teaching episode to enable the children to imagine replicates of the fractional whole, which in Protocol XIV was a pizza cut into eight slices. By working in this context, both Joe and Patricia were able to construct fractional connected number sequences (28th of April teaching episode, Protocol XVIII) as a coordination of their number sequences for one and their iterative fraction schemes.

Stages in the Construction of Fraction Schemes

The unit fraction composition scheme emerged in Joe in the 8th of March teaching episode (Protocol X), which was approximately 2 weeks after he was observed engaging in the splitting operation in the 22nd of February teaching episode and approximately 4 weeks after the emergence of the iterative fraction scheme in the 10th of February teaching episode. These cognitive events all point to the presence of the underlying operations that produce a unit of units of units. We have already argued that Patricia's adaptive power in producing recursive partitioning, the splitting operation, and reversibility in her partitive fraction scheme in the context of interacting with Joe and the children's teacher was made possible by such operations. So, rather than attempt to argue that, say, the splitting operation is first constructed in toto, and then children use this operation in constructing recursive partitioning, reversibility of the partitive fraction scheme, the iterative fraction scheme, the commensurate fraction scheme, or the fraction composition scheme, our argument is that these operations and schemes emerge in context and that their emergence is indicative of an underlying presence of operations that produce a unit of units of units as assimilating operations. For example, we argued that when Joe produced improper fractions for the first time, he also split the fractional whole and in doing so he produced a "whole-to-part" relation where the fractional unit (the part) was included in the whole (the 6/5-stick or 9/7-stick) as the result of his operating. So, a split of a stick can be constructed by unitizing the elements of a connected number and uniting those unitized elements together into a connected number at a new level of abstraction, where this abstracted connected number is thought of as a unit of connected units. This permits the splitter to be aware of the results of the operations of partitioning and iterating *without carrying them out*. An equipartitioning of a stick is made possible by the presence of a composite unit of discrete iterable units and a goal to share a continuous unit equally among, say, five individuals, but without the presence of the next level of containing unit. In this case, partitioning and iterating are sequential operations and are not realized in structural relation prior to operating.

The difference between two and three levels of units as assimilating structures is quite significant in that they produce distinctly different stages in the construction of fraction schemes. Two levels of units apparently produce only the partitive fraction scheme (cf. Jason's construction of the partitive fraction scheme in Chap. 5 and his limitations), whereas three levels of units as an assimilating structure produce the recursive partitioning, the splitting operation, the iterative fraction scheme, the unit commensurate fraction scheme, and the unit fraction composition scheme (cf. Jason's constructions in Chap. 6). The production of the unit fraction composition scheme apparently emerges upon the emergence of three levels of units, but a more general fraction composition scheme requires the construction of distributive partitioning operations, which neither Joe nor Patricia demonstrated (Protocol XVII).

Chapter 8
Equipartitioning Operations for Connected Numbers: Their Use and Interiorization

Leslie P. Steffe and Catherine Ulrich

During her fourth grade, Melissa had been paired with another child who had constructed only the tacitly nested number sequence. The teacher of the two children geared her activities to the other child and Melissa essentially served as the other child's interlocutor. The other child did not construct any fraction schemes during her fourth grade and, Melissa, whose role was to interact with the other child at her level, constructed at most the partitive fraction scheme. Melissa had constructed the explicitly nested number sequence, so we paired her with Joe in their fifth grade in order to investigate her constructive itinerary. Our purpose in the upcoming analysis is to investigate her construction of the iterative fraction scheme as well as the unit fraction composite scheme and compare her progress with that of Laura while she was in the fifth grade. Although Joe served as Melissa's interlocutor, we investigate whether the schemes that he established during his fourth grade were permanent as well as any accommodations he might make in them. In what follows, we analyze 17 protocols that were selected from the teaching episodes that started on the 20th of October and proceeded on through the 4th of May of the children's fifth grade.

Melissa's Initial Fraction Schemes

As a result of the first two teaching episodes held on the 20th and the 27th of October, it was clear that Melissa had already constructed the partitive fraction scheme. In the third teaching episode held on the 3rd of November, Melissa's use of her units-coordinating scheme in the context of the children partitioning already-partitioned sticks was analyzed to investigate whether she had constructed recursive partitioning. The first task involved the children making fraction sticks, beginning with a 4/4-stick.

Protocol I. Using a 4/4-stick to make new fraction sticks.
T: (Points to a 4/4-stick on the screen.) Each one of those pieces is how much of the whole stick?
J: One-fourth.
T: I want you to use this fraction stick to make a new fraction stick. What could you do to make this fraction stick into a new fraction stick?

M: You could use PARTS.
T: What would be the next fraction stick you could make?
J: One-half.
M: (Dials Parts to "10" and clicks on the right-most part of the 4/4-stick. PARTS partitioned just the selected part into ten parts.)
T: OK. Before clicking any more, Melissa, how many pieces would be in the whole stick if you kept that up?
M: Forty.
T: (To Joe.) How much of the whole stick would each tiny piece be?
J: One-fortieth.
M: (Completes partitioning the 4/4-stick into forty parts and pulls out one-fortieth of the stick upon being directed by the teacher. She then erases all marks and repartitions the stick into a 4/4-stick again upon being directed by the teacher.)
T: What would be the next fraction stick that you could make doing the same thing? What would be the very next one you could make?
J: Two!
T: OK, go ahead.
J: (Partitions the first of the four parts of a copy of the 4/4-stick into two parts.)
T: What is each piece of that one going to be?
M: (Immediately.) It would be eighths!
T: (After Joe finishes partitioning the four parts of the 4/4-stick.) You make the next one. (To Melissa). What is it going to be?
M: Umm..., eleven...no, umm, five!
T: (Looks at Melissa sharply.) You used two, right? How many pieces would be in the next one?
M: It would be twelve.
T: OK, go ahead. Copy the stick first.
M: (Partitions each part of the 4/4-stick into three parts.)

Melissa's choice to use PARTS was made independently of the teacher's language and actions after the teacher's question, "What could you do to make this fraction stick into a new fraction stick?" Further, after the teacher asked Melissa how many pieces would be in the whole stick if she kept up partitioning each fourth into ten parts, she immediately answered, "Forty." Independently choosing to use PARTS and answering "forty" together constitute solid indication of a generalizing assimilation of her units-coordinating scheme in the context of connected numbers. Assimilation follows from her independently choosing to use Parts and its generalizing nature follows from her using her number concept, ten, to partition a part of the 4/4-stick into ten parts.[1] Further, she at least anticipated partitioning each of the other parts of the 4/4-stick into ten parts, which is the basis for inferring that she coordinated her two number concepts, ten and four, by using ten to partition each of the four parts.

An anticipatory units-coordinating scheme is certainly basic to recursive partitioning because, to find how much one-tenth of one-fourth is of the whole stick, the child mentally partitions each fourth into ten parts to produce forty, exactly as

---

[1] In the discrete case, a units-coordination would consist of Melissa inserting the unit of ten into each unit of the four units of four.

Melissa demonstrated. Melissa's answer, "It would be eighths!" to the teacher's question, "What is each piece going to be?" is indicative of recursive partitioning, but her answer followed on from Joe partitioning one-fourth of the unit stick into halves and previously saying "one-fortieth" as the fraction of the unit stick constituted by one of the forty parts that Melissa would make had she completed partitioning the 4/4-stick. Consequently, it is uncertain whether Melissa could have independently answered in that way. On the other hand, Joe's answer of "one-fortieth" corroborates that, for him, recursive partitioning was a permanent operation (cf. Protocol X of Chap. 7).

Contraindication of Recursive Partitioning in Melissa

In further exploration of whether Melissa had constructed recursive partitioning, the teacher began by asking Joe to make the next one.

Protocol I. (First Cont)

T: Joe, you make the next one.
J: Five!
T: Five?!
J: You have already got four. (Pointing the cursor arrow at the 4/4-stick.)
T: Ooooh. You already have four there, but did you use four?
J: No.
T: All right, make a copy.
J: (After some discussion, the teacher asked Joe if he wanted to do his fraction stick.) Oh, I forgot! (Partitions each part of a copy of the 4/4-stick into four parts.)
T: OK, Melissa, its your turn now. What will the next one be?
M: Twenty. (Makes a 20/20-sick using the 4/4-stick.)
T: Each piece will be how much of the whole stick?
M: (Hesitates.)
J: One-twentieth.
T: OK, Joe, you make one more. What is this one going to be?
J: Twenty-fourths. (Makes a 24/24-stick using the 4/4-stick.)

Melissa's hesitation after the teacher asked the additional question, "Each piece will be how much of the whole stick?" indicates an uncertainty on her part about what the answer should be. It also indicates that she was still in the process of linking her units-coordinating scheme and her partitive fraction scheme in order to establish a way of operating that resembles a unit fraction composition scheme. If she established a link between the two schemes, it was after she answered "twenty." Before she answered, "twenty," Melissa's goal seemed to be to produce the next partition of her partitioned 4/4-stick. After she answered "twenty" and partitioned each part of the 4/4-stick into five parts, she closed off using her units-coordinating scheme because she had reached her goal. So, when the teacher asked her how much each piece would be of the whole stick, the task seemed to appear to her as a new task. If the question evoked her partitive fraction scheme she would use it to assimilate the results of using her units-coordinating scheme, and in this way establish a link between the two schemes. So, she may have been in the process of establishing using the results of her units-coordinating scheme as a situation of her partitive fraction scheme. Even if Melissa did establish a link between the two schemes, she would need to demonstrate recursive partitioning in order for us to judge the linked schemes as a unit fraction composition scheme.[2]

Reversibility of Joe's Unit Fraction Composition Scheme

Joe's answer, "One-twentieth," again corroborates the inference that he had constructed a unit fraction composition scheme while he was in his fourth grade (cf. Protocol X of Chap. VII). The teacher continued on and changed the question in such a way that answering it would seem to necessarily involve reversibility of the unit fraction composition scheme. What resulted was a chance to test if Joe's scheme was reversible and a new chance to test whether Melissa had constructed a unit fraction composition scheme and, hence, recursive partitioning.

Protocol I. (Second Cont).
T: Make one-fourth using that stick. (Points to the 4/4-stick.)
M: Pulls one part out of the 4/4-stick.
T: (Points to the 1/4-stick.) I want you to cut that up so each little piece is one-twentieth of the whole stick.
J: (Immediately partitions the 1/4-stick into five parts.)
T: (Asks Melissa to pull another 1/4-stick out of the 4/4-stick.) I want you to cut that up so each little piece is one-thirty-second of the whole stick.
M: (Immediately partitions the 1/4-stick into thirty-two parts.)
T: Would that be right?
J: (Shakes his head.) No, because thirty-two times four is not thirty-two!
T: OK. You can erase marks if you want to. (To Melissa.) (After Melissa erases marks.) What would you have to do, Melissa?
M: Cut it into eight pieces! (Partitions the 1/4-stick into eight parts.)
T: Why wouldn't each one of those little pieces be one thirty-second of the whole stick? (Referring to her partitioning the 1/4-stick into thirty-two parts.)
M: Because I thought you meant thirty-two pieces in all the parts.
T: Would you like to ask Melissa a question, Joe?
J: (After a long pause when Melissa was rearranging the screen.) Each little piece would be one-sixteenth.
M: (Partitions a 1/4-stick into four parts.)
T: How could you prove that each part is one-sixteenth?

---

[2] It would also require that she could solve tasks like, "Find how much one-fifth of one-fourth of a stick is of the whole stick."

M: Because four times four is sixteen.
T: You give Joe one.
M: (Puts up fingers apparently counting by seven four times. But the monitor partially hid her hands, so it was not possible to see all of what she did.) Each little part would be one-twenty-eighth!
J: (Immediately dials PARTS to "7" and clicked on a copy of the 1/4-stick.)

After a long pause, Joe then told Melissa that each part would be one-thirty-sixth in posing a problem to her. She immediately dialed PARTS to "9" and clicked on a copy of the 1/4-stick. She seemed to have abstracted how to use her anticipatory units-coordinating scheme to solve the situations that Joe presented to her. However, to infer that she could engage in recursive partitioning even though she performed flawlessly after Joe's corrective, "No, because thirty-two times four is not thirty-two!" was not plausible. The fact that she initially partitioned a 1/4-stick into thirty-two rather than eight parts after the teacher asked her to cut it up so each little piece would be one-thirty-second of the whole stick is contraindication of recursive partitioning in the context of solving the problem. At that point, "Thirty-two" referred to the result of a single partition rather than a result of a coordination of two partitions. Joe's corrective contained explicit multiplicative language and that would be sufficient to provoke her units-coordination scheme. It is reasonable to infer that her units-coordinating scheme was a reversible scheme because of the way she used it, for example, to pose the problem, "Each little part would be one-twenty-eighth!" to Joe.

When Joe presented the problem "Each little piece would be one-sixteenth." to Melissa, she partitioned a 1/4-stick into four parts and explained why she did it by saying, "Because four times four is sixteen." "One-sixteenth," then, apparently referred to partitioning the whole stick into sixteen parts, i.e., her partitive fraction scheme was a reversible scheme. Rather than attribute recursive partitioning to Melissa, the conjecture is that her reversible partitive fraction scheme became linked to her reversible units-coordinating scheme in that a result of the former was used as a situation of the latter. For example, when she posed the problem "Each little part would be one-twenty-eighth!" involved using her reversible partitive fraction scheme because one-twenty-eighth referred to partitioning the whole stick into twenty-eight parts. She could then reason, "Four times what is twenty-eight?" which is an indication of a reversible units-coordinating scheme. The linkage between the two schemes might seem sufficient to infer that she had constructed a reversible unit fraction composition scheme. However, the unit fraction composition scheme contains recursive partitioning as a subscheme whereas the linkage of the two schemes as we have explained it does not involve recursive partitioning.

The indication is solid that Joe had indeed constructed his unit fraction composition scheme as a reversible scheme when he partitioned the 1/4-stick into five parts after the teacher's request, "I want you to cut that up so each little piece is one-twentieth of the whole stick." His goal was to make a piece that is one-twentieth of the whole stick, whereas the most that can be inferred is that Melissa's goal was to make so many pieces of the whole stick using the 4/4-stick. These differences in the schemes the two children used are subtle. If the differences are consequential, they should be manifest in the adaptability of the two schemes.

After the tasks of Protocol I, the teacher proceeded to ask the children a more complex question that we regard as a test of the adaptability of the two children's schemes.

Protocol II. A test of adaptability.
T: Joe, make three-fourths of this stick.
J: (Pulls a 3/4-stick out of the 4/4-stick.)
T: I want you to cut this up so each little piece is one-eighth.
M: (Dials PARTS to "8" and clicks on each of the parts of the 3/4-stick.)
T: Is each little piece one-eighth of the big stick?
M: Uh, huh. (Yes.)
T: Well, pull a piece out and see.
M: (Pulls out one of the twenty-four parts she made by clicking on the 3/4-stick three times. She then activates MEASURE[3] but hesitates before clicking on the 1/32-stick she pulled out.)
J: (With urgency.) Click on it and see!!
M: (Clicks on the 1/32-stick and "1/32" appears in the number box.)
J: (Smiles knowingly.)
T: I want you to cut three-fourths up so each little piece is one-eighth of the whole stick.
J: (Takes the mouse and starts to dial PARTS. The teacher asks him to take another three-fourths out first, so he pulls a 3/4-stick from the 4/4-stick and, after the teacher reposes the question, he dials PARTS to "2" and clicks on each part of the 3/4-stick.)
T: How did you do that, Joe?!! Is that right?
M: (Counts the six parts on the now 6/8-stick.)
J: (Pulls a 1/8-stick out from the 6/8-stick and measures it. "1/8" appears in the number box.)
T: Can you tell us how you thought that out?
J: (Smiling.) I put it on two parts, and two right here. (The 3/4-stick.) And six and eight in this one. (The 4/4-stick.)
T: (To Melissa.) Did you hear what he said?
M: (Nods.)
T: (Asks Melissa to pull out another 3/4-stick and cut it up into pieces so that each little piece would be one-twelfth of the whole stick.)
M: (Dials PARTS to "4" and clicks on each of the three parts of the 3/4-stick.)
T: What do you think, Joe?
J: That would be one-sixteenth!

Having three rather than one part of the 4/4-stick to partition to make a stick that was one-eighth of the 4/4-stick introduced a constraint that Melissa did not eliminate by making an adjustment in her way of operating in the second continuation of

---

[3] MEASURE works by designating the 4/4-stick as a unit stick and then using the mouse to click on the stick to be measured relative to the unit stick.

Protocol I. From the observer's perspective, Melissa regressed to the way she operated at the beginning of the second continuation of Protocol I. Her way of operating in Protocol II proved to be nonadaptive, and when contrasted with Joe's insightful comments, this nonadaptability serves as contraindication that she had constructed a reversible unit fraction composition scheme. Joe, on the other hand, operated in a way that was compatible with the hypothesis that he had in fact constructed his unit fraction composition scheme as reversible.

Based on Joe's partitioning each part of the 3/4-stick into two parts and on his explanation, Melissa did modify her initial partitioning activity where she partitioned each of the three parts into eight parts. Rather than partition each of the three parts into twelve parts upon being asked to cut the three parts into pieces so that each piece would be one-twelfth of the whole stick, she partitioned each of the three parts into four parts to produce twelve parts. However, partitioning each of the three parts into four parts to produce twelve parts did not constitute the insight that stems from using a reversible unit fraction composition scheme.

A Reorganization in Melissa's Units-Coordinating Scheme

The goal in the analysis of the teaching episode held on the 1st of December is to explain the schemes the children used to transform a unit fraction into a commensurate fraction. To begin the teaching episode, Melissa made a long and rather narrow bar. The teacher asked Joe to partition it into five parts and to pull out one-fifth of the bar. Joe then made unmarked copies of the bar at the teacher's request and, after the teacher explained that he was to partition the first copy into a different number of parts so that one-fifth could be still pulled out, Joe immediately partitioned it into ten parts and pulled out two after the teacher asked him to pull out one-fifth of the bar. He also said "two-tenths" after the teacher asked him for another name for that fraction.

As Joe spoke, Melissa was counting the number of parts in the 10/10-bar by twos. After she was done, she agreed with Joe that another name would be two-tenths. This seemed to evoke Melissa's units-coordinating scheme because she immediately dialed PARTS to "15" and partitioned the next copy into fifteen parts and pulled out three of the parts after the teacher asked her to pull out one-fifth. She also immediately said that another fraction name for the part she pulled out would be three-fifteenths. Protocol III starts with Joe's next turn.

Protocol III. Joe and Melissa produce fractions commensurate with one-fifth.

T: OK, it's your turn Joe. Make it go as far as you can. (Three unmarked copies of the 5/5-bar remained.)

J: (Dials Parts to "25" and clicks on the next copy. He then drags a rectangle over the first five parts, one at a time, while intently looking at what he was doing. He did not appear to look at the previous one-fifth fractional parts. Melissa looked intently at the computer screen while Joe was working silently.)

M: (Dials Parts to "30" and clicks on the next copy. She then drags a rectangle over the first six parts, one at a time, while intently looking at what she was doing. She did not appear to look at the previous one-fifth fractional parts.)
T: (To Joe.) What was the name for your fraction?
J: Five-twenty-fifths.
T: (To Melissa.) What was the name for your fraction, Melissa?
M: (Immediately.) Six-thirtieths.

Melissa knew that she was going to pull out the first six parts of the 30/30-bar that she made before she began dragging the rectangle over the parts. Even though it is quite likely that she knew that Joe pulled out five parts because she watched him intently as he was doing so, she did proceed with confidence when she pulled out six parts. Moreover, she knew immediately that she had made six-thirtieths of the bar. For these reasons, it is plausible that she understood that she needed to pull out six parts because five times six is thirty. My hypothesis is that after Joe partitioned a copy of the unit stick into twenty-five parts and pulled out five, she conceived of the stick as partitioned into the original five parts where each part was partitioned into five parts using her units-coordinating scheme. She then proceeded to use her units-coordinating scheme to mentally partition each of these five parts into six parts each to produce thirty parts and then pulled out six of the thirty parts.

To illustrate the children's reliance on the counting-by-five patterns involved in producing the sequence of fractions commensurate with one-fifth, the teacher asked Melissa what the next one would be. She said thirty-five and seven, and then seven-thirty-fifths. Joe then said the next one would be eight-fortieths, and Melissa said the next one after that would be nine-forty-fifths. The two children continued without hesitation, taking turns, until they reached fifteen-seventy-fifths. The children did not know that, say, five times fifteen is seventy-five nor that five times fourteen is seventy without computing, but they had a way of producing a sequence of fractions, each commensurate with one-fifth, that avoided multiplicative computation. Nevertheless, we infer that each fraction of the sequence symbolized an invariant conceptual structure. For example, "fifteen-seventy-fifths" symbolized the 5/5-unit stick where each of the five parts was partitioned into fifteen parts as indicated by Melissa saying "one-fifth" when asked for another fraction name by the teacher. The children knew that fifteen-seventy-fifths was one-fifth, not because they could engage in the operations involved in reducing fifteen-seventy-fifths to one-fifth,[4] but because they maintained the original partition of the unit stick into five parts across their partitioning activity.

---

[4]These operations involve partitioning two composite units, one of numerosity seventy-five and the other of numerosity fifteen, into subunits of equal numerosity. It also involves an awareness of both unit structures produced in each case (e.g., a unit of five units of three and unit of twenty-five units of three) as well as awareness that five out of the twenty-five units of three determines fifteen out of the seventy-five units of one.

Given Melissa's strong performance in Protocol III, the teacher decided to change the task to one similar to the second continuation of Protocol I. He asked them how many parts the 1/5-stick would need to be cut into so that each part would be one-one-hundredth of the whole stick. In solving the task, it would have been possible for the children to continue coordinating the two number sequences for one and five beyond "fifteen and seventy-five." However, introducing reversibility led the children to abandon this way of proceeding and to using their units-coordinating schemes to make conjectures when trying to solve the task.

Protocol IV. Melissa's conjecture.
T: (Points to the 1/5-bar with the cursor.) This is the part you are going to break – break this piece up so that each little part of this is one-one-hundredth of the unit bar. (Running the cursor over the 5/5-unit bar.)
M: You would put one hundred in this bar. (The 5/5-unit bar.)
T: But what would you have to put in this one? (Pointing to the 1/5-bar.)
M: You would have to pull one out of this one? (The 5/5-unit bar.) You would have to pull ten out of this bar.
T: So, there would be ten in this? (The 1/5-bar.)
M: Yeah. There would be ten in this. (The 1/5-bar.)
T: And that would give you a hundred in the unit bar?
M: Uh-huh. (Yes.)

It is quite significant that Melissa's partitive fraction scheme was reversible when there was no bar in her visual field that was one-one-hundredth of the 5/5-bar: She knew if each little part of the 1/5-bar was one-one-hundredth, then the 5/5-unit bar would be partitioned into one hundred parts. She also seemed to understand that the unit bar, when partitioned into one hundred parts, was one hundred times one of its parts. "One-one-hundredth" referred to a partitive unit fraction and her comment, "You would put one hundred in this bar," seemed to refer to *splitting* the bar into one hundred parts. We emphasize "splitting" because to split a bar into one hundred parts means that the bar is conceived of as partitioned into one hundred parts, and of one hundred times any one of its parts. Given that one-one-hundredth was posited as a part of the 1/5-bar, Melissa would need to inject that 1/5-bar along with its part (one-one-hundredth) into the 5/5-bar and then mentally split the 5/5-bar into one hundred parts. When she said that "You would have to pull one out of this one (the 5/5-bar)?" and then "You would have to pull ten out of this bar." there was no bar in her visual field partitioned into one hundred parts. She was speaking hypothetically, and the ten parts she referred to were apparently conceived of as ten parts of one of the five parts of the 5/5-bar ["Yeah, there would be ten in this. (The 1/5-bar.)"]. She seemed to establish a connected number, one hundred, where the one hundred units were partitioned into five parts, which constitutes the structure of a unit of units of units in the context of connected numbers. Melissa performed these operations without considering how many parts each one of the five parts would need to be partitioned into. For this reason, and because it was legitimate to infer that she engaged in mentally splitting the 5/5-bar into one hundred parts, we infer that she engaged in reciprocal reasoning in the case of one-one-hundredth and one hundred.[5]

Her saying that there would be ten in the 1/5-bar indicates that she did not establish how many of the one hundred parts would be in each one of the five parts. Melissa was very precise in her calculations, and she would not have made such a blatant mistake had she actually counted by ten five times in a test to find if ten worked. So, the teacher decided to ask Melissa to check her estimate of ten.

Protocol IV. (Cont)

T: (To Melissa.) Make a copy of this. (The 5/5-bar.)
M: (Breaks the 5/5-bar into its parts.)
T: Do you think ten in each part is going to give you a hundred parts?
M: (Nods.)
J: I know how to do it!
M: Twenty!
T: OK! Go ahead and do it.
J: There is another way, too!
M: (Partitions each of the five parts into twenty parts and then joins the five parts together into a 100/100-bar.)
T: Is there a hundred parts in that?
J: (Uses the menu PARTS IN A BAR to verify that there are one hundred parts in the bar and "100" appears in the NUMBER BOX.)
T: Can you pull one out, Joe?
J: (Pulls one of the one hundred parts out from the 100/100-bar.)
T: Try measuring that.
J: (Measures the part he pulled out and "1/100" appears in the NUMBER BOX. Both children express pleasure at being able to measure such a small part of the bar.)
T: (Points to the original 1/5-bar.) So how many parts do you have to break this into to get one-one-hundredth?
M: Twenty.

It was not until the teacher asked, "Do you think ten in each part is going to give you a hundred parts?" and Joe said, "I know how to do it!" that Melissa doubted her answer of ten and made a corrective. In that she quickly said, "twenty," it is quite possible that her original answer of ten was based on her associations among the numbers five, ten, and one hundred. In fact, Joe's "another way to do it" was to partition the bar into ten parts and then partition each part again into ten parts as he demonstrated after the continuation of Protocol IV. By their fifth grade in school, children have learned well that "ten times ten is one hundred," so it is very plausible that Melissa relied on this knowledge to produce ten. In any event, this is the first time that we were able to infer that Melissa had constructed the opera-

---

[5] From this, I do not infer that she had constructed the concept of the reciprocal of a fraction. Nor do I infer that she had constructed more general reciprocal reasoning involving proper and/or improper fractions.

tions that produce a unit of units of units in the context of connected numbers and the splitting operation. It is also the first time that we were able to infer that she had constructed her partitive unit fraction scheme and her units-coordinating scheme as reversible schemes. These inferences are corroborated in Protocol V when the teacher changed the number of parts from one hundred to eighty.

Protocol V. Inferring equipartitioning operations for connected numbers.

T: How many parts would you have to break this up into, the one-fifth, if each little part was one-eightieth of the unit bar?
M: (After approximately 25 seconds) Twenty-five. Cause, um, three times twenty is sixty and you add two more tens to it and you would get eighty. (Three twenties and two tens!)
T: That would give you eighty parts, but would they all be the same size?
M: Uh-uh. (No.)
T: So that wouldn't work, would it?
M: No.
J: I know how many sticks. Put twenty in one and then thirty in another, and then twenty in another.
T: Would all those parts be the same size?
J: (Shakes his head.) I don't know.
M: (While Joe and the teacher are talking, Melissa intently enacts a computational algorithm on the table by tracing numerals with her right forefinger. She then sits back in her chair and appears to be in deep concentration.)
J: (Continues on working while Melissa is attempting to find a result. The children do not communicate verbally. Joe proposes a possible answer, and the teacher casts doubt on it, so he keeps working.)
T: Maybe I have forgotten the problem. What is it that we are trying to do?
M: How many eightieths can you get to fit into one-fifth!
T: That's right! Very good! How many eightieths can you get to fit into one-fifth?
M: (Asks for a paper and pencil and leaves her seat to retrieve her own supply of paper and pencil.)
J: (Makes two copies of the 1/5-bar and partitions the original 1/5-bar and the two copies into thirty parts each. He then indicates to the teacher that he is trying to make ninetieths!)
T: Melissa, what are you looking for? We have a calculator on the computer that you can use. (Joe activates the calculator and Melissa retrieves a pencil and pad and returns to her seat. The teacher re-poses the problem to each of the two children. Each child works separately, Melissa using her pencil and paper and Joe mentally without using the calculator. She tries five times twelve, and then others. She then says that she got it. Joe then says, "sixteen.") Sixteen!! How did you get that? (Checks Melissa's answer before he allows Joe to explain.) What did you get?
M: I got fourteen! (Explains that she divided five into eighty. She then checks Joe's answer by computing five times sixteen.)
T: How did you get sixteen, Joe?
J: I know five times twelve, so I did thirteen would be sixty-five, fourteen would be seventy, fifteen would be seventy-five, and sixteen would be eighty!
T: Fantastic! (Guides the children to partition the 1/5-bar into sixteen parts, pull out one part and measure it to check.)

Due to the difficulty of finding sixteen mainly, both Melissa and Joe tried to fit eighty-eighties into five-fifths without maintaining an equal number of the eightieths in each fifth. When the teacher asked the children what they were trying to do, Melissa answered, "How many eightieths can you get to fit into one-fifth!" In that Melissa used fraction language in her answer, it is clear that she was aware that each of the eighty parts was one-eightieth of the unit stick, and that it was implicit in her answer that finding how many eightieths would fit into one-fifth would inform her about how many eightieths would fit into each of the one-fifths. Her clearly stated problem is solid indication of equipartitioning operations for connected numbers with the proviso that the five-part stick was one of her givens. In that she computed to check whether five times sixteen is eighty, there is indication that she understood that it was a matter of necessity that the number of eightieths in each fifth could be iterated five times to produce eighty. Melissa still focused on portioning in her language. Nevertheless, her equipartitioning operations permitted her to work at the level of re-presentation and there was no necessity for there to be material in her visual field to stand in for the eightieths. Moreover, she worked symbolically using her division algorithm to achieve her goal.

Joe's strategic reasoning, starting with five times twelve is sixty and coordinating incrementing twelve by one and sixty by five until he reached eighty, is also an indication of equipartitioning operations for connected numbers. The coordination corroborates that he regarded it as necessary for five iterations of the numerosity of the items placed into one of the fifths to yield eighty-eightieths. His reasoning warrants the inference that Joe, like Melissa, assimilated the situation using three levels of units, i.e., he constructed the situation as a unit containing five connected units each of which contained an unknown numerosity of units, where his goal was to find how many parts he should partition each of the five connected units into so that, together, there would be eighty one-eightieths in the 5/5-stick. This situation and goal was a situation and goal of his *reversible* units-coordinating scheme that he used strategically. Using his reversible units-coordinating scheme in this way can be thought of as a reversible unit commensurate fraction scheme if the goal is to find how many eightieths is equal to one-fifth, which was Melissa's explicitly stated goal. This goal, along with the equipartitioning operations for connected numbers that induced a modification in the units-coordinating scheme, is what differentiates a reversible unit commensurate fraction scheme from a reversible units-coordinating scheme. It is quite significant to realize that a reversible unit commensurate fraction scheme is a *multiplicative* scheme. But it is yet to be constructed as an *equivalence* scheme.

Melissa's Construction of a Fractional Connected Number Sequence

Joe already constructed a fractional connected number sequence in the teaching episode that was held on the 28th of April of his fourth grade as a functional accommodation of his iterative fraction scheme. So, the primary interest was in analyzing

Melissa's language and activities in Protocol VI, which involved the production of improper fractions. It was important to know whether the operations that she used to produce an equipartitioning of the composite unit eighty-eighths into five connected composite units were permanent operations or whether they were specific to the situation of Protocol V. In Joe's case, whether his iterative fraction scheme was permanent and ready-at-hand for him to use, or whether he had to reconstruct it using the operations that were available to him was investigated.

Protocol VI was extracted from the teaching episode held on the 12th of January. The children had made an 11/11-bar and pulled out several fractional parts, one of which was a 6/11-bar.

Protocol VI. Melissa's construction of the iterative fraction scheme.

T: Tell me, is that (The 6/11-bar.) more or less than a half?
M: More.
T: (To Joe.) Can you make one that is a little bit less than that, that would be less than a half? What would be left of the candy bar if you took six-elevenths out? (Pretending that the bar was a candy bar.)
M: Five.
T: Five what?
M: Five-elevenths.
T: OK. Can you do that one, Joe?
J: Five-elevenths?
T: Yeah.
J: (Pulls a 5/11-bar from the 11/11-bar in such a way that it is the complement of the 6/11-bar that Melissa pulled out.)
T: If you had two pieces like that, how much of the bar would you have?
M: Ten-elevenths! (Joe also answered "ten-elevenths.")
T: Would that be the whole bar?
J&M: Less than the whole bar.
T: How much less?
J&M: One-eleventh.
T: OK, now put out the six-elevenths.
M: (Drags the 6/11-bar from the side of the screen and places it and the 5/11-bar end-to-end.)
T: If each of you have a bar like that… (The 6/11-bar.)
M: It wouldn't be any more of the bar. (The 6/11-bar and the 5/11-bar together made a bar commensurate with the 11/11-bar.)
T: What if you both had six-elevenths?
M: It would be one more than the bar.
J: Twelve-elevenths.
T: Twelve-elevenths! How much more than a bar would that be?
J&M: One-eleventh.
T: If you took eight of these two-elevenths… (Melissa dragged a 2/11-bar under the 11/11-bar.)
M: Sixteen-elevenths!
T: And how much of the bar would that be?
M: That would be…
J: Five-elevenths.
M: Five-elevenths more…

Protocol VI (continued)
T: Five-elevenths more than…
J&M: The whole bar!!
T: Referring to a 4/11-bar.) If you took eleven pieces like this, can you tell me how much you would have?
M: Forty-four!
T: Forty-four what?
M: Forty-four elevenths!
T: How many whole bars would that be?
J: Four!
T: Why would forty-four elevenths be four bars?
M: Because four times eleven is forty-four!

In this segment of the teaching episode, Joe and Melissa did not communicate directly with each other, but rather with the teacher. Nevertheless, the thinking of the two children seemed to be in harmony. Melissa's comment that the 6/11-bar was more than one-half of the 11/11-bar led the teacher to ask the question, "What would be left of the candy bar if you took six-elevenths out"? Melissa's answer of five-elevenths serves as confirmation that, for her, the six-elevenths when joined to five-elevenths constituted the whole of the 11/11-bar. This inference is corroborated by her subsequent comment that the 6/11-bar together with the 5/11-bar would not be any more of the bar, meaning that together, they would not be more than the bar. Not only did she enact pulling the 6/11-bar from the 11/11-bar using PULL PARTS, but also she mentally disembedded five-elevenths from eleven-elevenths, six-elevenths from eleven-elevenths, and integrated them together to reconstitute eleven-elevenths.

Both children convincingly demonstrated that they understood that a 12/11-bar contains the 11/11-bar and that it is one-eleventh more than the 11/11-bar. That is, they demonstrated a reversal in relation between the fractional part and the fractional whole in that what was before a fractional part now contained the fractional whole as a part. Because the children knew that the composite unit of numerosity twelve is one more than the composite unit of numerosity eleven, it would seem that it would be rather straightforward for them to understand that the fractional connected number twelve-elevenths is one-eleventh more than the fractional connected number eleven-elevenths. However, this relation between twelve and eleven has to be constructed anew in the case of fractions. The children's operations opened the way for their construction of the reversal in relation between the part and the whole and, as demonstrated in Protocol VII, for the construction of the relation between twelve-elevenths and eleven-elevenths as well as a more general iterative fraction scheme for composite fractions.

Both children knew that if they each had a 6/11-bar, then together they would have a 12/11-bar and that that bar would be 1/11 more than the whole bar. This was especially remarkable because they seemed explicitly aware of their reasoning and of the elements on which they operated: Melissa said, "It would be one more than the bar." In reply to the teacher's question, "What if you both had six-elevenths?"

and Joe knew that it would be twelve-elevenths as well. The children seemed aware of integrating the two composite units together, each containing six 1/11-bars.[6] In that one of these composite units was not in the children's visual field, the children worked in re-presentation when performing the integration operation. That is, the children operated on elements in visualized imagination, which also supports the inference of awareness.

Another indication of their explicit awareness of how they operated is that they mentally produced a 1/11-bar that was a part of the twelve-elevenths bar they mentally established but was not contained in the 11/11-unit bar. Mentally producing this 1/11-bar was a major achievement for the children and saying that the 12/11-bar was one-eleventh more than the 11/11-bar stood in for actually producing this 1/11-bar. My judgment is that the children's comments symbolized producing the 1/11-bar, which would entail an awareness of the operations involved.

The inference that the children were explicitly aware of the numerical whole-to-part relation between twelve-elevenths and eleven-elevenths as well as of the status of each 1/11-bar contained in the 12/11-bar as a unit fractional part of the 11/11-bar is corroborated by their knowing that sixteen-elevenths is five-elevenths more than the bar (the 11/11-bar). Producing five-elevenths further indicates that one-eleventh had become an iterable unit for the children that was on a par with their iterable unit of one. The children had constructed a fractional connected number sequence of which one-eleventh was the basic unit element, a sequence analogous to their explicitly nested number sequence where the basic unit element was one.

Further, the children seemed to operate with their fractional connected number sequence for one-eleventh in a way that was analogous to how they operated with their explicitly nested number sequence for one. This claim also finds corroboration in how the children operated after the teacher asked them the incomplete question, "If you took eight of these two-elevenths…?" Melissa almost immediately replied "Sixteen-elevenths!" and both children knew that this result was five-elevenths more than the unit bar. The children's way of operating indicates not only that two-elevenths was an iterable composite fractional unit for the children, it also indicates that the children engaged in a generalizing assimilation[7] of their units-coordinating scheme for composite units. They could now iterate two-elevenths eight times and produce sixteen-elevenths as the result.

It is quite impressive that the children also knew that eleven 4/11-bars would yield forty-four-elevenths. This knowledge is another corroboration of the inference

---

[6] To integrate the two 6/11-bars together means to unite them into a composite unit containing the two bars as component units and, further, to disunite the two bars into their elements and then unite these elements into a composite unit while maintaining an awareness of the two component units.

[7] Recall that an assimilation is generalizing if, from the point of view of the observer, the scheme involved is used in situations that contain elements that are novel for the scheme and if there is an adjustment in the scheme without the activity of the scheme being implemented.

that the children had engaged in generalizing assimilation of their units coordinating scheme for composite units. It is especially impressive that both Melissa and Joe knew that there were four unit bars in a 44/11-bar, "Because four times eleven is forty-four!" They were aware that if they iterated a 4/11-bar eleven times, they would produce a 44/11-bar. They were also aware that they could produce this 44/11-bar by iterating the 11/11-bar four times, "Because four times eleven is forty-four!" Although this knowledge may have been based on a functional interchange of the number of iterations and the number of elements in the composite unit being iterated rather than on operations that produce an awareness of commutativity, it still indicates a generalization of their units-coordinating scheme for composite units. Not only was the 44/11-bar eleven times the 4/11-bar, it was also forty-four times the 1/11-bar. That is, the fraction 44/11 now stood in multiplicative relation to one-eleventh in that it was forty-four times one-eleventh.

It could be said that both Joe and Melissa had constructed *fractional numbers* because, after making the 12/11-bar using the 6/11-bar, they knew that the 12/11-bar was one-eleventh more than the 11/11-bar. They did not resort to constituting one part of the 12/11-bar as one-twelfth because the 12/11-bar had twelve parts. This indicates that the children regarded a 1/11-bar as contained in the 11/11-bar but also as a unit fractional number that could be iterated enough times to produce improper fractions. Maintaining an awareness of both of these aspects of the 1/11-bar was essential for the children to designate each part of the 12-part 12/11-bar as one-eleventh. But it is not sufficient. The children also regarded the 11/11-bar as both a composite unit (i.e., a single entity) and as consisting of eleven times one of its unit parts (Melissa said that the 12/11-bar would be one more than the bar). Speaking of "the bar" indicates that she regarded the 11/11-bar as an entity, and by saying that the 12/11-bar was one more than the bar, she indicated an awareness of "the bar" as being eleven times one of its parts. Moreover, both children said that a 16/11-bar would be five-elevenths more than the whole bar. Implicit in this statement is the understanding that the whole bar is also an 11/11-bar, i.e., a bar that is eleven times one of its parts. A fraction was no longer simply a part of a fractional whole for the children, because the relationship between the fractional numbers twelve-elevenths, sixteen-elevenths, and forty-four elevenths and the fractional whole had changed in that the fractional whole was now contained in the fractional numbers. This is a defining characteristic of fractional numbers and clearly differentiates the iterative fraction scheme from the partitive fraction scheme. The splitting operation is lacking in the partitive fraction scheme, and it is this operation along with the operations that produce three levels of units that permits a child to reorganize the partitive fraction scheme into an iterative fraction scheme. These operations are also the operations that permit the construction of the unit fraction composition scheme.

The question concerning whether Joe's iterative fraction scheme was permanent and ready-at-hand for him to use is answered in the affirmative. Further, establishing the iterative fraction scheme and a fractional connected number sequence for elevenths in the context of Protocol VII is indication that Melissa's equipartitioning operations for connected numbers, which were used to find how many eightieths

8 Equipartitioning Operations for Connected Numbers: Their Use and Interiorization

you can fit into one-fifth, were permanent operations[8] that she could at least use in activity. It was indeed surprising that Melissa could operate so powerfully the first time that we presented improper fraction tasks to her and it is testimony to the adaptability of children who can use three levels of units in assimilating improper fraction situations.

Testing the Hypothesis that Melissa Could Construct a Commensurate Fraction Scheme

Given that Melissa had constructed equipartitioning operations in Protocol V and the iterative fraction scheme in Protocol VI, the hypothesis that she could construct a commensurate fraction scheme seemed plausible. The test of this hypothesis turns on whether she could transform a proper fraction into a commensurate fraction. In Protocol VII, which served in testing the hypothesis, the unit bar was still the 11/11-bar and the teacher began by asking the children to partition the bar so they could still pull out elevenths.

Protocol VII. Making six-elevenths using a 22/22-bar.
T: Joe, and Melissa, both of you, can you think of another number of parts you could put in this bar (A copy of the 11/11-bar.) and still pull your elevenths out of it?
M: (Partitions the bar into twenty-two parts using PARTS.)
T: Pull three-elevenths out for me.
M: (Pulls a 3/22-bar from the 22/22-bar.)
J: (Watches Melissa intently.) That's three twenty-two.
T: (To Joe.) Could you pull out three-elevenths?
J: (After Melissa erases the 3/22-bar, mistakenly pulls out a 5/22-bar, but tries again this time pulling out a 6/22-bar.)
T: OK, now Melissa, why is that three-elevenths?
M: (Drags the 6/22-bar to the end of the 5/22-bar, and doesn't respond to the teacher's inquiry.)
T: (After asking Melissa to explain four more times.) OK, explain to Melissa why that's three-elevenths.
J: Because...I know if you pull out two twenty-twos it will be one-eleventh. So, if you pull out six of those it will be three-elevenths!
T: (To Melissa.) Does that make sense to you?
M: (Nods.)

Because it was Melissa who independently made the decision to partition the unmarked copy of the 11/11-bar into twenty-two parts, it was a surprise that she pulled out a 3/22-bar instead of a 6/22-bar after the teacher asked her to pull out three-elevenths. It was especially surprising in view of her clearly stating in

---

[8]These operations are the same operations that produce three levels of units that both children used to produce the iterative fractional scheme.

Protocol V, "How many eighties can you get to fit into one-fifth!" when it was her goal to find how many parts she would need to break a 1/5-bar into so that each part was one-eightieth of the unit bar. In contrast to Joe, she seemed to operate as she did in Protocols III and IV; she produced the 22-part bar without establishing, prior to activity, the 11/11-bar as a unit containing eleven connected units each of which could be partitioned into two connected units using her equipartitioning operations. That she could not offer an explanation for why the six-part bar Joe pulled out was three-elevenths of the bar contraindicates that she used her equipartitioning operations to produce the 22-part bar at the very beginning of the protocol.

Joe, on the other hand, not only pulled out a 6/22-bar when asked to pull out three-elevenths of the bar, but also explained that, "I know if you pull out two twenty-twos it will be one-eleventh. So, if you pull out six of those it will be three-elevenths!" This explanation indicates that he iterated a 2/22-stick three times where the 2/22-stick and the 1/11-stick were identical, which corroborates Joe's use of equipartitioning operations applied to connected numbers as the key operations that permitted Joe to act so powerfully. So, equipartitioning operations were assimilating operations of Joe's, but not Melissa's, units-coordinating scheme for connected numbers.

The question now is whether Melissa's equipartitioning operations that she used in Protocols V and VI could be activated as assimilating operations in the context of finding fractions commensurate to proper fractions. The continuation of Protocol VII provides an occasion to further analyze Melissa's assimilating operations.

Protocol VII. (Cont)
T: (To Melissa.) You give Joe an elevenths fraction to pull out of your bar. (The 22/22-bar)
M: Umm, let me see. Ten-elevenths.
T: Joe, can you pull out ten-elevenths?
J: (Using PULL PARTS, slowly drags a rectangle around twenty parts and pulls them out of the 22/22-bar. He counts by twos in the process.)
M: (Counts the twenty parts as Joe counted them.)
T: (To Melissa.) How many parts is that?
M: (Immediately.) Twenty.
T: (After Joe drags the 20/22-bar directly underneath the 11/11-bar with left endpoints coinciding.) Can you check it, Melissa?
M: (Counts the ten parts of the 11/11-bar that are spanned by the 20/22-bar.)
T: Joe, you give Melissa one.
J: Six-elevenths.
M: (Counts off six of the eleven parts of the 11/11-bar starting with the left-most part. She then drags the 22/22-bar directly beneath the 11/11-bar.)
T: (Laughing.) How many parts is that, Melissa?
M: Twelve.

Melissa's decision to ask Joe to pull out ten-elevenths seemed to be based on the 11/11-bar rather than on the 22/22-bar. To justify why Joe's answer of twenty was related to 10/11-bar, she counted the ten parts of the 11/11-bar that corresponded to the 20/22-bar. She did not give an explanation based on the logical necessity that

8 Equipartitioning Operations for Connected Numbers: Their Use and Interiorization

two one-twenty-seconds in each one-eleventh implies that twenty-twenty-seconds would be ten-elevenths because two times ten is twenty. Rather, she made a direct reading using the bars to justify Joe's answer and proceeded in a very similar way when Joe asked her to find six-elevenths of the 22/22-bar. For these reasons, we infer that when she produced the 22/22-bar in Protocol VII, she simply added eleven and eleven, which is to say that she counted by eleven. By counting this way, she proceeded in a way that was similar to how she produced the sequences for five in the teaching episode held on the 1st of December (cf. Protocol III). In order to continue probing Melissa's assimilating operations, the teacher gave Joe the problem of making another bar using the unit bar with a different number of parts so that he could still pull out elevenths.

Protocol VIII. Melissa produces a 33/33-bar by inserting three into each part of a 11/11-bar.
T: (To Joe.) Go ahead, Joe. Copy the unit bar and make a different one now.
J: (Makes a copy of the unit bar and wipes the marks off from it using WIPE BAR. He then partitions it into twenty-two parts.)
T: That is the same as Melissa's bar!
J: Oh, I know. (Wipes the bar and tries to dial Parts to "33," but the greatest numeral on the dial is "32.")
T: You are trying to get to what number?
J: Thirty-three.
T: Is there a way to use the unit bar to get a bar with thirty-three parts in it?
M: Make it eleven parts!
T: So you can use a bar with eleven in it. OK.
J: (Dials PARTS to "11" and clicks on the copied, but blank, bar.)
M: Then put three in each one. (Pointing to the first three parts of the 11/11-bar that Joe made.)

Melissa's decision to "make it be eleven parts" and her comment, "then put three in each one," are the first indications that she made a units-coordination to produce a different number of parts in the unit bar so she could still pull out elevenths. The teacher in the continuation of Protocol VIII explored the consequences of her perhaps momentary and contextual insight.

Protocol VIII. (Cont)
J: (Joe breaks the 11/11-bar and partitions each part into three parts. He then joins the eleven 3/33-bars together to form a 33/33-bar and indicates to the teacher that he wants to pull out six-elevenths. He then slowly drags a rectangle over a few pieces but stops after he realizes that the rectangle did not contain the first part he wanted to pull out.)
T: Is there a way you could pull one-eleventh out? How many times would you have to repeat it?
J: Six.
T: OK. Do that.
J: (Pulls a 3/33-bar out.)
T: How many of them will you need – how many thirty-thirds?
J: Eighteen.
T: Do you think he is right, Melissa?

Protocol VIII (continued)

M: Yep.
T: Why?
M: No! He needs twelve!
J: (Makes the 18/33-bar and drags it directly underneath the 11/11-bar with left endpoints coinciding.)
M: (Counts the parts in the 18/33-bar and agrees that there are eighteen.)
T: How did you get twelve?
M: I was using the twenty-seconds!

Even though Melissa conflated using twenty-seconds and thirty-thirds, the fact that she said that Joe would need twelve of what she considered to be twenty-seconds indicates that she regarded each 1/11-bar as containing a 2/22-bar. There was at least a temporary modification in Melissa's units-coordinating scheme in that equipartitioning operations were used in constituting the situation. This local modification opens the possibility that Melissa could construct a commensurate fraction scheme, but it does not establish it at this point in the teaching experiment.

Whether Joe could produce a plurality of fractions commensurate to one-eleventh[9] is doubtful. He did seem to be aware that, for any number of elevenths up to and including eleven-elevenths, he could produce a commensurate fraction using the 33/33-bar. But he seemed unaware that he could simply use his number sequence in systematically generating fractions commensurate to one-eleventh. There is no reason to believe that either child could produce a plurality of fractions commensurate to one-eleventh in the sense that they produced a fractional connected number sequence (cf. Protocol VI).

To test the hypothesis that Melissa's use of equipartitioning operations as assimilating operations in the first part of her units-coordinating scheme in Protocols VIII and its continuation was an accommodation, Protocol IX is selected from the teaching episode held on the 2nd of February. The teacher asked the children to use one-ninth, one-eighteenth, and one-twenty-sevenths to produce a fraction commensurate with a multiple of one of the other fractions.[10] The solution of these tasks would involve a coordination of the iterative fraction scheme and commensurate fraction scheme. These tasks were chosen to test the hypothesis that Melissa's equipartitioning operations for connected numbers were assimilating operations for her units-coordinating scheme because, in the case of her iterative fraction scheme, Melissa apparently took the results of her equipartitioning operations

---

[9] By a plurality of commensurate fractions, I mean an unbounded sequence of fractions. Such a sequence could be realized as an unbounded sequence where the next fraction of the sequence entails partitioning the 11/11-stick, say, using the next number in the number sequence for one. The children "produce" this sequence only in the sense that they are aware that they could continue on making fractions of the sequence an indefinite number of times.

[10] For example, use one-eighteenth to produce four-ninths.

8   Equipartitioning Operations for Connected Numbers: Their Use and Interiorization   245

as material in further operating when she justified why forty-four-elevenths (which she produced as eleven times four-elevenths) was four whole bars (cf. Protocol VI). Hence, if her construction of equipartitioning operations in Protocol VIII was on a par with her construction and use of operations that produce three levels of units to assimilate fraction situations in Protocol V, then she should be able to use her equipartitioning operations to establish how many of one unit fraction are in another, which would open the way for her to coordinate her scheme to find a particular commensurate unit fraction with her iterative fraction scheme in order to solve this novel task. Joe had already established equipartitioning operations as permanent operations in the first part of his units-coordinating scheme and the operations that produce three levels of units as the first part of his iterative fraction scheme, so a comparison of how the two children operated in the coordination tasks can be used in testing the hypothesis. There were three congruent bars in the children's visual field prior to Protocol IX: a 9/9-bar, an 18/18-bar, and a 27/27-bar. There was also a 1/9-bar, an 1/18-bar, and a 1/27-bar available.

Protocol IX. Melissa attempts to use the 1/18-bar to make a fraction that is not made up of eighteenths.

T: You have to make a different kind of fraction from the one you are using. If you make ninths, you can't use ninths, and if you make twenty-sevenths, you can't use twenty-sevenths. (To Melissa.) You pose a problem for Joe, now. Choose which units to use. (Pointing to the 1/9-bar, the 1/18-bar, and the 1/27-bar.)

M: (Chooses the 1/18-bar. She drags the 1/18-bar to a position beneath the three partitioned whole bars on the screen; the 9/9-bar, the 18/18-bar, and the 27/27-bar.) I want you to make.... (Sits looking intently at the screen for approximately 23 seconds) I know what I want to do, but I don't know how to say it!

T: What kind of a fraction do you want him to make? You have the eighteenth, and you want him to make what?

M: (Looks intently at the screen for approximately 20 seconds) Um, one-fourteenth.

T: (Clasps his hands to the side of his head.) Make one-fourteenth? Of the unit bar?

M: Use this (The 1/18-bar.) to make one-fourteenth of that. (Runs her hands along the 27/27-bar.)

T: You mean make fourteen parts out of this bar? (Pointing to the 27/27-bar.) Show me, I am not sure I understand.

M: (Repeats the 1/18-bar into an 8/18-bar and drags it directly underneath the 27/27-bar with left endpoints coinciding. The 8/18-bar is adjacent to twelve parts of the 27/27-bar. Shakes her head.)

J: Two-eighteenths makes three-twenty-sevenths!

T: (To Melissa.) Did you hear what he said?

M: Two-eighteenths makes twenty-sevenths.

T: (Asks Joe to repeat what he said.)

J: Two-eighteenths makes *three*-twenty-sevenths!

M: Oh.

T: Can you show us that, Joe?

J: (Fills three parts of the 27/27-bar directly beneath two parts of the 18/18-bar at the rightmost end.)

M: (After the teacher asks her if she sees that, nods.) Mm-hmm.

Melissa sitting and looking intently at the screen for approximately 23 seconds coupled with the comment, "I know what I want to do, but I don't know how to say it!" indicates that she experienced a perturbation that she could not resolve. She could iterate the 1/18-bar as many times as she wanted, so iteration or lack thereof was not a source of her experienced perturbation. To make a 1/9-bar using her 1/18-bar certainly would entail establishing two-eighteenths as commensurate to one-ninth prior to activity. So, her comment indicates that she could not establish this relation as she sat intently looking at the screen, although she had used the operations that produce three levels of units to produce a fractional connected number sequence (Protocol VI). Her choice of one-fourteenth as the bar that she wanted Joe to make is corroboration that she could not establish the two fractions as commensurate. She certainly did not coordinate her iterative fraction scheme with making commensurate fractions. It is important to note that if such a coordination had occurred, it would have also occurred prior to activity as it did for Joe when he said that two-eighteenths makes three-twenty-sevenths. That is, an inference that Melissa made a coordination of the two schemes would need to be based on her choice of an appropriate fraction that she wanted Joe to make, prior to observable activity.

Her use of the operations that produce three levels of units as constitutive operations of her iterative fraction scheme in Protocol VI apparently was not accompanied by an explicit awareness of herself as an operating agent, which means, in other terms, that she did not willfully execute the operations. Although she did know that sixteen-elevenths was five-elevenths more than the whole bar, this reasoning can be carried out without being aware of the operations involved in producing the unit of units of units on which it is based. Further, after Melissa produced forty-four-elevenths as eleven times four-elevenths and after Joe said that there were four unit bars in the forty-four-elevenths, she explained that there were four because four times eleven is forty-four. Although her explanation seems to imply that she was aware of the involved equipartitioning operations on which transforming eleven times four-elevenths into four times eleven-elevenths are based, her adding and multiplying schemes for whole numbers were evoked and used throughout the protocol. My interpretation is that the reason these whole number schemes were evoked was because she used her equipartitioning operations for connected numbers in assimilating the tasks and these operations evoked her adding and multiplying schemes for whole numbers. Such massive transfer, although it is solid corroboration of the reorganization hypothesis, does render opaque the issue of whether she was aware of the operations to which her explanation pointed. These operations entail reassembling a composite unit that contains eleven units each of which contains four connected 1/11-units into a composite unit containing four units each of which contains eleven connected 1/11-units. Joe, on the other hand, did willfully use equipartitioning operations in Protocol IX, which is compatible with his explicit awareness of the operation of recursive partitioning and its inverse (cf. the continuation of Protocol I) because equipartitioning operations contain the operation of recursive partitioning.

Melissa's Use of the Operations that Produce Three Levels of Units in Re-presentation

How Melissa might willfully use her operations that produce three levels of units in re-presentation was of primary interest in the analysis. In Protocol V, Melissa had constructed two composite units, a composite unit of eighty-eightieths and a composite unit of five connected units comprising five-fifths, and asked how many eightieths would go into each fifth. She produced an equipartitioning of eighty-eightieths and engaged in finding if fourteen of the eightieths would fit into each fifth. Although she did not explicitly iterate fourteen five times, she did use her multiplying algorithm to test whether five times fourteen is eighty. Her complete activity, when considered together, implies that she formed an image of eighty-eightieths, whatever it might have been, and was aware of partitioning re-presentations of a part of the 5/5-bar in order to produce eighty-eightieths. So, it was a puzzle to us why recursive partitioning was not ready-at-hand for her in those cases where there was no material in her visual field on which she could operate because recursive partitioning operations are operations that produce three levels of units. In the next section, we analyze protocols selected from teaching episodes where the children were asked to repeatedly make fractions of fractional parts of bars. The children's engagement in the tasks presented an opportunity to analyze Melissa's interiorization of recursive partitioning operations and Joe's construction of a scheme of recursive partitioning operations.

Repeatedly Making Fractions of Fractional Parts of a Rectangular Bar

In the teaching episode held on the 9th of February, the children made repeated partitions of a bar. At each step, the children were to find the fractional part of the whole bar that one of the parts comprised.

Protocol X. Joe and Melissa find a half of a half of a half of a....

T: (The children had made a rather large rectangular bar on the screen.) Melissa, you go first. You cut it into halves. Make it into two parts. Then Joe, you take one of those pieces and cut it into halves.

M: And then cut it and cut it and cut it....

T: And then take one of those parts and cut it again, and then keep on. At each one, I want you to tell me how much of the unit bar it is and why did you say that.

M: (Uses VERTICAL PARTS to partition the rectangular bar into two parts, breaks it upon the request of the teacher, and fills one of the parts a different color. She then says the part she filled is one-half.)

J: (Partitions the filled part into two parts using HORIZONTAL PARTS, breaks the two parts and fills one of them.) One-fourth. (Explains that it takes four of them to cover the whole piece.)

M: (Repeats the action using VERTICAL PARTS on one of Joe's parts.) One-sixth – (Shakes her head "no.") one-eighth! If you would do it to all of them, you would have eight. (With Joe, counts the parts that would have been made if a complete partitioning had been made at each step in verification after the teacher asked Joe if he thought it would be one-sixth or one-eighth.) You have four sets of two! (Fig. 8.1)

J: (Repeats the actions on the lower right rectangle using VERTICAL PARTS.) One-sixteenth.

T: Why do you say its one-sixteenth?

J: Cause that's the way we were doing it.

M: Cause that's the way we were doing it by twos. Two, four, six, eight, ten, twelve, fourteen, sixteen. (Pointing to eight places on the original rectangle where pairs of rectangles would have been made if a complete partitioning was made at each step.)

T: (Asks Joe to measure the part he made, and "1/16" appears in the NUMBER BOX.)

M: (Repeats the action on the right most one-sixteenth of the rectangle using HORIZONTAL PARTS.) One-eighteenth. (Fig. 8.2)

T: How much do you think it is, Joe?

J: (Counts from the bottom to the top of eight imagined columns of four rectangles starting from his right to proceeding to his left.) One-thirty-two.

T: (To Melissa.) How did you get one-eighteenth?

M: Well, I used doubles. (After the teacher asks her how she used doubles, she counts the imagined parts as did Joe and agrees with Joe's answer.)

J: (Repeats the action on one-thirty-second.) One-sixty-fourth. (Immediately.)

T: Wow! How did you get one-sixty-fourth?

J: Thirty-two and thirty-two. (A witness asked him how he knew to double thirty-two.) Because the first time we did it we got two. Then we doubled it and got four. Then we doubled that and got eight, and then doubled that, sixteen, and then kept on going!

T: Do you know why those numbers double like that?

J: Uh-uh. (No.)

M: One one-hundred twenty-fourths. (In anticipation of the fractional part of the rectangle she will make next.) (Repeats the actions on the lower right most rectangle.) (Fig. 8.3)

T: One one-hundred twenty-fourth, and you say? (Pointing to Joe.)

J: One one-hundred and twenty-eighth.

M: Oh, I guessed! (Repeats Joe's answer.)

J: (In explanation.) I took one four off from sixty-four. And then I added sixty plus sixty is a hundred twenty and I added the fours.

T: (To Melissa.) Do you know why we are going from sixty-four to one hundred twenty-eight?

M: Because we are counting by.... (Stops and looks confused.)

J: Double it.

M: We are doubling the numbers we get.

T: OK. Let's keep going.

J: (Repeats the action on the lower rightmost one one-hundred twenty-eighth. Looks into space.) I can do it! Forty (After a pause.) two fifty-six! One two hundred-fifty-sixths.

T: Wow!

J: I added the hundreds first, then I added the twenties and got forty, and then added the eights.

M: (Requests paper and pencil, and uses a computational algorithm.)

J: (Measures the part and "1/256" appears in the number box. Both children express pleasure that the computer confirmed their answers.)

8 Equipartitioning Operations for Connected Numbers: Their Use and Interiorization 249

Fig. 8.1. Making one-eighth by partitioning one-fourth.

Fig. 8.2. Making one-thirty-second by partitioning one-sixteenth.

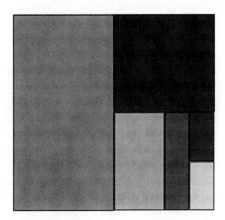

Fig. 8.3. Making one-one-hundred twenty-eighth by partitioning one-sixty-fourths.

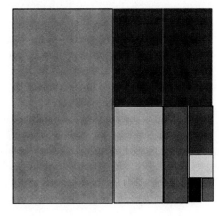

Melissa's initial comment, "And then you cut it and cut it and cut it…" indicates that she imagined the steps of cutting into one-half. This is crucial because when engaging in recursive partitioning in re-presentation, the child regenerates the results of an immediately prior partitioning as input for further partitioning. However, Melissa's anticipation of cutting into one-half proceeded forward in sequence rather than recursively – "And then cut it and cut it and cut it…" Nevertheless, the fact that she could anticipate sequential partitioning acts is encouraging because it is fundamental in constructing a scheme of recursive partitioning operations.

After Melissa cut one-fourth of the rectangular bar into two pieces, she momentarily said that one of the parts was one-sixth of the rectangular bar. This same way of proceeding occurred after she cut one-sixteenth of the rectangular bar into two parts and said, "one-eighteenth." In both cases, she operated sequentially and added the number of parts she made to the preceding number of parts. In the first case, she made a self-correction and proceeded to count the parts that would have been made if a complete partitioning had been made at each step. Although we would not refer to these counting actions as partitioning prior partitions in re-presentation, they certainly constituted a modification in her way of judging how much a part of a current rectangle was of the whole rectangular bar. She definitely was aware of the history of the partitioning actions in that she organized their potential results as "four sets of two." In doing so, she projected partitioning into two parts into the unpartitioned parts that were in her visual field. She also projected partitioning into two parts into the unpartitioned parts in explaining why Joe said, "one-sixteenth" – "Cause that's the way we were doing it by twos. Two, four, six, eight, ten, twelve, fourteen, sixteen (pointing to eight places on the original rectangular bar where pairs of rectangles would have been made if a complete partitioning had been made at each step)." These were definitely enactments of partitioning the unpartitioned rectangles into two parts.

Joe's recursive partitioning operations enabled him to independently produce the parts of the rectangular bar that would be produced if partitioning into two parts had been completed at each step. His abstraction of the sequence of doubling acts – "Because the first time we did it we got two. Then we doubled it and got four. Then we doubled that and got eight, and then doubled that, sixteen, and then kept on going!" – indicates that he was aware of his acts of partitioning. Repeatedly engaging in recursive partitioning was not simply an operation that Joe carried out without an awareness of how he proceeded. Rather, he seemed aware of recursive partitioning in activity. If he took the results of recursive partitioning operations at each step as a unity and set those results at a distance, this would enable him to abstract that partitioning each part into one-half doubles the number of parts. This is a major step in the construction of a *scheme of recursive partitioning operations* because, in such a scheme, the child does not need the results of the preceding partitioning operations in his or her visual field to produce the next partition. But Joe would need to generalize it to other partitionings before we would judge that he had indeed constructed such a scheme. In addition, when the teacher asked him "Do you know why those numbers double like that?"

Joe did not know. This is a contraindication that he had explicitly correlated numerical doubling with doubling the number of parts made in the rectangular bar.

Melissa assimilated Joe's way of doubling the preceding numerical result to produce the next partition and guessed what the double of sixty-four would be before she partitioned the rectangle that was one-sixty-fourth of the original rectangular bar into two pieces. Her guess, "One one-hundred-twenty-fourths," was close, but she did not engage in strategic reasoning of the kind in which Joe engaged to produce "One one-hundred and twenty-eighths" and "One twohundred-fifth-sixths". Joe's strategic reasoning is a corroboration of his ability to take partitioning operations as a unity[11] and set their results at a distance and operate on the material he set at a distance using powerful numerical operations.[12] We take Melissa's lack of strategic reasoning as an indication of her more or less general way of organizing her experience into definite and knowable structures. Whether this orientation to mathematical activity constrained her progress in the creative construction of recursively partitioning the re-presented results of a prior partitioning will be addressed in the analysis of the remaining teaching episodes.

Melissa Enacting a Prior Partitioning by Making a Drawing

The modifications Melissa made in her partitioning operations in the teaching episode held on the 9th of February, where she repeatedly projected units of two into the partially partitioned portions of the rectangular bar, reemerged as a drawing in a similar task in TIMA: Sticks in the teaching episode held on the 16th of February. But prior to making the drawing, she enacted partitioning the parts of a 16-stick into two parts each by making cutting motions with her hand. This enactment preceded enacting partitioning by making drawings. The teacher used TIMA: Sticks during the teaching episode because of the way in which PARTS was programmed. In TIMA: Sticks, the children did not need to break a stick in order to partition a part of the stick. Rather, if a stick was partitioned into, say, eight parts, each of the eight parts could be partitioned into as many parts as desired up to and including thirty-two parts. At the beginning of Protocol XIII, Joe had partitioned a stick into eight equal parts and said that each part was one-eighth of the stick, and Melissa took her turn without comment from the teacher.

---

[11] Taking a partitioning operation as a unity means that the child abstracts the relation between the input and output and can use a current output as input for further partitioning. Abstracting the relation means that the child is on the "outside" of partitioning and is able to generate and analyze relations between intermediate states with the proviso that the relation is reversible and can be regenerated at will.

[12] For this reason, I believe that he was very close to constructing a scheme of recursive partitioning operations and, hence, the operations necessary to construct, for example, the cyclic nature of the Hindu-Arabic numeration system as a result of productive thinking.

Protocol XI. Melissa enacts partitioning the parts of a 32-stick.

M: (Partitions the parts of the eight-part stick into two parts each and then counts the sixteen parts.) One-sixteenth.
J: (Partitions the seventh one-sixteenth part into two parts and fills one of the parts gray.)
T: How much is that little gray piece in there?
J: (After a short pause.) One-thirty-second.
T: Will you tell me how you got one-thirty-second?
J: I had one-sixteenth, and I just cut that in half and added sixteen and sixteen.
M: (Partitions the 1/32-part Joe made into two parts.) One-sixty-fourth. (Immediately.)
T: How did you get one-sixty-fourth?
M: I doubled two times thirty-two.
W: Why did that work, Melissa?
M: Because you would do it by counting by eights, oh not by eights, by sixteens.
T: You did not say anything about sixteen, did you?
J: You would make them in halves.
M: You would make one-half (Making cutting motions with her hand.) in all of them.

Melissa's comment that, "You would make one-half in all of them" while making cutting motions with her hand indicates awareness, however unarticulated, of partitioning each of the thirty-two parts into two parts.[13] Her saying that she doubled thirty-two to produce sixty-four seemed to be based on Joe's explanation of his answer of one-thirty-seconds rather than on the results of using her units-coordinating scheme because she also said that they counted by sixteens, which was a recapitulation of Joe's previous explanation. Still, we can deduce that she was aware that her action of partitioning one of the thirty-two parts also partitioned each of the other parts because she made cutting motions with her hand while saying, "You would make one-half in all of them." This enactment of imagined acts of partitioning into two parts in the presence of the 32-stick, which is indeed a units-coordination of thirty-two and two, will prove to be a crucial step in her ability to recursively partitioning a stick in re-presentation.

Joe saying, "I just cut that in half and added sixteen and sixteen," was another example of strategic reasoning like that in Protocol X. It again corroborates his ability to take partitioning operations as a unity, set their results at a distance, and operate on the material he set at a distance using powerful numerical operations, because the result of cutting each of the sixteen parts into two parts was reorganized into two composite units of sixteen. Such a reorganization involves an awareness of a composite unit of sixteen units of two, splitting each unit of two into two separated parts, and integrating these parts into two composite units of sixteen elements.

Before judging whether Melissa's imagining of partitioning acts and her enacting them by cutting motions with her hand constituted an accommodation in her units-

---

[13] In the first part of the protocol, the fact that Melissa counted the sixteen parts of the stick she just made by partitioning each of eight parts into two parts before she answered, "one-sixteenth," should not be interpreted as meaning that she had constructed a fractional composition scheme. Retrospectively, I will interpret it as an important step in such a construction.

8 Equipartitioning Operations for Connected Numbers: Their Use and Interiorization

coordinating scheme, a recurrence of them in the case of a partitioning of the elements of a partition into three or more parts would need to be observed. Fortunately, the teacher turned to a similar task that involved making thirds. We pick up the task just after Joe partitioned each of the three parts of a stick into three parts and said that one of these parts was one-ninth of the whole stick.

Protocol XII. Melissa's drawing of a partition of a partition of a partition.

M: (Partitions the first part of the 9/9-stick into three parts and fills one of the parts.) (Fig. 8.4)
T: OK. How big is that?
M: One-eighteenth!
T: What do you think, Joe?
M: Oh, no, no. One-twelfth!
M: (Before Joe replies, she tries to explain, and then starts to make a drawing, but her drawing is hidden by the monitor.)
J: One-twenty-seventh!
T: Melissa is drawing a number line.
M: (Looks up from her drawing.) It would be about one-twenty-seventh!
T: Joe says it is about one-twenty-seventh. Oh, show them how you figured that out.
M: (Holds her drawing over the screen of the monitor and it was structured something like the diagram that follows.) (Fig. 8.5)

Fig. 8.4. Melissa's partition of the first part of a nine-part stick into three parts.

Fig. 8.5. Melissa's drawing of a partition a partition of a partition.

Melissa's drawing completed each step of the original partitioning – she first partitioned the stick into three parts, then Joe partitioned each one of the three parts into three parts, and then Melissa partitioned the first of the nine parts of the stick into three parts. Her initial two answers of "one-eighteenth" and "one-twelfth" preceded her drawing so these answers were not based on her drawing. After the teacher cast doubt on her answer of "one-eighteenth" by asking Joe what he thought, Melissa turned inwardly and enacted the complete partitioning of the 9/9-stick by making her drawing. Her enactment of the partitioning solidly indicates that she was in the process of interiorizing the involved operations. It was an indication because, while in the process of drawing, it would be necessary for her to monitor her activity. Should her enactment of partitioning a partition by making a drawing recur, the making of a drawing could be considered as an accommodation in her units-coordinating scheme in the context of fractional connected numbers.

A Test of Accommodation in Melissa's Partitioning Operations

If Melissa's enactment of partitioning activity constituted an accommodation, then it should reoccur in situations similar to the situation of Protocol XII. In the teaching episode held on the 23rd of February, the teacher returned to using TIMA: Bars. This time the teacher asked the children to first partition a bar into three parts, then one of those parts into two parts, then one of those parts into three parts, etc. This change was designed to break the process of always doubling, or always tripling, the last number of parts to produce the current number of parts, a process that had been abstracted by Joe and that Melissa had appropriated without engaging in a similar abstractive process. If the children were to produce the sequence; 1/3, 1/6, 1/18, 1/36, 1/108, 1/216, 1/648, 1/1,296,..., it would be based on at least partitioning the whole of an immediately preceding partitioning. We pick up the conversation where Joe partitioned one-sixth of the bar into three equal parts.

Protocol XIII. Melissa partitions unpartitioned parts of a bar in re-presentation.

J: (Partitions the lower 1/6-bar into three horizontal parts and fills one of them a different color.) (Fig. 8.6)

T: Now, the question is, what's that piece? Figure it out before you say it.

M: (Counts the three parts Joe made from the bottom to the top, and then continues on, pointing to three more places on the 1/6-bar immediately above the three 1/18-bars. She then starts at the top of the adjacent 1/3-bar and points to it six times from the top to the bottom. She then completes her counting episode by starting at the top of the leftmost 1/3-bar and points to it six times from the top to the bottom. Each time she pointed, she subvocally uttered a number word.)

T: Do you know Joe?

J: One-eighteenth.

T: (To Melissa.) What do you say?

M: (Nods.)

T: How did you get that?

M: Because there are three sets of six! (Partitions the bottommost 1/18-bar into two vertical parts, breaks the bar and fills the right most part yellow.) (Fig. 8.7)

T: OK, we've got that little yellow piece out there now. How much is that little yellow piece?

M: (Looks at the screen.) That little yellow one?

J: (Starts to answer, but is hushed by the teacher.) Wait for Melissa to get hers.

M: (Looks downward while she is subvocally uttering number words.) One-thirty-second.

T: One thirty-second. And what did you say, Joe?

J: One-thirty-sixth.

T: (To Melissa.) Show me how you did it.

M: Well, I went by twos. (Points to the two 1/36-bars, then to the 1/18-bar immediately above those two bars, then to the next 1/18-bar, then three times to the 1/6-bar.) Two, four, six, eight, ten, twelve,....

T: How did you do it, Joe?

J: I just doubled it.

T: Why did you double it? Why didn't you triple it, or quadruple it?

J: Because I tried thirds, and that didn't work out, so I went to halves and doubled it.

Fig. 8.6. Joe partitioning one-sixth of the bar into three parts.

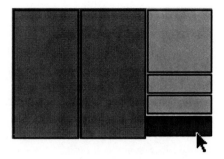

Fig. 8.7. Melissa's partition of one-eighteenth of a bar into two parts.

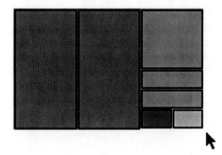

In the first step of partitioning in the protocol, Melissa's counting of the parts that would be made had the partitioning been completed at each step and structuring her activity as "three sets of six" constitutes a structuring of the results of completing the prior partitions even though she did not make a drawing as she did in Protocol XII. Melissa was very precise in structuring her activity as three sets of six – as a unit containing three units each containing six units. This comment is a solid indication that she structured her experience as a unit of units of units and that she was aware of her unit structure. The whole of the counting episode constitutes an enactment of recursive partitioning. Further, when Melissa looked downward while she subvocally uttered number words before answering "one-thirty-second," she explained, "Well, I went by twos." Counting in this way would entail making a visualized image of the partially partitioned bar and projecting units of two into parts of the bar as she counted by twos. Partitioning a visualized image of a prior partitioning is precisely the operation that is involved in recursive partitioning at the level of re-presentation given that it was her goal to find how much one-half of one-eighteenth was of the unit bar. We take this to be her goal, and to eliminate the discrepancy between her expectation of establishing a fraction for one-half of one-eighteenth, and not knowing how many parts this partitioning act implied for the unit bar, she completed the partitioning activity of the prior two steps at the level of re-presentation. So, making the drawing in Protocol XII was an accommodation in her recursive partitioning operations.

A Further Accommodation in Melissa's Recursive Partitioning Operations

If she had not monitored "going by twos" in Protocol XIII, Melissa probably would have become lost in her activity. Her monitoring became more explicit in the following protocol that was extracted from the teaching episode held on the 2nd of March. In that teaching episode, Melissa independently introduced a notational system as a correlate of her drawing. Her notational system was a surprise and it seemed to serve two functions for her, as we explain below In Protocol XIV, Joe was to make fourths and Melissa was to make thirds of the resulting bars.

Protocol XIV. Melissa's notational system.
- J: Partitions the bar into four horizontal parts and fills one part with a different color. (Both children agree that one-fourth of the bar was filled.)
- M: (Partitions the filled part into three vertical parts and fills the middle part with another color.)
- T: (Joe starts to say something.) Hold on, I want you both to get the answer. (After both children have the answer, Melissa says she got one-twelfth.) Why did you say its one-twelfth?
- M: Because if you put all in those squares, you would get twelve pieces. You would have four sets of three.
- J: (As explanation, counts across the uppermost 1/4-bar three times, then across the next 1/4-bar three times, etc.)
- T: Joe, I believe it is your turn now.
- J: Partitions the middle 1/12-bar into four parts and fills the uppermost part (Fig. 8.8).
- M: (Points to three places on the uppermost 1/4-bar as she moves her hand horizontally from the left to the right side. She then points to four places on the uppermost 1/4-bar as she moves her hand downward along the left side. Continuing, she points to four places on the second 1/4-bar, and then to four places on each of the two remaining 1/4-bars as she moves her hand downward.)
- J: (Sits silently in deep concentration while Melissa is counting. As Melissa is almost done.) One-forty-eighth!
- T: Could be. Wait for Melissa to get her answer, too.
- M: (Repeats the four modules of four counting acts downward.)
- J: (While Melissa is counting downward the second time, uses MEASURE to measure the 1/48-bar he filled. He tries to hide the answer in the NUMBER BOX using his hand. His activity seems to distract Melissa.)
- T: (To Melissa.) What did you get?
- M: (Joe removes his hand.) I can see it!
- T: What do you think it was before you saw it?
- M: I was doing fifteen times four.
- T: (Hands Melissa a piece of paper upon her request.) Where did your fifteen times four come from? (Hands Joe a piece of paper.) I will give Joe one so he can use it later.
- M: (Writes on her paper which is hidden by the monitor. Looks intently at the screen for approximately 20 seconds. She then returns to writing on her paper.) And the next one was…He did three, so…
- T: (After about 15 seconds.) So you're starting to make a sequence of pieces here. (After about 30 seconds.) Now you're drawing a picture of the pieces. (Indicates to Joe that he should show what he did using his paper and pencil.) How did you get your answer, Joe, while she's working on hers?

8   Equipartitioning Operations for Connected Numbers: Their Use and Interiorization   257

J: I did twelve times four.
M: (After about 45 seconds.) And the next one was one-twelfth. And the next one was one-forty-eighth.
T: How did you get one-forty-eighth?
M: First, he did his fourths. Then I did my thirds, then he did his fourths.
T: OK, let's hold this up here to see what you have been doing. (Holds Melissa's paper up to the monitor.) Joe started out with a fourth, Melissa thirded it and got a twelfth, Joe fourthed it and got a forty-eighth. Then she drew the picture at the bottom so they could see what's going on.

Fig. 8.8. Joe's partition of one-twelfth of a bar into four pieces.

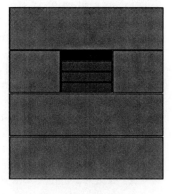

Fig. 8.9. Melissa coordinating her drawing and her notational system.

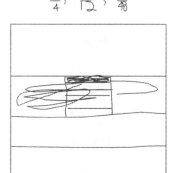

It is quite significant that Melissa recorded both the number of parts into which a particular bar was partitioned and the fraction of the whole bar produced by that partitioning. For example, the "4" in third step in her notational schema [^{4}1/48] referred to partitioning one-twelfth of the whole bar into four parts and the "1/48" referred to both forty-eight parts in the bar and the fraction one of those parts was of the forty-eight parts. Her notational schema not only recorded the history of her partitioning acts, but also indicates that she took the output of each prior partitioning act as input for the next partitioning act. For example, after producing the 1/12-bar in her drawing by partitioning a 1/4-bar into three parts, she partitioned the 1/12-bar into four parts and produced the 1/48-bar. She only completed the first partition in her drawing, so her notation $^3 1/12$ stood in for partitioning each 1/4-bar into three parts in order to partition the whole bar into twelve parts. That is, she seemed well on the way to constructing her partitioning operations as recursive at the level of re-presentation. To make such an inference, of course, does not require building a notational system because Joe said "one-forty-eighth" after sitting in deep concentration while Melissa was counting over the computer graphics.

The effect of Melissa's abstractive activity was that, when it was her goal to find, say, how much one-third of one-fourth of a bar was of the bar, she established this situation as a situation of her units-coordinating scheme. It is no exaggeration to say that these partitioning operations had become the operations of her units-coordinating scheme and that the activity of her scheme – finding how many parts was in the partitioning so produced – was recorded in these operations. So, if Melissa produced the goal of finding how much one of the three equal parts of one of the four equal parts was of the unpartitioned bar, this activated the abstracted operations of partitioning into three parts and distributing these operations over the remainder of the four parts while monitoring how many parts were produced as a result of the distribution. The implementation of these operations, along with comparing the part to the partitioned whole, constitutes the activity of a unit fraction composition scheme. The continuation of Protocol XIV contains corroboration that Melissa used her units-coordinating scheme to produce the number of parts of the partition implied by partitioning a 1/48-bar into three parts.

Protocol XIV. (Cont)

T: All right. What's next, the thirds? Who takes the thirds?
M: I do. (Partitions the uppermost 1/48 bar into three vertical parts and fills the middle part.) (Fig. 8.10)
T: Think about it before you actually do it.
M: Count by twelve.
T: Can you tell me what the answer is going to be if you do that? OK, how big is that yellow piece there?
J: One one-hundred forty-fourth.
T: Melissa, what did you get?
M: One one-hundred forty-fourth.
T: OK, let's hold this up. (Melissa's paper.)

$$\frac{12}{1} \qquad \frac{12}{48}^{\frac{4}{3}} \qquad \frac{^2 48}{144}^{\frac{3}{3}}$$

$$\frac{^4 1}{4)} \qquad \frac{^3 1}{12)} \qquad \frac{^4 1}{48)} \qquad \frac{3}{144}$$

T: (Asks Joe to hold up his paper.)
J: $\qquad \frac{48}{144}^{\frac{3}{3}}$

Fig. 8.10. Melissa filling one-eighth of the bar.

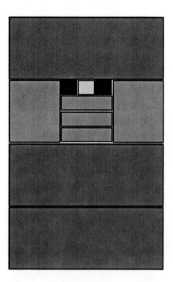

Melissa extended her sequence of fractions beyond one-forty-eighth by multiplying forty-eight by three, which corroborates that she used her units-coordinating scheme in the production of one hundred forty-four.

Melissa's independent production of her notational system was indeed a surprise and it was not preceded by any intentional action or comment by the teacher. She had already constructed one- and two-digit numerals as symbols in that they referred to at least an iterable unit of one that could be iterated the number of times indicated by a particular numeral to produce a composite unit containing the units produced by iteration, and she could use her numerals to stand in for these operations.[14] Her numerals had also taken on new meaning in that now "4," say, could stand in for partitioning a bar or a stick into four parts. Further, she used her fraction numerals, such as "1/48," to stand in for the operations of her iterative fraction scheme because she did not need to carry out all of those operations to give meaning to the shaded bar in Fig. 8.8. Her drawing that is replicated as Fig. 8.9 solidly indicates that she could give meaning to the numeral prior to making the drawing. So, one source of her creative act of producing the notational system resided in the symbolic nature of her whole number and fraction numerals. Another, of course, was her use of recursive partitioning at the level of re-presentation.

Given that it was Joe who began the teaching experiment in fifth grade having already constructed a reversible unit fractional composition scheme, it was indeed surprising that Melissa seemed to be the stronger of the two students in symbolizing a sequence of recursive partitioning operations using drawings and notation. Joe's abstractive power is well illustrated in the second continuation of Protocol XIII when he partitioned a 1/18-bar into two parts and said that one of the three parts

---

[14] In this case, I consider Melissa's numerical concepts as multiplicative concepts.

was one-thirty-sixth of the whole bar after sitting silently in deep concentration. Melissa, on the other hand, resorted to doubling and said that the part was one-thirty-second of the whole bar. In this case, Joe used the results of his prior recursive partitioning operations in his current recursive partitioning operations. This opens a possibility that Joe had constructed a scheme of recursive partitioning operations. However, it must be remembered that Joe was not asked in Protocol XIII to explain how he could operate. Had he been able to make such an explanation, this would be a solid indication that he was becoming aware of how he could operate on any but no particular turn,[15] which is essential in inferring a scheme of recursive partitioning operations. The notational system that Melissa generated in Protocol XIV and its continuation is more of an indication of a scheme of recursive partitioning operations than was Joe's ability to produce the next fraction in the sequence of fractions being produced by the children's partitioning actions by multiplying the number of parts in the current partition and the number of parts produced by the preceding partitions. But Melissa's notational system is still not sufficient to infer a scheme of recursive partitioning operations, although it is a strong indication. Making such a notational system is based on an awareness of taking the results of current operation as input for the next operation, so she definitely became aware of how she was operating. This is the basis for my inference that she had finally constructed recursive partitioning. But to infer a scheme of recursive partitioning operations, Melissa would need to use her notational system as input for further operating without resorting to making drawings.

A Child-Generated Fraction Adding Scheme

Both Joe and Melissa had constructed a unit fraction composition scheme that they used recursively in the sense that they used the results of a prior use of the scheme as input for another use of the scheme. During the course of the teaching experiment, we did not emphasize written notation with Joe and Melissa because we were primarily interested in exploring the children's operations and the schemes that they could use to supply meaning for written notation. For example, neither child construed finding how much of the whole bar was constituted by one of the three equal parts of one-forty-eighth as fractional multiplication in the continuation of Protocol XIV, even though their ways of operating were constitutively multiplicative in the fractional sense. Given Melissa's independently generated notational system, however, we became keenly interested in analyzing her construction of a symbolized fraction adding scheme in the context of her recursive use of her unit fraction composition schemes.

In the teaching episode held on the 6th of April, the teacher presented situations like those in which Melissa's drawings appeared to explore the children's creative production of the first few terms of the series $1/2 + 1/4 + 1/8 + 1/16 + 1/32 + \ldots$.

---

[15] This is akin to operating with a variable.

8 Equipartitioning Operations for Connected Numbers: Their Use and Interiorization   261

Protocol XV. Finding partial sums of the series, 1/2 + 1/4 + 1/8 + 1/16 + 1/32.

T: OK, we are going to do some fraction addition today. What we are going to do is take half, then half again, then make half again. I'll show you what I mean. (To Melissa.) Go ahead and make the first one.
M: (Partitions the bar horizontally into two parts, breaks the bar, and fills the lower part.)
T: (To Joe.) OK, you are going to take half of the unfilled part. Go ahead.
J: (Partitions the upper part into two parts vertically, breaks the bar and fills the rightmost part.)
M: Now we've got three-fourths!
T: Now, the question is, how much of the unit bar is filled?
M: Three-fourths.
J: One-half plus one-fourth.
T: Yeah, but what does that work out to?
J: (Picks up his pencil and writes on his paper.) I got three-fourths.
T: OK, let's see how you did it. (Holds Joe's paper up to the monitor)

$$\frac{1}{2} = \frac{2}{4}$$
$$\frac{1}{4} = \frac{1}{4}$$
$$\overline{}$$
$$\frac{3}{4}$$

Joe's use of his computational algorithm for adding fractions was a complete surprise. Apparently, the teacher's comment, "OK, we are going to do some fraction addition today." Oriented Joe to use his computational algorithm. Melissa, on the other hand, knew almost immediately that the filled portion of the bar was three-fourths of the bar. Joe replied, "One-half plus one-fourth," but he did not use his recursive partitioning operations to partition the bottom one-half of the bar into two parts and then disembed the three parts from the four parts to produce three out of the four parts. These operations were available to him, but the use of his computational algorithm apparently closed off the activation of relevant operations.

Protocol XV. (First Cont)

M: (Horizontally partitions the leftmost 1/4-bar into two parts and fills the bottommost part as shown below. The unfilled part of the bar is red.) (Fig. 8.11)
T: OK!
M: (Points her pencil to the unfilled part and then to the part to its right. She then repeats this pointing action at the filled 1/8-bar and then to the right of that bar. She then points to the 1/2-bar twice and then twice more. She then writes something on her paper.)
J: (While Melissa is pointing.) One-half plus one-fourth plus… (Points to the lower 1/8-bar but does not know the fraction for that bar. He then points in pairs to the bar downward four times as if he is counting 1/8-bars.) One-eighth!
M: (Points to the filled portion of the bar an indefinite number of times while Joe was pointing to the bar eight times. She then again writes something on her paper.)
J: One-half plus one-fourth plus two-eighths!
T: (To a query by a witness.) Two-eighths, he said.
J: One-eighth! (Continues writing on his paper.)

Protocol XV. (continued)

M: (In explaining to the teacher, she points over the bar eight times and then counts over the filled part seven times, where the pointing actions indicate regions of the bar that would be 1/8-bars had the whole bar been partitioned each time.)

J: I got it – I got it.

T: (Places Melissa's paper up to the monitor.)

$$1, \tfrac{1}{2}, \tfrac{3}{4}, \tfrac{7}{8}$$

T: (Places Joe's paper up to the monitor.)

$$\tfrac{1}{2} = \tfrac{4}{8}$$
$$\tfrac{1}{4} = \tfrac{2}{8}$$
$$\tfrac{1}{8} = \tfrac{1}{8}$$
$$\tfrac{7}{8}$$

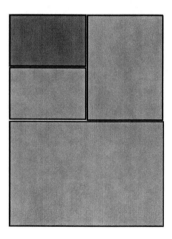

Fig. 8.11. Melissa's partition of one-eighteenth of a bar into two parts.

When Joe counted over the bar eight times to produce one-eighth, at that point he did not realize that he had essentially solved the problem of finding how much of the whole bar was filled. His goal was to find how much the filled 1/8-bar was of the whole bar. That goal, when coupled with his goal of finding the fractional part each filled bar was of the whole bar so he could use his computational methods, apparently excluded him from asking himself how many eighths could be made from the filled portion of the bar. Because his paper and pencil methods were separated from his recursive partitioning operations that he so powerfully demonstrated in earlier teaching episodes, his goal also apparently excluded finding how many eighths could be made from each of the 1/4-bar and the 1/2-bar using the operation of partitioning the results of prior partitioning, i.e., using reasoning. Melissa, not being in a computational frame of mind, worked insightfully and produced a sequence of partial fractional sums of the bar. However, Melissa had only begun to construct a fraction adding scheme when operating to find seven-eighths. In later tasks, she insightfully produced "15/16" and "31/32," also without explicitly using a fraction adding scheme. It is important to note that in the task where Melissa produced "31/32," when Joe tried to find the sum of whatever fractional meaning he gave to "1/32," "1/16," "1/8," "1/4," and "1/2," he became very despondent and made computational errors in attempting to change each of the last four fractions into thirty-seconds. Apparently, Joe's fraction adding scheme was procedural in nature and it excluded him from reasoning insightfully when he was capable of doing so.

An Attempt to Bring Forth a Unit Fraction Adding Scheme

The current teaching episode held on the 4th of May was conducted for the purpose of investigating the operations involved in finding the sum of two given unit fractions that are not produced as a part of a sequence of recursive operations. The teacher's goal was for the children to find what fractional part a 1/3-bar and a 1/4-bar together is of the whole bar. Because the reversible unit fraction composition scheme is involved in solving this task, the teacher began by presenting a task designed to reinitialize Joe's reversible fraction composition scheme.[16] In the task, the teacher began by asking the children, starting with a 3/3-bar, to partition the bar so that one-ninth could be pulled out.

Protocol XVI. Producing sequences of fractions using the reversible unit fraction composition scheme.

T: (After Joe made a 3/3-bar.) OK, Melissa. Use PARTS so you could pull a ninth out using the thirds.
M: (Dials PARTS to "3" and clicks on the leftmost one-third of the 3/3-bar. She then uses PULL PARTS to pull a 1/9-bar from the unit bar.)
T: (Asks Joe to erase the marks Melissa made to restore the bar to a 3/3-bar.) Joe, you make it so you can pull out a twelfth.
J: (Dials PARTS to "4" and clicks on the leftmost one-third of the 3/3-bar. He accidentally clicks twice so that the second part of the four is again partitioned into four parts. He then erases the extraneous marks and pulls out one of the four parts he made using PULL PARTS.)
T: (To Melissa.) OK, Melissa, it is your turn. What would you like to do now, Melissa?

---

[16]In the second continuation of Protocol I, the teacher asked Joe to cut up a 1/4-stick so that each part is one-twentieth of the whole stick and Joe immediately partitioned the stick into five parts. Based on this protocol and Protocol II, where Joe partitioned each part of a 3/4-stick into two parts to make eighths of the whole stick, I inferred that he had constructed his unit fractional composition scheme as a reversible scheme. That is, given a result of the scheme, he could recursively partition a partition to produce the result.

M: One-fifteenth. (After erasing the marks Joe made to pull out twelfths, dials PARTS to "5," clicks on the leftmost one-third of the 3/3-bar, and pulls out one part.)
J: I will pull out an eighteenth!
T: You guys work out all of the ones that you can do.
J: (Dials Parts to "6" without erasing the five parts Melissa just made. He moves the cursor back and forth between the five parts that Melissa made and the middle one-third of the 3/3-bar without clicking. He then clicks on the middle one-third of the 3/3-bar, erases three of the four marks Melissa made, and then uses PULL PARTS to pull out one part of the six parts he made.)
T: (Asks the children to write down on their paper all of the ones they could pull out.)
J: (Writes the sequence of fractions along the top of his paper without difficulty until the teacher interrupts.) 1/3, 1/6, 1/9, 1/12, 1/15, 1/18, 1/21, 1/24, 1/27, 1/30, 1/33, 1/36.
M: (Writes the sequence of fractions across her paper also without difficulty.) 1/3, 1/6, 1/9, 1/12, 1/15, 1/18, 1/21, 1/24, 1/27, 1/30, 1/33, 1/36, 1/39.
T: That's really neat, too. OK, you guys, let's do fourths! Can you do that for fourths?
J: (Nods. He then writes the sequence of fractions without difficulty until the teacher interrupts.) 1/4, 1/8, 1/12, 1/16, 1/20, 1/24, 1/28, 1/32, 1/36, 1/42, 1/48, 1/52. (Holds his paper up to the monitor.)
T: (Reads the first few fractions.) Fourths, eighths, twelfths, sixteenths, twentieths, twenty-fourths, ooooh! (To Melissa.) Hold yours up there.
M: 1/4, 1/8, 1/12, 1/16, 1/20, 1/24, 1/28, 1/32, 1/36, 1/42, 1/48, 1/52, 1/56.

The children could have also produced a sequence of fractions each *commensurate* with one-third or any other reasonable unit fraction had the teacher presented such a task. Because it was the goal of the teacher to investigate whether the children could construct a unit fraction adding scheme involving one-third and one-fourth, after Protocol XVI, however, rather than actually ask the children to produce sequences of commensurate fractions, he asked the children to partition a 3/3-stick so that they could pull out a one-fourth of the whole stick.[17]

Protocol XVII. Attempts to pull a 1/4-stick out from a 3/3-stick.
T: OK, Melissa. I want you to think about this too, Joe. I want you to use PARTS and make it so you can pull out a fourth of this 3/3-stick.
J: (After about 20 seconds.) I know how to do it.
T: OK, I want you to write on your paper what you are going to do. (Both children write on their paper what they plan to do. After the children are done, he asks Melissa to hold her plan up to the monitor.)
M: (Writes: "I need to pull out 1/3 and change it to 1/4.")
J: (Writes: "I will clear the marks and cut it into four parts and pull one out.")

---

[17] To find how much of a fractional whole 1/3 + 1/4 comprises involves finding a unit fraction for which both one-third and one-fourth are multiples. To produce this unit fraction, both one-third and one-fourth must be partitioned into a sufficient number of parts to produce such a unit fraction.

8 Equipartitioning Operations for Connected Numbers: Their Use and Interiorization

T: OK. Let's pretend, Joe, that you can't clear the marks. So, you have to revise your plan. (To Melissa.) Let's pretend that you can't pull a third out. But you can pull a fourth out. You can't pull a third out first. (Encourages the children to write their plans down rather than act using the TIMA: Sticks.) You have to pull one-fourth of the bar out. You have to partition it so you can pull one-fourth of the bar out.

J: (Writes, "I will erase the marks.")

T: (Laughingly reminds Joe that he cannot erase the marks.) I want you to use PARTS.

M: (Writes, "I will make 1/3 into a 1/4 without pulling anything out [she meant without pulling one-third out].")

T: OK, let's see if Melissa can use PARTS first.

M: (Activates PARTS, dials it to "4," and clicks on the leftmost part of the 3/3-stick.)

T: OK, now pull one-fourth out.

M: (Activates PULL PARTS and pulls the leftmost part out of the four parts that she made.)

J: Don't you have to make one-fourth of the whole stick?

M: (Covers the two remaining unmarked parts of the 3/3-stick with her hand.)

T: That is a fourth of what?

J: One-third!

T: That's a fourth of a third, isn't it? I want a fourth of the whole bar.

J: (Erases all of the marks on the stick, including the original two marks that marked the stick into a 3/3-stick.)

T: You've got to leave those third marks in there.

J: (Continues on in spite of the teacher's admonition. He erases all marks and uses MARKS, freehand, to subdivide the stick into four parts. This action circumvented the constraint that he was not to erase the hash marks he made using PARTS.)

T: Oh, I see what you are going to do!

J: (Tries to measure the last part he made, but there was not a unit stick in MEASURE. So, after he uses PULL PARTS to pull the last part out of the four-part stick he made freehand using MARKS, the teacher helps him copy a unit stick into MEASURE. Joe then measures the part he pulled out, and "11/39" appears in the number box.)

T: Is that a fourth?

J: (Hangs his head and laughs.)

T: Wow. Your plan didn't work (To Joe.) and your plan didn't work. (To Melissa.) Make a new plan. Go back and put in thirds. (Melissa wipes the stick clear of marks and uses PARTS to make another 3/3-stick.) Using PARTS, leaving the thirds there, I want you to use PARTS so that after you are done using PARTS, you can pull one-fourth of the bar out.

M&J: (Sit silently for approximately 60 seconds, so the teacher abandons the situation.)

Melissa did not enact her initial plan, "I need to pull out 1/3 and change it to 1/4" because the teacher introduced another constraint that she was not to pull out a 1/3-stick. In fact, the basic purpose of the teacher in asking the children to write their plans out was to eliminate certain actions which, if executed, might close out the possibility of the children engaging in those actions that would solve the problem. If the teacher had judged that Melissa could have indeed changed a 1/3-stick into a 1/4-stick after she had pulled out the 1/3-stick, there is no question that he would have encouraged her to do so. However, it was the judgment of the teacher that Melissa could not engage in such transformative actions because such actions would imply that Melissa had constructed fractions as rational numbers of arithmetic.[18] In fact, when Melissa executed her second plan, which was, "I will make 1/3 into a 1/4 without pulling anything out," she partitioned the leftmost part of the 3/3-bar into four parts and pulled out one part. This does indicate what she meant by making "1/3 into a 1/4" without pulling anything out. Presumably, had she pulled a 1/3-bar out from the 3/3-bar, she would have made it into a 1/4-bar in a similar way.

Joe eventually enacted his initial plan, which was, "I will clear the marks and cut it into four parts and pull one out." He planned to do this by erasing the marks on the 3/3-bar and using MARKS to mark the bar into four parts. This occurred even after the teacher attempted to induce the constraints of not erasing the marks and using the 3/3-bar in making a 1/4-bar. Rather than being arbitrary, the teacher permitted Joe to use MEASURE in verifying if he had indeed made a 1/4-bar using MARKS. Fortunately, the part Joe pulled out measured "11/39" and not "1/4" so that Joe realized that his way of proceeding did not work. That permitted the teacher to again restate the situation that they were to use PARTS to partition the 3/3-bar without erasing any marks. At this point in the protocol, both children sat silently and made no plans for how they could proceed.

Had the children been successful, we would infer that they had constructed their unit fraction composition scheme as a distributive scheme. That is, to find one-fourth of three-thirds, the children would find one-fourth of each third and unite these three parts together into three-twelfths. So, we close the case study by advancing the hypothesis that the construction of a unit fraction adding scheme entails constructing the unit composition scheme as a distributive scheme.

Discussion of the Case Study

Melissa made rapid progress from the 20th of October to the 1st of December in a way that is reminiscent of the progress that Patricia made during the first part of the teaching experiment when she worked with Joe in her fourth grade (cf. Chap. 7). During this time, however, Melissa seemed to experience internal constraints characteristic of children who have constructed only the partitive fraction scheme when engaging in attempts to solve the tasks of Protocols I and II. In her attempts to solve these tasks, she was dependent on Joe's independent solutions of the tasks in a way that was similar to how Laura was dependent on Jason's independent solutions of similar tasks. In essence, Melissa assimilated Joe's language and actions in Protocol I using her units-coordinating scheme in such a way that we characterized her use of

---

[18]Operating on a 1/3-bar to make a 1/4-bar would entail partitioning the 1/3-bar into four parts, pulling out one part and iterating it three times to produce a 3/12-bar or a 1/4-bar. This kind of operating implies that the child has abstracted fractions as an ensemble of operations of which the child is explicitly aware, which is what I mean by the rational numbers of arithmetic.

the scheme as generalizing assimilation. But she could not modify this scheme to remove the constraint that she experienced in the second continuation of Protocol I. In that protocol, the contrast between Joe partitioning each of four parts of a 1/4-stick into five parts so he could pull out one-twentieth of the 4/4-stick and Melissa partitioning a congruent 1/4-stick into thirty-two parts to pull out one-thirty-second of the stick was quite pronounced and served as contraindication that she could modify her units-coordinating scheme to construct recursive partitioning operations at the level of re-presentation at this time in the teaching experiment.

It soon became apparent that there was a distinction between Melissa and Laura in the fall of their fifth grade in that Melissa had constructed the splitting operation. In fact, in the teaching episode held on the 1st of December (cf. Protocol IV), Melissa knew that she would need to put one-hundred parts in a 5/5-bar as a consequence of breaking a 1/5-bar so that each little part would be one-one-hundredth of the 5/5-bar. To know that she needed to put one hundred parts in the 5/5-bar, we inferred that she conceived of the 5/5-bar as partitioned into one hundred parts and that the whole bar was one hundred times any one of its parts. That is, we inferred that she conceptually split the 5/5-bar into one hundred parts and established a unit of five units, each of which contained, mistakenly, ten parts. It was in this same teaching episode that Melissa posed the question, "How many eightieths can you get to fit into one-fifth?" after the teacher changed the task from one hundred parts to eighty. On the basis of her question, we inferred that she was aware of a unit containing the eighty parts that she could partition into five equal parts, each of which contained an unknown but equal number of the eighty parts. Although essential, this inference is not sufficient to infer equipartitioning operations. It must be also possible to infer that each of the five parts is an iterable composite unit. We were able to make this second inference on the basis that she actually tried to find the number of eightieths that would fit into one-fifth using her computational algorithm. So, we inferred that she had constructed equipartitioning operations for connected numbers in the context of operating.

In Joe's case, the second of the two inferences was based on his strategic reasoning to produce sixteen as how many of the eighty parts could be fit into one-fifth. The first of the two inferences hinged on the indicators that he had constructed a reversible unit fraction composition scheme in Protocol II. For this scheme to be reversible means that the results of the scheme are taken as input for further operating. In that these results constitutively involve a unit of units of units, reversibility of the unit fraction composition scheme implies equipartitioning operations for composite units.[19] Joe's equipartitioning operations were more or less permanently constructed and available for him as assimilating operations that he used to constitute and independently solve situations that involved direct or inverse reasoning. However, we soon became aware that had Melissa not been engaged in operating,

---

[19]In Protocol II, Joe knew that he had to partition each part of the 3/4-stick into two parts in order to pull a 1/8-stick out from each part, so before he acted he had already mentally partitioned each part of the 4/4-stick that contained the 3/4-stick into two parts.

she would not have so clearly posed the question, "How many eightieths can you get to fit into one-fifth?" to herself, nor would she have explicitly formed the goal implied by the question.

The Iterative Fraction Scheme

In retrospect, Melissa's construction of the iterative fraction scheme and a connected number sequence in the teaching episode held on the 12th of January (cf. Protocol VI) might be regarded as an anomaly because, at that point in the teaching experiment, she was yet to use operations that produce three levels of units for connected numbers as assimilating operations. Melissa did, however, mentally split a bar into eighty parts in the context of producing three levels of units in operating in Protocol V. So, in the production of relations among parts and wholes, as indicated by her reasoning that sixteen-elevenths was five-elevenths more than the whole bar in Protocol VI, she could mentally split the whole bar into eleven parts. Therefore, to explain Melissa's construction of the iterative fraction scheme, it was sufficient that she split the fractional whole into eleven parts where each part could be iterated eleven times to produce eleven-elevenths. She could then disembed one-eleventh from the eleven-elevenths and use it as if it were an iterable unit of one. In fact, when she integrated a 6/11-bar with another 6/11-bar, this produced a 12-part bar that she said was *one* more than the 11/11-bar. She then said it was one-eleventh more upon the teacher asking, "How much more than a bar would that be?" Melissa reinterpreted the 12-part bar as containing the 11/11-bar, so each part of the 12-part bar was one-eleventh. She was quite capable of producing three levels of units in operating using discrete quantity, so her interpretation of the iterable unit of one as one-elevenths evoked generalizing those operations in the context of fractional numbers. For example, she construed forty-four-elevenths as forty-four times one-eleventh just as forty-four was forty-four times one. She interpreted forty-four-elevenths as if it were forty-four discrete, rather than connected, segments.

That Melissa was not aware of the operations she used to produce three levels of units[20] when constructing the iterative fraction scheme was corroborated in Protocol IX where she chose to make a fraction that was not made up of eighteenths using a 1/18-bar. To make such a bar entails coordinating the commensurate fraction scheme and the iterative fraction scheme prior to activity, as indicated by the way in which Joe produced three-twenty-sevenths as a fraction that could be made from two-eighteenths. This coordination is based on equipartitioning operations because, when it was Joe's goal to produce a fraction that was not made up of eighteenths using the 1/18-bar, he based his solution on establishing a 9/9-bar in visualized imagination and partitioned each ninth into two or three parts, whichever suited his goal. The operations that produce three levels of units were assimilating operations for Joe's commensurate fraction scheme and his iterative fraction scheme as well as

---

[20]This is another way saying that she did not use the operations as assimilating operations.

his unit fraction composition scheme. Melissa, on the other hand, produced the operations whose results are a unit of units of units in the context of actually operating. But she was yet to interiorize these operations in such a way that she could produce their results mentally without actually engaging in the operations.

Melissa's Interiorization of Operations that Produce Three Levels of Units

As analysts, we were constrained to the affordances of the teaching episodes when exploring Melissa's interiorization of operations that produce three levels of units and focused on her interiorization of recursive partitioning operations instead. This was justified because the latter operations are constitutively involved in the former. Focusing on recursive partitioning operations led in turn to investigating Melissa's use of her units-coordinating scheme in the context of three levels of units because units-coordinating is the mathematical activity in recursively partitioning a partitioned continuous unit.

Making drawings to complete prior partial partitionings was the key element in Melissa's interiorization of recursive partitioning. Melissa independently contributed her drawings and they were a surprise to the teacher. In explaining the emergence of Melissa's drawings, a reconsideration of how she used her units-coordinating scheme in completing her prior partitions is essential. In Protocol XIV, to explain how she arrived at one-twelfth, Melissa said, "Because if you put all in those squares, you would get twelve pieces. You would have four sets of three." We consider her saying, "because if you put in all those squares," as indicating that she mentally partitioned each one of the four parts into three parts, and her saying, "you would have four sets of three," as indicating that she structured the result into a composite unit containing four units, each of which contained three units. What is left implicit in Melissa's comments is her use of three as a partitioning template that she used to partition each of the four parts into three parts each. That is, the unit of three served as a template to partition the four units of the connected number four when it was her goal to find how much 1/3 of 1/4 of the whole bar was of the whole bar.

So, the change in Melissa's units-coordinating scheme consisted of extending the situation of the scheme from numbers containing discrete, separated units to include numbers containing continuous, connected units. The operations of units-coordination still consisted of two programs of operations, but with an alteration. When activated, the first program of operations produced, say, a connected number, four, and a partitioning unit, three, that could be used to partition each unit of the connected number, four. The second program of operations consisted of partitioning each of the four units into three units, uniting each trio of units together into a composite unit, three, and uniting these units of three into a composite unit containing four units of three, which is an accommodation in the operations of units-coordinating. She still inserted a unit of three into each unit of four, but the meaning of insertion changed from filling each unit of four with a unit of three to partitioning each unit of four into three equal parts.

The activity of the scheme was to progressively integrate the units of three with those preceding and increment the numerosity of the preceding units by three in order to find the numerosity of the parts produced by the partitioning. From an observer's perspective, Melissa's implementation of the activity of the scheme consisted of her counting how many parts would have been made had the partitioning activity been completed at each step (cf. the first continuation of Protocol XIII).

The making of a drawing in Protocol XII introduced a modification in the second program of operations in that, rather than using the stick in her visual field to carry out the units-coordination, she re-enacted units-coordinating activity by making a drawing. Such re-enactment involved sufficiently re-presenting the prior partitioning activity to enable her to make the drawing. However, what she was aware of was the drawing that she was making rather than an image of a stick on which she was operating in re-presentation. One might say that she was *in* the re-presentation and that the re-presentation was found in the activity of making the drawing. There was definitely visualization involved, kinesthetic as well as visual, but the visualization was an activity that was implemented as a drawing.

Had Melissa performed the visualizing activity mentally without actually making a drawing, this would involve monitoring the activity while it was being carried out in a recursive way using her concepts of four and of three. It is this monitoring of the activity which interiorizes the activity. Once interiorized, the child can execute the activity willfully and it can be said to be available to the child without the child actually engaging in the visualizing activity. The visualizing activity is produced by means of operating, so what the child has available is a program of interiorized operations that produce the visualizing activity.

For Melissa, the interiorization of the visualizing activity did not occur in one fell swoop. The interiorization process was also observed in the first continuation of Protocol XIII when she looked downward and away from the computer screen to count all of the parts that would have been produced had partitioning the whole bar been completed after partitioning the whole bar into three parts, then one of these parts into two parts, and then one of these parts into three parts, and then one of these parts into two parts.

After completing this partitioning activity, she partitioned the visible 1/18-bar into two parts and then continued on *in visualized imagination* (cf. Fig. 8.7). So, the material on which she operated was figurative. Metaphorically, she partitioned each visualized unit of her concept of three into two parts, and in this process, recorded this figurative material in operations of partitioning into two parts.[21] If the figurative material is "dropped out," leaving only its records, the figurative material becomes interiorized. So, what became recorded in the units of her concept of three were not

---

[21] The material on which operations operate become recorded or registered in the operations. The records are interiorized records to the extent that the results of operating – a partitioned bar – can be produced without actually executing the operations. All that is necessary is that the operations be evoked rather than implemented.

simply two unit items. Rather, the operations of partitioning into two parts were recorded into the units of her concept of three by means of the interiorized figurative material.[22]

On the Possible Construction of a Scheme of Recursive Partitioning Operations[23]

Given that it was Joe who began the teaching experiment in fifth grade having already constructed a reversible fraction composition scheme, it was indeed surprising that Melissa seemed to be the stronger of the two students in symbolizing a sequence of recursive partitioning operations using drawings and notation. Joe's abstractive power is well illustrated in the second continuation of Protocol XV when he partitioned a 1/36-bar into three parts and said that one of the three parts was one one-hundred-eighth of the whole bar after sitting silently in deep concentration. Melissa, on the other hand, resorted to doubling 36 and said that the part was 1/72 of the whole bar. In this case, Joe used the results of his prior recursive partitioning operations in his current recursive partitioning operations. So, this opens a possibility that Joe had constructed a scheme of recursive partitioning operations.

There is no doubt that Joe operated in a way that opens the possibility that he had indeed constructed such a scheme. However, it must be remembered that Joe was not asked in Protocol XV, or in either of its continuations, to explain how he could operate. For example, he was not asked how he would find how much the part would be if he and Melissa took, say, two more turns apiece. Had he been able to make such an explanation, this would be a solid indication that he was becoming aware of how he could operate on any but no particular turn,[24] which is essential in inferring a scheme of recursive partitioning operations.

The notational system that Melissa generated in Protocol XVI and its continuation is more of an indication of a scheme of recursive partitioning operations than was Joe's ability to produce the next fraction in the sequence of fractions being produced by the children's partitioning actions by multiplying the number of parts in the current partition and the number of parts produced by the preceding partitions. However, Melissa's making of her notational system is still not sufficient to infer a scheme of recursive partitioning operations even though we did infer that Melissa had constructed the operation of recursive partitioning based in part on her notational system. She definitely became aware of how she was operating because making such a notational system is based on an awareness of taking the results of

---

[22] A unit has two meanings. First, it is a unitizing operation, and, second, it is a proverbial slot in which records of experience are recorded.

[23] A scheme of recursive partitioning operations is different from a recursive partitioning scheme.

[24] This is akin to operating with a variable.

current operation as input for the next operation. This is the basis for my inference that she had finally constructed recursive partitioning.

To infer a scheme of recursive partitioning operations, it would be sufficient to be able to infer that Melissa was able to use her notational system as input for further operating without resorting to making drawings. However, in Protocol XVII, after both children were asked to find how many 1/144-pieces would fit into a 1/44-piece, Melissa resorted to using her drawing instead of her notational system. In that same protocol, she also made a drawing to find how many 1/144-pieces would fit into a 1/4-piece instead of simply resorting to her notational system that she had made and which we repeat below.

$$\frac{1\ 2}{4} \qquad \frac{1\ 2\ \frac{3}{4}}{48} \qquad \frac{^2 4\ \frac{3}{8}}{144}$$

$$\frac{^4 1}{4)} \qquad \frac{^3 1}{12)} \qquad \frac{^4 1}{48)} \qquad \frac{3}{144}$$

Melissa never realized that all she had to do was to "read" the numeral "3" from her notational system to find how many 1/144-pieces would fit into a 1/48-piece, etc., and to find the product of three, four, and three to find how many 1/144-pieces would fit into a 1/4-piece. Of course both "readings" of her notational system would entail reasoning reciprocally. In fact, an awareness of how she operated on any but no particular turn would be based on reciprocal reasoning – understanding that if a bar is partitioned into three parts, then each part is one-third of the original bar, and the original bar is three times any one of its three parts.[25] But, it involves more. She would also need to reason that if a 1/48-bar is partitioned into three equal parts, then each of the forty-eight 1/48-bars would be partitioned likewise, and so that would produce three times forty-eight, or one-hundred forty-four parts, and thus each one of the three equal parts would be a 1/144-bar. This reasoning is indicated by her notational system.

What her notational system does not indicate is an awareness that if she started with a 1/144-bar, to find how many of these bars are in a 1/48-bar, all she needed to do was reverse the steps she took to find how much of the whole bar each one of the three equal parts of the 1/48-bar was (creating a unit of units). This is the specific context of the reciprocal reasoning in which she needed to engage. To find how many 1/144-bars in a 1/12-bar, she needed to use the knowledge that there were three 1/144-bars in a 1/48-bar when finding how many 1/48-bars was in a 1/12-bar. This simply entailed finding the product of three and four had she correlated using her notational system with her reasoning, creating a unit of units of units. To finish the problem, she would need to reason reciprocally and use the numerosity of this unit structure, twelve, to find how many 1/144-bars were

---

[25] Such reciprocal reasoning is based on the operation of splitting.

in a 1/4-bar by finding the product of three and twelve. So, developing a scheme of recursive partitioning operations not only involves making units within units within units, which Melissa could do, and symbolizing these operations, but also involves using the most elemental unit produced through partitioning in uniting to make a unit of units and then a unit of unit of units. That is, partitioning and uniting must be constructed as reciprocal operations at three unit levels.

As necessary as it seems for partitioning and uniting to be constructed as reciprocal operations in the construction of a scheme of recursive partitioning operations, it seems necessary for there to be a notational system produced that symbolizes these operations and their properties. For a notational system to be a symbol system, the symbol system must stand in for the operations actually carried out to produce the notational system and for the child to be able to reason using the notational system without actually carrying out the symbolized operations. This frees the child from the need to actually operate for there to be a result, because the result is symbolized. A symbol system is especially crucial when it become necessary to produce more than three levels of units, because the child can symbolize the numerosity of the third level of units in a unit of units of units and use this symbolized numerosity as if it referred to a unit of units that can be operated on further. My assumption that human beings can learn to operate in such a way that they can produce three levels of units and then use those three levels of units in producing three more levels of units, and etc., is justified by the Hindu-Arabic numeration system. A scheme of recursive partitioning operations is definitely involved in the production of that system especially in that case where it is extended to include decimals and their symbolization. So, if Melissa's notational scheme was in fact a symbol system, then she should have at least given some indication of wanting to use her notational system in her reasoning especially after she just used her diagram to find how many 1/144-pieces would fit into a 1/44-piece when finding how many 1/144-pieces would fit into a 1/4-piece. Instead, she made a whole new drawing and seemed to be in the process of constructing a symbol system rather than having completed the process.

The Children's Meaning of Fraction Multiplication

Both children had constructed a unit fraction composition scheme[26] and learned to use this scheme in the embedded recursive partitioning tasks. In the "recursive partitioning" activities from Protocol X forward, we did not emphasize multiplica-

---

[26] In the latter part of the teaching experiment, there was contraindication that both children could find, say, how much one-third of four-fifths of a unit bar is of the unit bar. This is a nontrivial generalization of finding how much one-third of one-fifth of a unit bar is of the unit bar because it involves distributive reasoning which the children were yet to construct.

tive language nor did we emphasize written notation. Rather, it was our goal to bring forth Melissa's recursive partitioning operations and, hence, her fraction composition scheme and to provide Joe with situations in which he could use his fraction composition scheme and modify it in the possible construction of a scheme of recursive partitioning operations. In retrospect, it would have been very easy for us to encourage the children to develop multiplicative fraction language in both spoken and written contexts as they worked in the embedded recursive partitioning tasks. For example, in the continuation of Protocol XVI, it would have been appropriate to suggest to her that she was in fact finding the part of the unit bar indicated by "1/3 × 1/48" to aid her in interpreting what she was doing using fraction language after Melissa made her sequence of written fractions:

$$\frac{^4 1}{4}, \frac{^3 1}{12}, \frac{^4 1}{48}, \frac{3}{144}$$

It would have been possible then to ask her to write "one-third of one-forty-eighth" and help her in formulating the notation, "1/3 × 1/48." In finding how much this piece was of the whole bar, it would have been possible to take advantage of her product, $\frac{^{48} 3}{144}$, and to ask her what this product meant, how many pieces were there of size 1/3 × 1/48, and how she used it to produce 1/144, using standard numeric notation. Of course, the product of forty-eight and three was the result of recursive partitioning operations, and this product, along with her use of its result to find the fractional part of the whole bar that was constituted by the bar of size 1/3 × 1/48, would constitute a child-generated algorithm for finding the product, 1/3 × 1/48. This child-generated algorithm could have been brought forth in Melissa as follows. To find how much the piece of size 1/3 × 1/48 is of the whole bar, find the product of three and forty-eight to find how many such pieces are in the whole bar. Since 3 times 48 is 144, 1/144 is how much the piece of size 1/3 × 1/48 is of the whole bar. So, 1/3 × 1/48 = 1/144.

It also would have been possible to induce a child-generated computational algorithm in Joe. The key in his operating is that he found the product of three and forty-eight just as did Melissa. Encouraging Joe to explain why he found this product would have led to a rational explanation on his part because he was aware of why he operated as he did, which is essential in a child constructing a child-generated algorithm for finding the product of two unit fractions. Further, Joe could have explained that the yellow piece was one-third of one-forty-eighth, so he could have been encouraged to write "1/3 × 1/48" as a *record of his operating*. From this point on, his operating was quite similar to the operations in which Melissa engaged, so making records of them should have followed approximately the same path as that followed by Melissa. Child-generated algorithms as they are manifest in notation are nothing but records of operating, and these records serve the function of constructive generalization.

A Child-Generated vs. a Procedural Scheme for Adding Fractions

In the first continuation of Protocol XVII, Melissa's writing of $|, \frac{1}{j}, \frac{3}{4}, \frac{7}{8}$ is an example of a child-generated scheme for adding fractions. The sequence of numerals was a record of her production of combining a 1/2-bar and a 1/4-bar to produce a 3/4-bar, and then combining this configuration with a 1/8-bar to produce a 7/8-bar. Recursive partitioning was a key operation in her production of this sequence of numerals, and the sequence constituted a record of her operating. Joe also used recursive partitioning to produce the fractional part of the whole bar produced at each step of the partitioning. However, his paper and pencil algorithm for adding fractions, which was a procedural scheme, essentially excluded his conceptual solution of the task as well as his production of a child-generated scheme for finding how much of the whole bar was constituted by the 1/2-bar, the 1/4-bar, and the 1/8-bar combined when he was entirely capable of doing so. The investigation of a more general child-generated scheme for adding fractions than Melissa constructed was not possible because of the lack of distributive reasoning in the two children.

Chapter 9
The Construction of Fraction Schemes Using the Generalized Number Sequence

John Olive and Leslie P. Steffe

In this chapter, we trace the construction of the fraction schemes of two of the children in our teaching experiment, Nathan and Arthur, who apparently had already constructed a generalized number sequence before we began working with them. We interacted with these two children as we did with the other children in the sense that our history of the children along with their current mathematical activity served in creating possibilities and hypotheses that we continually explored in teaching episodes. In our interactions, we found that their construction of the operations that produce the generalized number sequence opened possibilities for their constructive activity that we did not experience with the other children. We did not decide a priori to use higher-order tasks in our interactions with these two children than we used with the other children. Rather, their ways of operating served as the basis for our constructions of tasks that we used. Nathan participated in the teaching experiment during his third, fourth, and fifth grades,[1] whereas Arthur participated only during his fourth and fifth grades.

The Case of Nathan During His Third Grade

In the first part of this chapter, Nathan's strategies for finding commensurate fractions, adding fractions with unlike denominators, and simplifying fractions that emerged during the first year of the experiment are discussed. Nathan worked with Drew for the first 12 of the 21 first-year teaching episodes using TIMA: Bars. It became obvious that although Drew had constructed the explicitly nested number seqeunce, he did not construct fraction schemes that were on a par with Nathan. For that reason, the decision was made in mid February to find a more compatible partner for Drew, so Nathan worked alone for the remainder of the first year and Arthur joined him at the outset of their fourth grade. We shall make some references to Drew's activities to illuminate the differences in children who have constructed the operations that

---

[1] Excepting Arthur, we worked with the other children on their multiplying schemes during their third grade.

produce the generalized number sequence and those who constructed the explicitly nested number seqeunce.

Nathan's Generalized Number Sequence

Nathan's strategic multiplicative reasoning in the teaching episode held on 15 November revealed his ability to coordinate two different iterable composite units. His coordination involved keeping track of the iterations of the two composite parts that he had united into a new composite whole. Using the computer tool TIMA Toys, Nathan and his partner Drew had created several strings of toys (linked toys) with from one to six toys in each string. The task was to use copies of the 3-string and the 4-string to make 24 toys. Nathan reasoned out loud as follows:

> Three and four is seven; three sevens is twenty-one, so three more to make twenty-four. That's four threes and three fours!

In solving the task, Nathan integrated a unit of three and a unit of four into a unit of seven, iterated the unit of seven three times to produce 21, increased 21 by three to produce 24, disunited 21 into three threes and three fours, integrated the additional three with the three threes, and produced four threes and three fours.

Nathan's strategic reasoning was multiplicative because not only were his composite units of three, four, and seven iterable units, but also the units served as material for the operations he carried out – integrating and disuniting. He definitely had constructed his units-coordinating scheme as an iterable scheme, and his strategic reasoning further indicates that he had constructed the operations that produce the generalized number sequence. Prior to operating strategically, it would be necessary for Nathan to posit a composite unit of units of three of unknown numerosity, and a composite unit of units of four also of unknown numerosity, whose units of one when integrated together would, if counted, equal twenty-four.[2] But rather than attempt to operate sequentially, counting by four and then by three in a test to find how many threes and how many fours would be needed, Nathan integrated these two units into a unit of seven to establish a unit of units of seven, and iterated the unit of seven until reaching a number just less than 24. These operations are operations of a GNS because in a GNS, any composite unit can be taken as the basic unit of the sequence and the composite unit implies the sequence just as the composite units of three, four, and seven implied units of composite units. Critically, Nathan established two number sequences, a sequence of units of three and a sequence of units of four "side-by-side," as it were, and combined the basic units of each

---

[2] This is more than sufficient indication that Nathan had constructed a splitting scheme for composite units. This scheme permits a child to mentally mark off one composite part of a composite unit as a fair share (e.g., to mark off five elements out of, say, 30 elements as one of six equal shares) and iterate that share six times in a test to find if it is a fair share.

sequence together to produce another sequence. Nathan used the two basic units, three and four, to establish another basic unit of seven, and used the results of operating with seven to establish results in the two original sequences. This way of operating indicates a GNS.

Developing a Language of Fractions

It was in the microworlds Nathan created using the operations of the TIMA: Bars software that he generated fraction schemes during the first year of the teaching experiment that excelled the fraction schemes the ENS children constructed throughout the 3 years that we worked with them. He developed schemes of operations with fractions that enabled him to add fractions with unlike denominators, find a fraction of a fraction, rename fractions, and simplify fractions to lowest terms. These operations with fractions were all constructed by Nathan as he operated in TIMA: Bars using natural language. No written fraction notation was used. When we first started working with Nathan and his partner, Drew, on simple sharing tasks using the TIMA: Bars software (November of their third grade), both children demonstrated a naïve use of fraction language to refer to the shares they created.

Partitioning and sharing situations. The first four teaching episodes using TIMA: Bars (the 22nd of November to the 12th of December) were situated in the context of sharing a given number of equal-sized bars among a given number of mats, the number of bars being less than the number of mats. The first session (the 22nd of November) illustrated the lack of a meaningful language for describing a fractional part of a bar for both Nathan and Drew. The results of operating on the tasks did, however, cause some cognitive dissonance for Nathan that forced him to reconsider his naïve use of fraction words. Drew, however, experienced no cognitive dissonance and did not question the inconsistencies in his use of fraction words.

In one of the tasks, the children were presented with three identical bars each of which was partitioned into 24 equal parts (each bar was partitioned into four rows of six). The children were to share the bars equally among four mats and find out how many small parts each mat would receive.[3] Nathan began by using an experimental strategy to find the number of parts per mat, starting with three, then four, and five. Drew watched and then said, "I know! It's eighteen small pieces." He took one row of six from each bar and assigned these three rows to one mat. He explained that he could do this three more times for the remaining three mats (four rows in each bar, four mats). Drew had organized the three bars into four rows of eighteen. In doing so, it would be necessary for him to produce four units of units of units – four units each consisting of a unit containing three rows where each row contained six parts. So, Drew definitely had constructed the operations that produce a unit of units of units. In partitioning, he used this insight to allot one row of

---

[3] The task was structured to find if the children would simply take one row of six from each bar for each mat. The number of rows and the number of mats were equal.

18 parts to each mat. In doing so, he decomposed each row of 18 parts into its constituent rows of six parts (three rows with six parts per row) and distributed these three rows to one of the four mats.

Nathan had a hard time following Drew's reasoning but eventually agreed that it was a correct solution. So the teacher/research asked the question: "How much of one candy bar does each mat get?" Both children had trouble understanding the question. The following interchange illustrates the cognitive dissonance experienced by Nathan (N stands for Nathan, D for Drew, and T for Teacher):

Protocol 1: Sharing three partitioned bars among 4 mats.

N: A fourth! (Possibly indicating one row from each bar.) You mean if you put the three green rows together in one bar?
T: Yes!
N: One-third.
T: Why?
N: Because there are three rows. (Drew agrees. Nathan looked puzzled at this point. He was obviously feeling some doubt about his answer.) But we have four rows in a bar.
D: It's a third!
T: If each person gets a third of a bar, how many people can share the bar?
N: Three.
T: (Pointing to the three green rows of six.) Could three people each get this much of one candy bar?
N: No! This can't be a third – there are four pieces (In the bar.)!

The fraction words for both children were associated with the number of visible parts in a share, not to a part-whole relation ("one-third" referred to three parts). They also appeared to only have words for unit fractions (a fourth and a third). Nathan eventually related the unit fraction number word to the number of parts in the whole rather than the number of parts in one share.

During the next two teaching episodes (the 5th and 6th of December), Nathan and Drew began to use a language of parts that made sense to them. In the 6th of December episode, the teacher decided to explore whether the children had constructed distributive partitioning operations. So, he asked them to share two bars among three mats. The children partitioned each bar into three parts vertically and two parts horizontally (3×2). They then shaded four of the six parts in one bar to indicate the share that one mat would receive (cf. Fig. 9.1).

When asked why that would be the share for each mat, Drew's response was, "Take four away from six; that leaves two" – an additive view of the situation. Nathan's response was, "It's four pieces out of six pieces" – a result of part-whole operations. He then added: "I guess you could call that "four sixes" or something like that." The teacher then provided the conventional language for the fraction: "four-sixths." Nathan connected this conventional language for proper fractions with the result of using his part-whole operations. However, the teacher's test of

9 The Construction of Fraction Schemes Using the Generalized Number Sequence

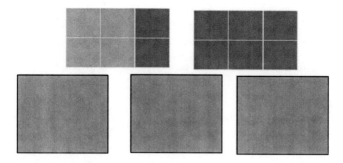

Fig. 9.1. Sharing two bars among three mats.

the presence of distributive partitioning operations was inconclusive because the children, by means of solving a sequence of tasks like that of Fig. 9.1 had learned to find the product of the number of bars and the number of mats as the number they were to use to partition each bar. Had the children shared the two bars equally among three mats by partitioning each bar into three parts and then distributing one part from each bar to each mat, then that activity would indicate distributive paritioning operations.

Nathan's generalized fraction language. In the first teaching episode following the winter school holidays (the 16th of January), Nathan was able to make sense of a reversible fraction task. Given an unmarked bar, he was told that this was two-sevenths of a whole candy bar and he was to make the whole candy bar. He immediately partitioned the given bar into two equal parts and copied one of these parts five times to give him a total of seven parts, which he then joined together to form the whole candy bar. We varied the task by presenting a partitioned part of a bar as the given quantity (e.g., a 6-part bar as three-eighths of a whole bar). Nathan had no problem with taking two of the parts as one-eighth and copying this 2-part piece five times to complete the whole bar (cf. Fig. 9.2).

We imputed a reversible partitive fraction scheme for connected numbers to Nathan. Evidently, the single encounter with four-sixths in the previous teaching episode on the 12th of December was sufficient for Nathan to establish meaning for at least proper fraction number words. It is remarkable that his fraction number words referred to the results of a reversible partitive fraction scheme without his establishing such a scheme in any encounter in the teaching experiment. It is even more remarkable that he partitioned the six parts into three units with two parts in each unit, took one of the three parts as one-eighth of a hypothetical fractional whole, and iterated one-eighth five more times to produce eight-eighths (cf. Fig. 9.2). That is, "three-eighths" implied a unit "eight-eighths" of which it was a constitutive part as well as a unit of units in its own right. If six parts was three-eighths, then two of the parts referred to a unit that was one of eight parts, a unit that he could produce by iterating the unit consisting of the two parts eight times. This way of operating was made possible by Nathan embedding his splitting

Fig. 9.2. Nathan makes five copies of two parts of a 6-part bar to complete eight-eighths from three-eighths.

scheme for composite units within his partitive fraction scheme, so that the elemental units were unit fractions rather than units of one.[4] It was an immediate embedding without observable antecedents, and it is a part of what made his use of fraction language so generative.

Improper fraction language: A temporary constraint. During the next task in the teaching episode, it became clear that Nathan's reversible fraction scheme was limited to starting with proper fractions of a bar. His partner, Drew, had created a bar with 25 parts and said that his bar was "twenty-four sixths" of a whole bar. Nathan erased the 25 parts and made 24 parts in the bar. He then cut off six parts. At this point it would seem that his reversible fraction scheme did extend to improper fractions if the 6-part bar was considered as the unit bar. However, when the teacher asked him what the 6-part bar was Nathan was unsure. He rejoined the 6-part piece with the other pieces to reconstitute the 24-part bar. He then copied this bar five times to make six of them and joined these six bars together to make what he claimed was the "whole bar."

The teacher posed a new problem: he drew a bar and called it "seven-fifths" of a whole bar and asked the two children to create the whole bar. Drew partitioned the unmarked bar into seven parts and then made five copies of the partitioned bar and joined them together. Nathan agreed with Drew's solution, but when the teacher asked if "seven-fifths" was more or less than a whole bar, Nathan responded that it was more but then added: "That can't be right." His initial response indicates an interpretation of "seven-fifths" as "seven of one fifth of a bar" but this was in conflict with what was visible on the screen: a bar that was five times the "seven-fifths" he started with.

Nathan's initial action to remove six of the 24 parts was an attempt to use his reversible fraction scheme for proper fractions in improper fraction situations. The result, however, was in conflict with his notion that the "whole" must be bigger than the given fractional part, or with his "inclusive" part-whole relation – he could not conceive of the 24 parts in the six-part whole, so he reversed the relation and interpreted the situation as representing "one-sixth partitioned into twenty-four parts" (a 24-part sixth rather than twenty-four sixths), which constitutes an assimilation of the situation using his reversible partitive fraction scheme. We never observed any of the other children use their reversible partitive fraction scheme to establish the fractional whole given that a 6-bar was three-eighths. In fact, such a task is not

---

[4] "One-eighth" could be a unit containing two parts out of a unit containing six parts that could be iterated eight times to produce eight-eighths, whereas "one" could refer to one unit of two parts out of six parts that could be iterated eight times to produce 16 parts. Both of these meanings were present in Nathan's activity.

within the scope of a partitive fraction scheme because such a scheme is limited to operating with two levels of units.

Construction of meaning for improper fraction language. We do not attempt to specify the operations of Nathan's "reversible partitive fraction scheme" that does not include the operations that produce improper fractions because Nathan did remove the constraint of having more parts in a bar than constitute the whole bar in the next teaching episode (the 17th of January). We decided to alter the task slightly and asked the children to make a bar that would be the same size as (or as much as) say, two and a half bars. To solve this problem, Nathan made three copies of the original bar and broke the third copy into two equal parts. He then joined two whole bars and one half bar together to form a bar (cf. Fig. 9.3).

Nathan was also successful with making a bar two and one-fifth as much as his bar. We then asked him to make a bar that was "as much as five-fifths" of his bar. He immediately said "That would be a whole bar!" We then posed the problem (in the restructured form) from the previous day: "Make a bar that is as much as seven-fifths of this bar." Nathan initially said that you could not have seven-fifths in a bar (This response is similar to Joe's initial response to making six-fifths of a bar in Chap. 7). Once Nathan realized that he was making a bar that was as big as seven-fifths of the original bar, he had no problem and explained that "it was two more fifths added to the whole bar!" Drew was then asked to make a bar that was as much as "one and two fifths" of his bar. Nathan immediately interrupted and said, "That's the same as mine!" and explained to Drew that "one and two-fifths" was the same as "seven-fifths" because "seven-fifths was two more fifths added on to a whole bar." Nathan was successful with all of the tasks that followed in this session, making a bar that was as much as thirteen-ninths of a bar, and calculating that there would be thirteen-sixths in two and one-sixth of a bar.

Nathan's comment that "you can't have seven-fifths in a bar" indicated the problem he had in the previous day's task in which a bar was presented as being "twenty-four sixths" of a bar. The change in the structure and language of the problem situation enabled Nathan to step outside his inclusive part-whole relation and to make whole-part comparisons. Drew had not yet constructed inclusive part-whole relations in fraction contexts, so he had no hope of making such comparisons. It should also be

Fig. 9.3. Nathan makes a bar that is two and a half times as much as a unit bar.

noted that, unlike Joe the following year, Nathan's concept of a fraction greater than the whole was quite stable following this one episode. In the course of two months and seven teaching episodes in his third grade, Nathan was in the process of constructing at least a reversible iterative fraction scheme. This scheme, however, when more or less completed, may contain operations that were not available to the other children in the teaching experiment that had constructed a similar scheme.

We reasoned that Nathan had at this point in time constructed a meaningful language of fractions that included both mixed numbers and improper fractions. Nathan used his operations to operate on subdivided or subdivisible regions in his visual field or images of these regions. It seemed necessary for him to actually operate on such figurative material to engage in fractional operating. By means of actually carrying out the operations, we hypothesize that aspects of the figurative material would become recorded in the operations so that, if the operations were evoked, figurative material would be generated that symbolized the operations.[5] If this occurred, it would be possible to speak of abstracted quantities and operative figurative material (cf. Chap. 8, *Melissa's Use of the Operations that Produce Three Levels of Units in Re-presentation*). That Nathan was still in the process of interiorizing[6] the figurative material he operated on is exemplified in the teaching episode on the 31st of January. He was asked to find out how much of a bar he would have if he joined a half of one bar with a third of another, congruent bar. His first response was "a whole bar!" He saw that he was wrong after carrying out the actions with TIMA: Bars and eventually reasoned that a whole bar would be a half plus a half. His strategy for finding a fraction to describe the half plus a third was, again, based on actually operating on figurative material. He tried different partitions and made visual comparisons. Three-fourths was not quite right so he then tried sixths. Four-sixths was not big enough, so he tried five-sixths and found that it was the same size as the half plus a third. He described the amount, however, as "one-sixth less than a whole bar!" The next series of episodes were designed to help Nathan construct a scheme for making commensurate fractions (Olive 2002).

Reasoning Numerically to Name Commensurate Fractions

In the 7th of February teaching episode, Nathan was set the task of making as many different fractions as possible from a 12-part bar. He was asked to copy one of the 12 parts and asked what fraction this was; he responded with one twelfth.

---

[5] When a bar is partitioned mentally into four equal parts, for example, the partitioning template consists of the operations that produce a unit containing four unit items. The operations are unitizing operations used in the service of making parts. Because of the nature of a unitizing operation, moments of focused attention register the sensory material on which they are focused (cf. Chap. 3).

[6] Recording figurative material in the operations that operate on the material constitutes part of the process of interiorization.

He was then asked to make a half and a third with this one-twelfth-part. He made sixth-twelfths and four-twelfths, respectively, by iterating the one-twelfth-part. He also named them in terms of the twelfths. He was next asked how many twelfths were in two-thirds. This question confused him initially as he had already said that four-twelfths was one-third. Eventually he said: "Oh! Now I've got it – one-third is four-twelfths so two-thirds would be eight-twelfths!" He went on after this to work with fourths: "one-fourth is three-twelfths, nine-twelfths is three-fourths, six-twelfths is a half which is two-fourths." The teacher asked Nathan if he could do the same thing for fifths (make one-fifth using the one-twelfth part). Nathan responded: "That would be five-twelfths, ten-twelfths." The teacher did not respond immediately and Nathan eventually asked: "Is it possible with fifths?" The teacher asked Nathan what he had done. Nathan indicated the five-twelfths piece that he had created and said "five-twelfths in one-fifth." The teacher suggested he check that. Nathan copied the original bar, wiped it clean, and partitioned it into five parts. He made visual comparisons between the twelfths and the fifths and eventually said: "You can't make one-fifth out of one-twelfth because it's not even."

Nathan continued with his task, making all the possible sixths and skipped sevenths "because it's not even!" He was not sure about eighths. The teacher asked him: "What would eight pieces be?" Nathan replied "eight-twelfths." When asked to think of another fraction word for eight-twelfths Nathan reasoned that: "four pieces is one-third, two-thirds is eight pieces, so two-thirds is eight-twelfths." He then reasoned that nine-twelfths would be three-fourths because "three pieces is one-fourth and three plus three plus three is nine-twelfths." When asked what ten-twelfths would be, he responded that "two pieces would be one-sixth so would it be five-sixths?" He was later asked what three-sixths plus one-third would be. He reasoned as follows: "three-sixths would be six and one-third would be four, so six and four is ten – ten-twelfths." The teacher asked what it would be in sixths? Nathan replied: "Oh! two, four, six, eight, ten – so it's five-sixths!" Nathan had used the equivalence of two-twelfths in one-sixth to count by two's up to ten (a unit-segmenting activity) and produced the commensurate fraction of five-sixths (a units-coordinating activity between units of one-twelfth and one-sixth). The teacher had helped Nathan make the renaming of fractions explicit through this activity.

This episode serves as corroboration of our hypothesis that Nathan was in the process of *interiorizing* the figurative material on which he operated when producing fractions commensurate to composite unit fractions,[7] because to operate as he did require that he monitor operating as it occurred. For example, when he was asked to think of another fraction word for eight-twelfths, Nathan would need to establish the complement of eight pieces in 12 pieces and use that unit of four pieces in splitting the 12 pieces into three parts. The relations among four, eight, and 12 (eight is four two times and 12 is four three times) were apparently available to Nathan prior to engaging in the problem, which means that he had already constructed the splitting

---

[7]When one-third, for example, refers to one unit out of a unit containing three units each of which contains four units, it is called a composite unit fraction.

operation for composite units. But because of the deliberate way in which he established that "two-thirds is eight-twelfths," we infer that not only was he aware of the intermediate steps, but also he was aware of how he composed them. For example, when he said that "four pieces is one-third," we infer that he was aware of a unit containing four pieces that he had disembedded from a 12-bar split into three parts, a one out of three relation between this unit and the split 12-bar, and the potential results of iterating the unit containing four pieces two times (two-thirds or eight-twelfths) and three times (three-thirds or twelve-twelfths). What served as a sentinel in the monitoring activity was his goal to make another fraction for eight-twelfths.

Corroboration of the Splitting Operation for Connected Numbers

The next teaching episode with Nathan took place after a 4-week gap.[8] This teaching episode, on the 6th of March, suggested that Nathan's fraction language did not always refer to figurative material that was produced by his interiorized operations in that he sometimes made associations between a fraction number word and the number of parts rather than the part-to-whole relation (as he did with the five-twelfths in one-fifth situation above). This episode, however, is also considered important for two other reasons: Nathan made further clarifications in his fraction language and he was introduced to COPYPARTS in TIMA: Bars. This action provides the user with a means of pulling part of a whole bar out of the bar to create a new bar while leaving the original bar intact. It provides users with an action to implement their disembedding operation. The action proved to be very powerful for Nathan and, we believe, helped him to construct an algorithm for simplifying fractions.

Nathan was asked to give two-thirds of a 15-part bar to a big mat and one-fifth of the same bar to a small mat. Nathan gave ten of the 15 parts to the big mat, which left five of the 15 parts in the original bar (the original unit bar was no longer visible as a whole – cf. Fig. 9.4).

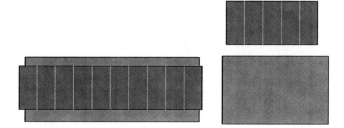

Fig. 9.4. Nathan gives two-thirds of a 15-part bar to the big mat.

---

[8] We worked with Nathan alone for the rest of the teaching episodes while he was in his third grade. Drew was partnered with another student who was more compatible in terms of his fractional schemes.

9 The Construction of Fraction Schemes Using the Generalized Number Sequence 287

Nathan reasoned to himself out loud as follows: "One-fifth? Five people could not get five pieces each. Three pieces? – three, six, nine, twelve, fifteen – so the little mat gets three pieces!" Nathan's solution corroborates that he had constructed a *splitting operation for connected numbers*. *Splitting* the 15-bar into five equal parts means that any one of the five parts produced by the partitioning of the bar into five parts, by necessity, when iterated five times must produce 15 parts.

An interesting event occurred after Nathan had found that the little mat would get three pieces. Two of the original 15 parts were left in the original bar at this point in the episode and the teacher asked Nathan "How much of the bar is left over?" He replied "One part less than a fifth." The teacher then asked him "Can you think of another way to describe that?" and he replied "Two-thirds of a fifth." This interchange indicates that Nathan could use a fraction as an operation (two-thirds of a fifth) as well as the result of the operation ("One part less than a fifth"). This double aspect of a fraction corroborates the concept of a fractional number (cf. Chap. 8) because, in a fractional number, a child establishes a whole-to-part relation where any one of the parts can be iterated to produce the whole. Nathan's comment, "One part less than a fifth" indicates that he regarded the three parts that constituted a fifth as a unit whole that could be produced by iterating any of its parts three times. He also established two parts as a result of an operation on the unit whole without explicitly producing the operation until he was asked, which he then expressed as "Two-thirds of a fifth." Perhaps more importantly, he established a relation between the two parts and the three parts in which the two parts were both contained and disembedded, which is crucial in establishing a fractional number.

Nathan did not always interpret the teacher/researcher's fraction language in ways that we expected. When Nathan was asked, "How much of the *whole* bar is left over?" he replied: "Two doesn't go into fifteen; two goes into fourteen – one-seventh?" For Nathan, "How much?" indicated the search for a unit fraction to describe the quantity, whereas, "two-thirds of one-fifth" indicated the operation of taking two of the three parts, which constituted the quantity "one-fifth." Still, his interpreting the two parts as necessarily being one unit of two out of a plurality of such units corroborates the inference that he had constructed a splitting operation for connected numbers and that this operation was ready-at-hand for him to use. But Nathan experienced perturbation, so we asked him if he could reconstruct the original bar from one of the two parts that were left. He did so by copying it 15 times and joining the copies together. This action led to describing subparts of the bar in terms of fifteenths: one-fifteenth, two-fifteenths, three-fifteenths, etc. Nathan was then able to describe the left over amount as two-fifteenths. If, in the future, Nathan were to make a distinction in the two situations[9] and produce an appropriate fraction word for each one, the episode could be considered as a functional accommodation of a scheme.[10]

---

[9]That is, between those situations when the numerosity of the fractional part is a divisor of the numerosity of the fractional whole and when it is not.

[10]A functional accommodation of a scheme is an accommodation of the scheme that occurs in the context of the scheme being used.

The scheme that Nathan was in the process of constructing transcends the iterative fraction scheme constructed by the other students in the teaching experiment. In that scheme, the assimilating operations consisted of (1) the splitting operation where the unit to be split was an unpartitioned continuous unit and the split was into individual units, (2) the disembedding and iterating operations, and (3) the operations that produce three levels of units used in the context of the other operations. What was novel in the scheme that Nathan constructed was that the first operation had been transformed into a splitting operation for connected numbers. For example, given a bar, he could split the bar into a unit containing five units, disembed one of the five units, and split it into three units where the latter split implied splitting the rest of the five units into three units to split the whole bar because it was an iterable unit. Rather than replace the splitting operations of the other students, Nathan could use whichever splitting operation was appropriate in a given situation. That is, the splitting operations for connected numbers superseded the original splitting operation. We refer to this scheme, when completed, as an iterative fraction scheme for connected numbers and distinguish it from the iterative fraction scheme using the following notation: (IFS: CN).

Renaming Fractions: An Accommodation of the IFS: CN

At the time of the teaching experiment, we had not hypothesized the IFS: CN, but the next task illustrates how Nathan used this scheme as well as the functional accommodations he could make while using it. The goal of the task following the task in Fig. 9.4 was to provide a situation for Nathan to build nonunit fractions as fractional numbers by taking multiples of a unit fraction. It was in this task that COPYPARTS was introduced. Nathan was asked to create a 16-part bar and was shown how to use COPYPARTS to make new bars consisting of one part, two parts, three parts, etc. of the 16 parts. In this situation, the original 16-part unit remained intact, thus providing a perceptual referent unit for all fractional parts created in this way. The following protocol illustrates how Nathan constructed a method for renaming fractions.

Protocol II: Renaming fractions made from a 16-part bar.
T: How many fractions will you have when you have finished?
N: Sixteen, with different names.
T: Can you name them?
N: One-sixteenth, two-sixteenths. I can name all of them by the sixteenths, so I'll just give you the other names. Two goes into sixteen eight times, so that would be one-eighth three-sixteenths, four-sixteenths – four would go into it four times. That would be two-eighths and it would be one-fourth. Five-sixteenths (No other names.). One-sixth (Pointing at the 6-part piece.) – no, yes, no! Six-sixteenths or three-eighths! Seven-sixteenths – no more names. Eight would be four-eighths or one-half, two-fourths. Nine would be nine-sixteenths. Ten would be ten-sixteenths and five-eighths and that's it! Eleven-sixteenths – that's it. Twelve would be two-sixths because six goes into twelve two times, or four-fourths.
T: (Asks Nathan to explain how he came up with two-sixths. Nathan pointed to the 6/16-piece on the screen and said that this was one-sixth. Note that he initially called this piece one-sixth and then corrected himself in the above protocol.)

T: Copy the twelve-sixteenths. (Nathan makes the 12/16-piece using COPYPARTS and visually compares it to the 6/16-piece.)
T: (Pointing to the 6/16-piece.) What was that?
N: Three-eighths – Oh! So it wouldn't be two-sixth! (Referring to the twelve-sixteenths) – It's six-eighths or four-fourths.
T: How many fourths?
N: Oh! Three-fourths – four-fourths would be sixteen (parts).
T: How did you do this?
N: I worked out how many times the number of parts goes into sixteen!

Initially, Nathan simply used his IFS: CN to generate the fractions from one-sixteenth to sixteen-sixteenths. What followed constitutes a functional accommodation of the scheme that was made possible by his splitting operation for connected numbers. He could conceptually transform, say, four-sixteenths into one-fourth by refocusing his attention on the units of four units comprising 16 or on the units of one comprising 16. What Nathan appeared to have done in the above protocol was to find a way to symbolize the equality of the two fractions "one out of four" and "four out of sixteen" by using the common divisor of the number of parts and 16, although he had no language for expressing this process. His expression "how many times the number of parts goes into sixteen" is the closest he gets to describing his symbolic method. The confusion over the 12-part piece illustrated that his IFS: CN was undergoing a modification-in-action (a functional accommodation).

The question remains, at this point in the teaching episode, whether Nathan had developed a numerical method for simplifying fractions or if he was superimposing the higher level partitions in his imagination over the partitioned pieces that were visible (e.g., did he imagine the 10-part piece as consisting of a unit comprising five 2-part pieces, which he had called eighths)? His confusion over the twelve-sixteenths and his explanation that "six goes into twelve two times" suggests that he had taken the visible six-sixteenths piece and mentally partitioned a nonvisible twelve-sixteenths into two of these, but had reverted to his initial label for the six-sixteenths (one-sixth) to come up with his answer of two-sixths for twelve-sixteenths. The visual comparison of the six-sixteenths and twelve-sixteenths pieces helped him realize his own inconsistency (in labeling) and to rename the twelve-sixteenths as twice the fraction he correctly used to rename the six-sixteenths (six-eighths from twice three-eighths). Monitoring his operations is solid indication that he was operating at the level of the splitting operation for composite units.

Construction of a Common Partitioning Scheme

Our goal in the next sequence of episodes was for Nathan to construct a scheme for finding a common partition of a bar so that he could make two different fractions from the same bar. The reason for developing such a collection of tasks is that when finding, say, the fractional part of a bar comprised by one-half of the bar and one-third of the bar, a unit fraction of the bar must be found so that both one-half and one-third are both multiples of the unit fraction. To produce such a unit fraction, it is sufficient to produce a partition of the bar so that both one-half and one-third of the bar are contained in the partition. It would seem that the IFS: CN would be sufficient to solve such tasks because the splitting operation for connected numbers is one of its assimilating operations. An adjustment in how this splitting operation is used would be necessary, however, because the goal of the splitting operation for connected numbers is to partition only one partition of a connected number into a lesser number of parts, such as to partition a 15-part bar into three parts. Finding a common partition of two given partitions involves finding a greater number of parts that contains both of the original partitions. One way to solve the task is to recursively partition each of the two partitions with the goal of producing an equal number of parts.

The first task that we used proved to be very confusing for Nathan. In the teaching episode on the 11th of March, he was given two bars of the same size: one partitioned into five parts and the other into three parts. He was asked to cut the two bars into equal-sized parts so that a child could have the same amount from each bar. Nathan at first tried to match two parts from the fifths-bar to one part from the thirds-bar. He eventually came up with a solution to the problem by repartitioning the 3-part bar into six parts and the five-part bar into ten parts and giving one-half of each bar to each of two children. The teacher made a further constraint in the situation at this point. He insisted that the pieces in each bar had to be the same size.

Protocol III: Finding a common partition for thirds and fifths.

N: How can they get equal pieces without using the same number?
T: That's right! – You've got to find out how to cut them into equal pieces.
N: Give me a sheet of paper. (The teacher did not supply a sheet of paper so Nathan proceeded to use the mouse to draw numerals on the computer screen! He made "10" and "6.")
N: I would go six, twelve, – fifteen. Ah! I have found it! (Nathan wiped both bars clean, made two mats and partitioned each bar into fifteen parts.)
T: (Pointing to the first bar.) How many fives are here?
N: Three fives.
T: (Pointing to the second bar.) How many threes are here?
N: Five threes. Oh! So there's three fives and five threes!

Once Nathan had realized that he had to use the original partitions of the bars to make further partitions he came up with the strategy of comparing multiples of the number of parts in each partition until he arrived at a common multiple (15). The act of forming the numerals for the second multiple of each partition (ten and six) appeared to have been a critical act for Nathan. The following episode (5 days later on the 16th of March) indicated that Nathan's strategy for finding a common partition was well established. Given a 5-part bar and a 7-part bar he compared multiples of five and seven, talking out loud, until he arrived at 35.

Protocol IV: Finding a common partition for fifths and sevenths.

T: We share these two candy bars among kids. We cut this candy bar (The fifths bar.) into so many smaller pieces, and give one piece to each kid; we do the same with the other candy bar. Each kid gets one piece of this candy bar, and each kid gets one piece of that candy bar, and those two pieces have to be exactly the same size.

N: Ten (With his finger, writes the numeral "10" on the table.) fourteen (Writes "14.") that wouldn't work (Scratches both numerals.), fifteen and twenty-one, twenty, add another ten, thirty and twenty-eight, that wouldn't work, so add another seven, thirty-five, ah – (Wipes both bars.)

Nathan mentally recursively partitioned each part of each bar into two parts and wrote down the result on the table ("10" and "14"). This indicates that the figurative material on which Nathan operated at this point was interiorized, or operative, figurative material in that he operated numerically. Using his concepts of five and seven, we would say that Nathan generated figurative material however minimal that he could imagine partitioning without actually carrying out partitioning acts. Doubling the two numbers symbolized the actual partitioning acts. From that point on, Nathan used his GNS to generate and compare the elements of two number sequences, one consisting of multiples of five and the other consisting of multiples of seven. In the section, *Nathan's Generalized Number Sequence*, we used how Nathan found how many strings of three and how many strings of four would be needed to make 24 toys to infer his GNS. This operational capacity of the GNS served Nathan in generating the common partition, 35, into which he could partition the 5-bar and the 7-bar to meet his goal. It was only after he operationally established 35 as the appropriate partition did he implement his operations in TIMA: Bars.

Constructing Strategies for Adding Unit Fractions with Unlike Denominators

In the teaching episode held on the 6th of April, the task situation was changed slightly. Three congruent bars were created. One bar was partitioned into thee parts and one into five parts. The third bar was not partitioned. Nathan's task was to partition the third bar so that he could make both one-third and one-fifth from this one bar. He would then be asked how much of the bar would be the combination of the two unit fractions. Nathan partitioned the blank bar into 12 parts and pulled out four parts for the one third (placing this 4/12-piece on the larger of two mats drawn at the bottom of the screen). He then dragged one part from the 5-part bar on top of his 12-part bar. He realized that the parts did not exactly match up.[11] He then checked

---

[11] Note that Nathan found a multiple, 12, for the 3-partition that produced one-third, but he did not find a multiple for the 5-partition that produced one-fifth. Nevertheless, he tested whether the 12-partition would work for the 5-partition, indicating that it was his intention to find a multiple that worked for one-third and one-fifth.

his 4/12-piece with one part from the 3-part bar to verify that these did match. The following interaction then took place between Nathan and the teacher:

Protocol V: Realizing the invariability of one-third of a fixed quantity.
N: Can I change the number (of parts) so that I can get a fifth?
T: Yes – as long as you can also get a third with that number.
N: Do the one-third pieces have to be the same size?
T: What do you think?
N: I don't know.
T: Tell me why they do NOT have to be the same size.
N: (Nathan thinks for a while.) They do have to be the same size! – Because one-third is one of three pieces, and four of twelve pieces will match, so any other number has to match!

Nathan's comment, "Can I change the number (of parts) so that I can get a fifth?" does indicate that it was his goal to find a common partition. However, his question concerning whether the one-third pieces have to be the same size was unexpected. In retrospect, it indicates that his strategy for finding a multiple of two partitions served the goal of finding a partition that contained the two given partitions and was basically a modification of his GNS. It was not a strategy that served the fraction goal of finding a partition that served in transforming the two given unit fractions into fractions that were equal to the unit fractions, even though he could produce such fractions commensurate with the unit fractions. One could say that his commensurate fractions were produced as a result of operating and were yet to be taken as input for further operating. Once Nathan took one-third and the result, four-twelfths, as input for operating, he explicitly realized that "They do have to be the same size!"

The hypothesis that Nathan's strategy for finding a partition that contained two given partitions was essentially a modification of his GNS is corroborated by his efforts to find a common partition for one-third and one-fifth because he continued his search through guess and visual test. He tried six parts and found that two-sixths made the third but the one-fifth did not fit. He then tried ten parts and found that two-tenths matched the fifth but he could not match the third. He gave the two-tenths to the small mat anyway. He still had four-twelfths on the big mat. The teacher/researcher asked him if he had solved the problem and he replied: "No – I need to find one that works for both." He continued his trials with 25 parts and then 15 parts. He exclaimed "Yeah!" after checking the 1/5-part with three-fifteenths.

Even though Nathan realized the need for a common partition ["one that works for both (the three and five)"], he did not use his common partition strategy in this situation. He used two sequences of multiples but did not coordinate them to arrive at a common partition. The activity of finding multiples was generalized to this new situation, but the coordination of the two number sequences was not activated until *after* he had reached a common partition through his trials with multiples of five: ten, twenty-five, fifteen. We asked him to tell us about his solution and this led to Nathan reflecting on his results of acting:

Protocol V: (First continued)

T: Describe the pieces.
N: One-third is five-fifteenths, one-fifth is three-fifteenths.
T: How much altogether (On both mats.)?
N: One-third and one-fifth or eight-fifteenths.

Following this interchange, we asked Nathan to set up a new, similar problem for himself. He created a 4-part and a 6-part bar and said "That will be a fourth and a sixth." Nathan used his complete strategy for finding the common partition of 12 parts, coordinating his multiples of four and six. He described his solution as follows: "two-twelfths or one-sixth and three-twelfths or one fourth." When asked how much together he replied "five-twelfths."

The goal of Nathan's strategy for finding a common partition for two given partitions had changed to a strategy for finding fractions equal to two given unit fractions. The goal of finding a common partition for two given partitions was still an operative goal, but now served the goal of finding fractions equal to two given fractions. So we decided to modify the situation once again. Nathan set a new problem for himself using a 7-part bar and a 3-part bar, which he made using a separate bar. We changed the situation by eliminating the mats and going straight to the addition of the two unit fractions.

Protocol V: (Second continued)

T: Copy one part from each bar and join them. (Nathan does so.) How much of one bar is that (Pointing to the joined piece.)?
N: One-third and one-seventh.
T: Can you think of a single fraction name for that?
N: Half a bar? (He compares it to the original bar by moving it up to the 3-part bar.) Almost!
T: Can you find out exactly? Think about what you have just been doing.
N: (Nathan cut the joined piece apart, reestablishing the one-third and one-seventh, and then thought for a while. He then partitioned the original blank bar into twenty-one parts and checked each of the one-third and one-seventh pieces against this bar to verify his partition.)
T: How much together?
N: Three and seven – that's ten.
T: Ten what?
N: Ten twenty-firsts
T: Is ten twenty-firsts less than or more than a half?
N: Less! (After visually comparing the ten twenty-first to the 21-part bar.)
T: Can you give me a reason?
N: Ten and ten is less than twenty-one – its only twenty.

Although Nathan may have altered the goal of his strategy of finding fractions commensurate to two given unit fractions to finding a fraction equal to the sum of two unit fractions, what followed the above protocol indicated that the inverse relations between multiples of parts and parts of parts were not yet explicit for Nathan.[12] We set Nathan a similar problem to the last one using a 9-part bar and a 6-part bar. Nathan used his strategy for finding a common partition and made a 36-part bar (he missed 18 as a common multiple). The teacher asked him if he could make the one-ninth and the one-sixth from this bar without visually comparing his existing one-ninth and one-sixth to the 36 parts. He copied four thirty-sixths for the one-ninth but lined the 6-part bar up with the 36-part bar to find the 6/36 for one-sixth. He was then asked if he could make the whole bar from the 4/36-piece. Nathan counted by fours until he got to 36 but miscounted his fours and said "eight fours." He then attempted to make seven extra copies of the 4/36-piece using the REPEAT button, which would have given him a bar consisting of eight of the four thirty-sixths. He accidentally hit the REPEAT button one extra time, thus creating a bar equivalent to nine of the 4/36-pieces! This was, of course, the same as the original bar. Nathan did, however, realize that he now had nine of the 4/36-pieces in this bar.

The above actions suggest that, in this situation, Nathan did not transform four parts out of 36 parts into one part out of nine parts, and did not construe the same bar as containing a 36-partition, a 9-partition, and a 6-partition. We hypothesized that at this point in time Nathan would not be able to answer the question: "How much more is one-sixth than one-ninth?"

Nathan began to make the numerical relations between two different fractions (one-third and one-thirteenth) and their common partition (39) explicit in the next episode 2 days later on the 8th of April. The task situation was the same as the one above: to find the fractional part of a bar constituted by the join of two different unit fractions of the bar. After revisiting the last problem from the previous episode (one-ninth and one-sixth) that Nathan was able to solve immediately with a 36-part bar and later with an 18-part bar, the teacher set the problem of using a 3-part bar and a congruent 13-part bar. Nathan's first action was to align the two bars as if checking for an immediate match between the partitions. He then started counting by threes. When he got to 39 he had made a record of 13 threes with his fingers (he had run through the fingers of one hand twice for ten and had three fingers on the table to complete the 13). He then partitioned a third, blank bar into 39 parts. He copied his one-third piece and moved the copy over the 39-part bar to visually determine how many thirty-ninths were needed to produce one-third. He did the same for the 1/13-piece. When he had done this Nathan made the following comment:

Protocol VI: Finding the sum of one-third and one-thirteenth of a bar.

N: Thirty-nine and the three took thirteen times to get to thirty-nine, so there are thirteen pieces in it (The one-third.), and the thirteen took three times to get to thirty-nine so there are three pieces in it (The one-thirteenth.)!
T: How much altogether?

---

[12] For example, if it takes 13 units of three to make 39, then each unit of three is one-thirteenth of 39.

N: There are thirteen plus three equals sixteen thirty-ninths altogether! (An observer asks Nathan to make the thirteen parts in the 1/3-piece and the three parts in the 1/13-piece. He does so and then compares these partitioned pieces to the 39-part unit bar.)
N: The small pieces are the same size!
T: What are they?
N: Thirty-ninths!

Nathan appeared to have established an inverse relation between the multiples of three in 39 and the number of 39ths in one-third (and similarly for the number of 39ths in one-thirteenth). This relation, however, may have only been relations that he established among three, thirteen, and thirty-nine rather than relations among one-third, one-thirteenth, and one thirty-ninth as suggested by the fact that he was surprised that the resulting parts were the same size as the parts in the partitioned unit bar. Still, Nathan used the relations among three, thirteen, and thirty-nine to establish the relations among the fractional parts of the bar.

Multiplication of Fractions and Nested Fractions

In the 6th of April teaching episode, Nathan used recursive partitioning to construct a unit fraction composition scheme. For example, he produced one-fifteenth as one-third of one-fifth. He also produced two-fifteenths as two-thirds of one-fifth where the one-fifth had been made out of three-fifteenths as in the 6th of March teaching episode. In all of these situations, Nathan operated on bars that he had made using TIMA: Bars, and the results that he produced were visible to him. Even though operating on bars in his visual field may not have been a necessity, Nathan was yet to abstract a program of operations that could be judged as fraction multiplication as indicated in the teaching episode on the 27th of April. This teaching episode was focused on inducing modifications in his unit fraction composition scheme.

Nathan was asked to choose a fraction and he made three-tenths of a 10-part bar. He was then asked to find one-fourth of three-tenths of the bar. His first action was to partition the 3/10-bar into four parts *vertically*. The three tenths were also marked vertically so an unequal partitioning resulted (cf. Fig. 9.5).

Fig. 9.5. The results of partitioning three-tenths of a bar into four parts.

Nathan then wiped the bar clean and remade his 4-part partition. He copied one of these four parts and declared this part as the solution, which was indeed the part of the bar that was one-fourth of three-tenths. The teacher/researcher asked if he could tell him another name for the part he had just made. Nathan replied: "For every three (tenths) there are four of them." He moved his piece on top of the original 10-part bar, and saw that his new piece (one-fourth of three-tenths) was less than a tenth. He then took his three-tenths piece (partitioned into four parts) and moved it over the original bar, noting that he was able to fit three of these into the original 10-part bar, giving him 12 of the new parts, but there was a one-tenth part left in the bar. He declared that the new part was one-thirteenth of the bar (there were 12 of the one-fourth of three-tenths and one more part, the remaining one-tenth)! He then wiped the original bar clean and partitioned it into 13 equal parts.[13] He compared one-thirteenth to his one-fourth of three-tenths and found it matched exactly (due to the limitations of the computer graphics – cf. Fig. 9.6)!

Because of the limitations of the computer graphics, the teacher/researcher made a concerted effort to help Nathan construct this apparent contradiction. Nathan knew that 12 of these new parts matched 9/10 of the bar and that one part was less than 1/10, so the teacher focused on attempting to help him construct why one-thirteenth matched the one-fourth of three-tenths of the unit bar. In his attempt, the teacher asked Nathan what was one-fourth of three-tenths of the bar. Nathan said he did not know. He copied one-tenth and partitioned it vertically into four parts. He then declared that one slither would be one-fortieth of the original bar using recursive partitioning operations. The teacher then tried to encourage Nathan to make his one-fourth of three-tenths in a different way, hoping to generate a horizontal partition of the three-tenths into four parts.[14]

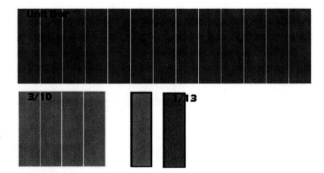

Fig. 9.6. Nathan compares one-thirteenth to one-fourth of three-tenths of a unit bar.

---

[13] Nathan relied on reasoning in terms of the visual configuration.

[14] Nathan could produce one-fortieth as one-fourth of one-tenth using recursive partitioning, so it is possible that he could have found one-fourth of three-tenths by finding one-fourth of each one-tenth and joining each of these three parts together into a unit, three-fortieths. However, this involves distributive reasoning, which he had not yet demonstrated.

9   The Construction of Fraction Schemes Using the Generalized Number Sequence   297

Nathan resisted all efforts to induce horizontal partitioning, so the teacher posed a new problem: Make a 20-part bar without using the number 20. Nathan was ingenious in finding ways to do this *without* using both horizontal and vertical partitions. He made a 5-part bar, copied one part, vertically partitioned this part into four parts, and then used one of these four pieces to create a 20-part bar. This way of operating is an indicator that he could indeed engage in recursive partitioning operations because, by iterating the part 20 times, he produced the bar that he would have produced had he partitioned each of five parts into four parts. Had he demonstrated that he could produce one-fourth of the five parts by producing one-fourth of each part and joining these five parts together, this would corroborate that he could engage in distributing reasoning.

In the last teaching episode of Nathan's third grade held on the 29th of April, we returned to exploring if Nathan could produce a general fraction composition scheme as an accommodation of his unit fraction composition scheme. Nathan was asked to make nine-sevenths of a given bar, which he did by copying the bar and then adding on two more sevenths. He was then asked to find one-fourth of the 9/7-bar. He did so by copying two of the 1/7-parts and then partitioning another one-seventh into four parts and joining one of these small parts onto the two-sevenths he had made (cf. Fig. 9.7).

Nathan's explanation was as follows: "I took this one piece and cut it into four and joined one bit to this (two-seventh) piece – then there would be another three like this (two-seventh) and one bit would go to each." He appeared to have divided the nine parts by four to arrive at "two parts and one left over" and then distributed a quarter of that remainder to each of the four 2/7-parts, which involves vestiges of

Fig. 9.7. Nathan makes one-fourth of nine-sevenths of a unit bar.

Fig. 9.8. Nathan makes one-third of a 7/7-bar.

distributive reasoning.[15] He was not able, however, to calculate how much of the original bar was the one-fourth of nine-sevenths.[16] Instead, he made a visual comparison of his constructed solution with the original bar and offered one-third as an estimate. He then constructed one-third of the 7-part bar by copying two parts and joining one-third of another part to these two (cf. Fig. 9.8).

Nathan's language and actions indicate that he had not made the necessary accommodations in his unit fraction composition scheme so that he could compose any two fractions at this point in the teaching experiment. During the second year of the teaching experiment, we worked extensively on composition of fractions with Nathan and his new partner, Arthur. Many of the above strategies reemerged for Nathan, along with his reluctance to use cross-partitioning.

Equal Fractions[17]

Arthur joined the teaching experiment as Nathan's partner in their fourth grade. In our interviews with Arthur prior to his work with Nathan, he was able to reason with composite units in ways similar to Nathan. His actions with discrete objects

---

[15]Had Nathan partitioned each of the nine parts into four parts and then reorganized them into four parts each containing nine parts, this would indicate that he anticipated that he could find one-fourth of nine parts by finding one-fourth of each of the nine parts and integrating these nine parts together, which involves distributive reasoning.

[16]Calculating how much of the original bar was one-fourth of nine-sevenths involves recursive use of distributive reasoning because the results in Fig. 9.7 have to be interpreted in terms of the seven-sevenths unit bar.

[17]This second part of the chapter is a revised portion of an article published in *Mathematical Thinking and Learning* (Olive 1999).

on the 1st of October indicated that Arthur had already constructed an iterative fraction scheme. Given six toys as two-fifths of all of the toys, he was able to determine that there would be nine more toys. Furthermore, when given two toys as one-fifth of a quantity of toys and asked to make seven-fifths of the toys, Arthur eventually added 12 more toys (in six groups of two) to the two given toys. Taken together, these two solutions indicate that Arthur had constructed a reversible iterative fraction scheme for composite units. Because he was operating with composite units, we infer that he had constructed the splitting operation for composite units, which is a constitutive operation of the IFS: CN if it is used in the context of continuous quantity. But these solutions are not sufficient to infer that Arthur had constructed the generalized number sequence, but there were good reasons throughout the time we worked with him to make that inference.

In the teaching episode with Nathan on the 20th of October, both children generated one twenty-fourth from a unit stick without using the number 24. Arthur partitioned the unit stick into three parts using PARTS and then partitioned one of those thirds in half and one of the resulting sixths in half, and eventually one of the resulting twelfths in half to generate a 1/24-piece. He correctly named all of the intervening parts. The certainty with which Arthur operated indicated that he could operate in a similar way with fractions other than one-half. If that were the case, then we could infer that he had constructed the splitting operation for connected numbers and that he could use it recursively. In any event, his partitioning actions seem sufficient to infer that the splitting operation for connected numbers was an operation that was available to Arthur.

Generating a Plurality of Fractions

In the teaching episode held on the 3rd of December, it was our goal to reinitialize the children's use of recursive partitioning by having them find fractions equal to a given fraction. In finding fractions equal to one third, both children made further partitions of the unit fractional part that had been pulled out of the original unit stick that was partitioned into three equal parts (using the computer tool TIMA: Sticks). Protocol VII begins after Arthur had drawn the original stick on the screen, and Nathan had used PARTS to mark the stick into three equal parts. He had filled the first part in a different color.

Protocol VII: Generating fractions equal to one third.
T: Arthur, can you pull that out? (Indicating the filled part.)
A: (Using PULL PARTS, pulls the colored part of the original stick out from the stick, leaving the original stick intact.)
T: Now, what is a way to represent that one-third by another fraction?
N: Two-sixths!
T: Arthur, how would you do that?

Protocol VII (continued)

A: (After approximately 15 s.) Cut the third in half! (He dials "2" with the PARTS button and clicks on the 1/3-stick pulled from the original stick.)

T: That would be?

A: Two-sixths. (Using LABEL, he selects the unit fraction "1/6" and the numerator "2" and labels the marked third-stick "2/6".)

Arthur's hesitation indicated that he experienced a constraint in that no operation was evoked that would allow him to produce another fraction. Nathan's comment "two-sixths" oriented Arthur toward finding a way to reconstitute one-third as two-sixths. "Sixths" had two meanings for Arthur. First, it meant six equal parts (of a unit whole), and second, it meant that one of the parts could be iterated six times to reconstitute the partitioned whole. He now faced the additional challenge of finding a way to transform three equal parts into six equal parts. This transformation involves a coordination of splitting the pulled part into two pieces, and the three parts still in the original stick into six pieces, while maintaining the relationship of the pulled part to the original stick. To do this, the child must partition the three pieces to produce six pieces without destroying the three pieces. It is even more demanding because the six pieces must remain embedded in pairs within the three pieces, so that one piece out of three yields two smaller pieces out of six. These operations constitute *recursive splitting* operations.[18] Arthur knew two results of applying his fraction scheme – one-third and two-sixths – and needed to transform the former into the latter. To eliminate the perturbation that constituted the goal, it suddenly occurred to him to cut the third in half, which is one reason why we referred to his operations as splitting rather than only partitioning. Moreover, cutting the third in half was not isolated to the material in his visual field. Rather, it implied partitioning each of the other thirds into two parts, not sequentially but simultaneously which is another indication of splitting. In essence, Arthur split each of the three parts into two parts simultaneously rather than sequentially.[19] For similar reasons, we also infer that Nathan engaged in recursive splitting operations.

The two children continued their recursive splitting activity by pulling out a third of the original stick and partitioning it with the next higher number. They did not need to enact the sequential partitioning of each of the three parts to obtain the resulting partition of the whole stick. They provided this result by applying their multiplication facts to the new situation. They also repartitioned copies of the whole stick to illustrate the denominator of their new fraction (and confirm their multiplication facts). They correctly labeled each of their new fractions (e.g., "2/6," "3/9," "4/12," "5/15") and eventually Nathan made a generalization that "All we've got to do is go up…This would be fifteenths (pointing to the one-third partitioned

---

[18] A split is indeed a partition.

[19] Apparently, recursive partitioning operations were available to Arthur prior to their evocation in the context of Protocol VII.

into five parts), the next one is eighteenths, then twenty-firsts, twenty-fourths." The teacher asked Arthur what the next two would be. Arthur responded with twenty-sevenths and thirtieths. He was asked: "How many thirtieths would make one third?" He responded "Ten…because ten times three is thirty." On the basis of his generalizing abstraction, we infer that Nathan established a plurality of *equal fractions* at least for one-third. But the concept of equal fractions seemed restricted in the sense that he had constructed the operations to produce a plurality of fractions equal to one-third *starting with one-third*. But there was yet no indication that he could judge whether any two fractions belonged to the same plurality. For example, given "13/39" and "18/51," there was no indication that he could judge the equality of the two fractions. Still, we infer that he could produce a conceptual plurality of fractions starting with one-third and continuing on indefinitely in the sense that he used his number sequence to guide the production of the next fraction in the sequence. But whether he could operate in such a way that he could judge the equality of any two fractions involves working at a symbolic level.

Working on a Symbolic Level

Encouraged by their apparent generalizations in the above task, we decided to engage the two children in symbolic activity in an attempt to transform the potential actions of producing different fractions equal to one-third into symbolic action. We established the meaning of the terms "numerator" and "denominator" in terms of the partitioning actions the children had been carrying out in the above tasks. We also related these terms to the two elements of the FRACTION LABELER that had been added to TIMA: Sticks (cf. Fig. 9.9). The denominator was determined by the choice of the unit fraction, and the numerator was chosen using the numeric keypad.

Protocol VII: (Cont)
T: (Selects "1/18" from the scrollable unit-fraction strip.) What should the numerator be?
A: Six! Because three times six is eighteen.
T: (Selects "1/66.")
N: Twenty-two! Because three times twenty-two is sixty-six.

Symbolizing the potential operations involved in making and coordinating partitions of the part and the whole was possible and almost immediate because both children abstracted a specific way of using their IFS: CN to produce fractions equal to one-third. It is this way of using their IFC: CN that we call an equal fraction scheme. It might appear that the equal fraction scheme was an equivalence scheme for fractions for the children. There were fundamental operations, however, for forming an equivalence relation that these children were yet to construct.

In the continuation of the above task, Arthur selected "1/69" for his denominator and Nathan eventually selected "23" for the numerator. The children were challenged

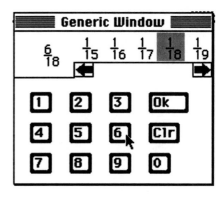

Fig. 9.9. The fraction labeler from TIMA: Sticks.

to make models of the last two fractions indicated by the numerals, "22/66" and "23/69," without using the denominator as input for the PARTS operation. Neither child used a one-third stick to produce their new fraction, even after the teacher had asked them if they could use that stick to help them. Nathan used a stick partitioned into 33 parts to produce the twenty-two sixty-sixths. He pulled out ten of the 33 and split each one into two. He then repeated this stick with 20 parts three times. He pulled out a copy of six of the 60 parts and joined this to the end of the 60-part stick to make his unit stick with 66 parts. For the twenty-three sixty-ninths, Arthur partitioned the unit stick into 23 parts and then partitioned eight of these into three parts each to produce twenty-four sixty-ninths! He pulled out ten of the resulting 24 parts (ten sixty-ninths), repeated the ten and pulled out another three to join onto the twenty to make twenty-three sixty-ninths.

Although Arthur's actions illustrate a coordination of the units of three and twenty-three to make sixty-nine parts (he used three to partition each of the 23 units into three parts each),[20] there was still no indication of an awareness of operating with the structure of his unit fraction scheme for one-third. The two schemes were combined sequentially when building up a sequence of fractions equal to one-third, but were not combined when trying to make twenty-three sixty-ninths using TIMA: Sticks. Nathan was also unable to make use of the structure of his unit fraction scheme for one-third when trying to make twenty-two sixty-sixths. Both children could transform one-third into equal fractions, but neither of them could reverse the operations. From our point of view, had the children constructed an equivalent fraction scheme, the most likely method for making twenty-two sixty-sixths and twenty-three sixty-ninths would be to partition a 1/3-stick into 22 and 23 parts, respectively. It appeared as though their equal fraction schemes were one-way schemes – they were not yet reversible.

---

[20] This is another way of talking about the splitting operation for connected numbers of the IFS: CN. However, Arthur did not operate sequentially as the use of "coordination of the units…" would imply. Rather, the coordination of units was implied by his multiplicative language and by his use of notation.

Construction of a Fraction Composition Scheme

It was evident from the preceding work that both children could create and find a unit fraction of a unit fraction (e.g., one-eighth of one-fourth was one thirty-second of the unit stick) but they had difficulties extending their unit fraction composition scheme to, say one-fourth of three-sevenths of a unit stick. One particular type of task, at first using the TIMA: Sticks and eventually transferring the task to TIMA: Bars, proved to be engendering for this construction. The task, given in the 8th of March teaching episode during their fourth grade, was to share *part* of a pizza (represented at first by a stick in TIMA: Sticks) among a number of friends and to find out how much of a *whole* pizza each friend would get. The task was:

> A pizza (stick) is cut into seven slices (parts of the stick). Three friends each get one slice. A fourth friend joins them and they want to share the three slices equally among the four of them. How much of one whole pizza does each friend get?

The typical approach that the two children took to this type of task was to partition the stick into seven parts, pull out three parts, partition each of those three parts into four parts and pull three of these parts out of one of the four parts for the share of one friend, which was distributive sharing because they distributed partitioning into four parts across each of the three parts and used three of these 12 parts as the share of one friend. Distributive fractional reasoning seemed not to be explicit in the way they pulled three parts out of one of the four parts rather than pull one part out of each of the three parts and join these three parts together as the share for one friend. That is, the children did not seem explicitly aware that to find one fourth of the three parts, they could find one-fourth of each one of the parts and join these three parts together as the fraction equal to simply taking one-fourth of the three parts together. Still, they acted as if the three parts they produced by distributive sharing was one of four equal shares, so fractional distributive reason was implicit in their actions. Following these partitioning actions, they would then use their IFS: CN to iterate this share four times to check that it matched the part of the pizza (stick) that they were given (cf. Fig. 9.10).

One difficulty for the two children was in naming the share as a fraction of the original whole pizza. They were not reversing their partitioning operations through the two levels of partitioning that had produced the share. For instance, in the above situation Arthur named the share as *three fourths of a seventh of the whole pizza*. He then attempted to measure the original stick with the share of one person by iterating it in an attempt to make a stick the same length as the unit stick.

Fig. 9.10. Four people share three-sevenths of a pizza stick.

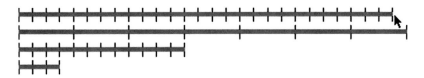

Fig. 9.11. Nine iterations of three-fourths of one-seventh of a pizza stick.

The stick produced by the iterations, of course, came up short (cf. Fig. 9.11). This unexpected feedback from the computer screen was a great occasion for learning. Arthur needed to make an accommodation in his current ways of operating.

Arthur's iteration of the share of one person to recreate the original stick produced a constraint in the results of using his IFS: CN. The goal of finding three-fourths of one-seventh of a stick was not attainable with his current operations that were based on his strategies for finding a unit fraction of a unit fraction.[21] Making distributive reasoning explicit was required. That is, he needed to decompose the three-fourths of one-seventh as three of (one-fourth of one-seventh) where each one-fourth of one-seventh was taken from one of the three parts. He could then use his recursive partitioning operations to find one-fourth of one-seventh, and use his uniting and unitizing operations to take three of these one twenty-eighths as one thing. The constraints Arthur encountered in these episodes and the modifications he made in the results of distributive sharing in overcoming these constraints provided the bases for the construction of his Fraction Composition Scheme.

Constraining How Arthur Shared Four-Ninths of a Pizza Among Five People

We hypothesized at this point in the teaching experiment that Arthur needed to refocus his attention on the component parts that constituted the one person's share so that he could then reflect on the results of his distributive sharing operations that he had used to produce that component part and project the involved operations back through the two levels of partitioning to establish what fractional part of the stick was constituted by the share of one of the four people. We decided to introduce a constraint into the situation that might provoke Arthur to refocus on the component part. When Arthur produced the three miniparts that constituted the share for one person, he did not select one mini-part from each of the four mini-parts that he made in each of the one-sevenths of the three-sevenths. Instead, he selected three mini-parts from one of the four miniparts in one of the sevenths. So, in the next teaching episode, the situation was modified to provide a constraint so that Arthur would need to select one minipart from each of the original pizza slices.

---

[21] After the children produced, say, one-eighth of one-fourth in TIMA: Sticks, rather than reason using recursive partitioning operations the children repeated the part 32 times to find what fraction it was of the whole (cf. the section Generating Fraction Families).

9 The Construction of Fraction Schemes Using the Generalized Number Sequence

Fig. 9.12. Sharing four-ninths of a pizza stick among five people.

The teacher in the next teaching episode held on the 10th of March posed the situation of a pizza stick with nine slices, each slice having a different topping (indicated by filling each of the nine parts in a different color). She asked Arthur to pull out four different slices, which he did (three of them attached and one separate piece). The task was to share the four slices equally among five people *so that each person gets a piece of each slice* (this was the constraint). Arthur partitioned copies of each of four slices into five miniparts and broke each slice into its five separate miniparts. He arranged the broken slices in five rows of four, one piece from each of the four slices in each row (cf. Fig. 9.12).

In the following protocol, the teacher asked Arthur to join the pieces that make the share for one person and to find out how much this was of the whole pizza. Arthur joined the four pieces in one row and compared this share to one-ninth of the original stick. He then thought for more than 1 min while looking intently at the screen. Protocol VIII begins at this point.

Protocol VIII: Finding four-fifths of one-ninth of a pizza.
T: What do you think?
A: I know it is four-fifths of a ninth of a pizza...
T: (Confirmed his response and asked if there was any way to find out how much that was of the whole stick.)
A: Yes there is, but...(Trails off and thinks some more.).
T: How many of these small pieces do you have in the whole thing (Pointing to one of the four parts of one share.)?
A: Forty-five.
T: Why is that?
A: There are nine pieces (In the whole stick.), five in each, so that's forty-five.
T: How much of the whole pizza is one share then?
A: Four forty-fifths. Because this is the shares of one person...and that's four... And in the whole thing there are forty-five, so the share of one person is four forty-fifths!

By focusing Arthur's attention on the relation between one minipart and the 9-part stick, the teacher provoked Arthur's recursive partitioning operations, enabling him to work out the unit fractional size for the smallest part. It was then simply a matter of uniting the four unit fraction pieces to establish the share as four forty-fifths of the 9-part stick. So, the teacher presented Arthur with a task of his unit fraction composition scheme, so Arthur might experience a recurrence in taking a proper fraction of a unit fraction.

Protocol IX: Finding two-thirds of one-seventh of a stick.
T: (Draws a stick.) The stick that I am thinking of is two-thirds of one-seventh of that stick.
A: (Uses PARTS to divide the stick into seven parts, then the first part into three, then pulls two of these three pieces – cf. Fig. 9.13).
T: How much of the whole stick is that?
A: There are twenty-one of that (The small piece.) in the whole stick so its two twenty-firsts.
T: How can you make sure?
A: Use the measure button.
T: Without measuring!
A: (Colors the two pieces red, pulls a third part, joins it with the two twenty-firsts, then uses the REPEAT function to iterate the 3/21-stick seven times to form a stick the same length as the original stick (cf. Fig. 9.14). He explains why it is two twenty-firsts.)
A: The two that are filled in are the two that I started with, and in the whole thing there are 21 (Referring to the subpartitions.).

Fig. 9.13. Two-thirds of one-seventh of a stick.

Fig. 9.14. Iterating three twenty-firsts to make a whole stick.

Arthur's actions of reestablishing a seventh by adding the third piece to his two twenty-firsts and then repeating this partitioned seventh seven times to make the whole were in contrast to the use of his IFS: CN in the teaching episode held on the 8th of March prior to Protocol IX in an attempt to produce a partition of the 7/7-stick using the newly established unit. In effect, it was an enactment of his recursive partitioning operations. Arthur seemed certain in how he operated, so in the next task he was asked to share five-elevenths of a stick among seven people.

Protocol X: Sharing five-elevenths of a stick among seven people.
T: (The teacher instructs Arthur to draw a stick, divide it into eleven parts, pull five parts out, and share those five parts among seven people.) How much of the whole stick does each person get?
A: (After pulling five parts out of an 11-part stick, looks at the screen, thinks for a few seconds and then partitions one of the five parts into seven parts; he pulls a minipart out of this partitioned part. He then erases the marks and thinks some more.)

T:	What were you going to do?
A:	I don't know…every body would get… If you divide each piece into seven pieces then everybody would get a piece from each piece, so five pieces. If there are seven in one-eleventh, then there would be seventy-seven in the whole and that would be five seventy-sevenths.
A:	(Arthur then repartitions the one-eleventh part into seven parts and pulls five of the parts out even though he had said "everybody would get a piece from each piece.")
Observer:	What's one-seventh of five-elevenths?
A:	I don't know.
Observer:	(Repeats the question.)
A:	(Thinks for several seconds, looking at the screen,) I think I have – It would be the same piece, I think. (He drives the 5/77-piece around on the screen.)

When Arthur said, "If you divide each piece into seven pieces then everybody would get a piece from each piece, so five pieces" he engaged in distributive sharing. But his intention to divide each piece into seven pieces and take one piece from each of the five pieces was markedly different than taking five mini-parts from the seven miniparts in one of the five parts in a way similar to how he operated in the 8th March teaching episode. He now was explicitly aware of how he operated and, had he been asked, we infer that he would have known that these five pieces together was one-seventh of the whole of the five pieces. But Arthur's goal was not to find one-seventh of five-elevenths as indicated by his saying "I don't know" when asked what is one-seventh of five-elevenths. Arthur seemed to reconceptualize the situation that he established when asked, "How much of the whole stick does each person get?" into fractional relations. That is, he seemed to establish five of the 11-part stick as five-elevenths and sharing these five parts among seven people as finding one-seventh of five-elevenths of the stick. We also infer that he now knew that the result was five-seventy-sevenths (I think I have – It would be the same piece, I think.). He seemed to assimilate the results of using his reorganized splitting scheme for connected numbers using his IFS: CN. It is our hypothesis that the assimilation reconstituted the IFS: CN in such a way that the splitting scheme for connected numbers that he had reorganized as manifest in solving the task of Protocol X became embedded in the IFS: CN and the reorganized IFS: CN became what we have called the fraction composition scheme. A test of the hypothesis would consist of Arthur finding a fraction of a fraction of a fractional unit in such a way that he used his fractional composition scheme. This hypothesis was tested in the teaching episode held on the 19th of April.

Testing the Hypothesis Using TIMA: Bars

In the teaching episode held on the 19th of April, we decided to introduce TIMA: Bars as a tool the children could use to solve problems like sharing four-ninths of a pizza among five people. The reason for our decision was that in TIMA: Bars, a bar could be partitioned both vertically and horizontally, or cross-partitioned. It was thought that Arthur might use cross partitioning when solving tasks like the Pizza

Task. If he did, this would be a corroboration of our inference that he could engage in distributive reasoning and that he had constructed a distributive partitioning scheme. It would constitute corroboration because, say, horizontally partitioning a bar into seven parts that is already vertically partitioned into four parts would at once constitute taking one of the horizontal partitions as one-seventh of the whole bar and as one-seventh of each of the four vertical partitions. The equality of these two partitions would constitute an identity.

Nathan did not spontaneously choose to use cross-partitioning when solving the problems presented him in his final episode of the first year of the teaching experiment. Whether by now he would spontaneously use cross-partitioning is investigated in Protocol XI along with the question whether he had constructed a fraction composition scheme.[22]

Protocol XI: Arthur's spontaneous use of cross-partitioning to find one-seventh of four-ninths.

T: Let's have nine pieces in our pizza to start with. Arthur, how many pieces shall we use?
A: Four. (Arthur makes a bar and partitions it into nine vertical parts.)
T: OK. Pull out the four pieces, Arthur.
A: (Arthur does so.)
T: You are going to share those four pieces among seven people.
A & N: Seven!?
T: Seven. Before you do anything, do you think you can figure out how much of one pizza each person will get?
N: I've got it! (Nathan reaches for the mouse.)
T: Wait. How much of a pizza do we have here?
A & N: Four-ninths.
T: OK. And how many people are sharing it?
A: Seven.
A&N: (Both children think for thirty seconds. Arthur stares intently at the screen, while Nathan stares off into space.)
N: It's easier to do it when you've got it done. (Meaning: It's easier to figure it out after you carry out the actions.)
T: Tell me what you would do.
A: If there are seven pieces in four then you have to think about how many in eight and then how many would be in the remaining one to make nine.
T: (To Arthur.) Share this among seven people, please.
A: Alert.
N: I've no idea! (My.) Head's busted!
A: (Uses PARTS to partition the four-part piece horizontally into seven rows of four.
N: You've done it! Each person gets one of those strips [pointing to a horizontal row of four. While Arthur is filling the share of one person (the top row – cf. Fig. 9.15), Nathan works out the number of small pieces in the whole bar and the fraction word for the share of one person.] Four times seven is twenty-eight, twenty-eight and twenty-eight is fifty-six, and seven more makes sixty-three. Each person gets four-sixty-thirds!

---

[22] We have focused on Arthur for clarity of presentation in his construction of the fraction composition scheme. But Nathan too was involved in the teaching episodes, so we take this opportunity to document his progress as well as Arthur's distributive reasoning.

Fig. 9.15. Filling one-seventh of four-ninths of a bar.

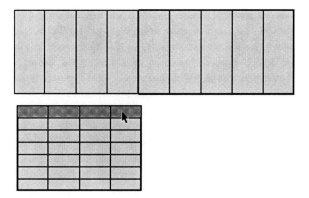

This was Arthur's first session using TIMA: Bars. So, that he independently used cross partitioning to partition a partition suggests that cross partitioning was an available operation for him prior to using the computer tool. During the 30 seconds, he sat staring at the computer screen, we infer that he generated an image of a 9-part bar and mentally partitioned each of the nine parts of this bar into seven parts – (If there are seven pieces in four then you have to think about how many in eight and then how many would be in the remaining one to make nine.). This comment preceded his cross partitioning the 4-part bar into seven parts horizontally, and it is the basis of our inference that he was aware of the results of partitioning each part of the 9-part bar into seven parts prior to his use of cross-partitioning. He apparently reorganized this result into an image of a structure consisting of seven rows and nine columns, where the first column comprised the first seven parts, the second column comprised the second seven parts, etc. This inference is based on his comment, "If there are seven pieces in four then you have to think about how many in eight and then how many would be in the remaining one to make nine," which indicates that he mentally engaged in recursive partitioning, which corroborates that he engaged in splitting the connected number nine. Further, his partitioning of the 4-part bar into seven horizontal parts corroborates that he could engage in distributive reasoning and that he had constructed a distributive partitioning scheme.

Try as he might, Nathan could not carry out recursive operations mentally [I've no idea! (My) head's busted!]. But once Arthur used cross-partitioning to partition the 4-part bar into seven horizontal strips, Nathan exclaimed, "You've done it! Each person gets one of those strips" and went on to find the number of subparts in the 9-part bar (63) and the fractional part that one of the strips was of the 9-part bar (four sixty-thirds). His exclamation and finding four sixty-thirds solidly indicates that it was his goal to partition each part of the 4-part bar into seven parts and to find the fractional part the share of one person was of the 9-part bar. In fact, he said, "I've got it!" prior to sitting approximately 30 seconds staring into space. He also said, "It's easier to do it when you've got it done." Indicating that he knew how to operate using the electronic manipulatives prior to Arthur's act of cross-partitioning had the teacher allowed him to do so (he did operate appropriately when finding four-sixty-thirds). So, his not using cross partitioning did not seem to constrain his reasoning when he sat for 30 seconds trying to engage mentally in solving the problem because he could vertically partition each part of the 4-part bar into seven parts when the 4-part bar was in his visual field. Our hypothesis for why he could not solve the problem without actually operating on material in his visual field is that, even though he could enact recursive partitioning operations using the electronic graphic items, he could not monitor their implementation in visualized imagination as did Arthur when he said, "If there are seven pieces in four then you have to think about how many in eight and then how many would be in the remaining one to make nine."

Discussion of the Case Study

Nathan constructed all of the fraction schemes while he was in his third grade that the other children (excepting Arthur) constructed while they were in their fourth and fifth grades. The key element in how Nathan made such rapid progress is the operations that generate the generalized number sequence. Using these operations, once he established meaning for four-sixths in the context of sharing two bars among three mats (Protocol I), he could independently solve the tasks in a way that went well beyond a partitive fraction scheme. In fact, later in his third grade, he independently solved tasks that went well beyond an iterative fraction scheme as well.

The Reversible Partitive Fraction Scheme

On the 16th of February of Nathan's third grade year, when he was given a 6-part bar and told that it was three-eighths of a whole bar, he partitioned the 6-part bar into three parts and copied the resulting 2-part bar five times to complete an 8/8-bar. These were the observable actions, and they indicate operations on which these actions were based that can only be inferred. Because he made an 8/8-bar where each 1/8-bar contained a 2-part bar (cf. Fig. 9.2), we infer that Nathan established the parts of the 6-part bar as belonging to a sequence of connected continuous units prior to action, where each unit of the sequence contained a part identical to the parts of the 6-part bar.[23] Each unit of the sequence was established at the level of the units of his generalized number sequence where the number of units of the sequence was yet unknown. Further, we infer that this sequence of continuous units was intended by Nathan to be the elemental units of a sequence of eight composite units that he also established at the level of the units of his generalized number

---

[23]Here, we are not using "bar" literally. Rather, we are using "bar" to refer to permanently recorded experiences of bars produced by the unitizing operation, experiences that can be in some form regenerated in visualized imagination.

sequence, where each composite unit contained a bar partitioned into an unknown number of equal-sized parts. So, Nathan had established two sequences of continuous units; a sequence of eight composite units whose elements were bars partitioned into an unknown number of parts, and a sequence of connected continuous units of unknown numerosity where the units were to be used to make the eight composite units. Establishing these two sequences side-by-side, as it were, where the sequence of eight composite units served as an assimilating structure as well as the structure that he implemented in operating on the sequence of individual bars, was based on his generalized number sequence.

It is crucial to note that had Nathan not generated images (however minimal) of these two sequences at the level of re-presentation, he could not have proceeded because he would have been embedded in his experience of the bars in the immediate here-and-now. As it were, one might say that Nathan's experience of the bars in his visual field was embedded in his visualized images of bars, where the images were produced using his two sequences in the production of the images. But that is not all, because based on his solving actions, the images constituted only a part of his experience of the situation. He also intended to act, which means that he formulated a goal to establish the number of parts of the sequence of eight composite units. When coupled with the actual operations in which he engaged and the results of his operations, the situation and goal constitutes a scheme that we called the iterative fraction scheme for connected numbers (IFS: CN).

There is no question that Nathan at least partitioned the six bars into three units with two units in each. The question is whether he *split* the six bars into three units of two units each to make the hypothetical fractional whole. Independently initiating the act of partitioning the 6-bar into three units with two parts in each unit means that he was explicitly aware of the relation between three-eighths and eight-eighths prior to his partitioning of the 6-part bar and that he was also explicitly aware that one-eighth iterated eight times completed eight-eighths, which is sufficient to infer that he split the 6-bar rather than simply partitioned it. It is also sufficient to infer that his splitting of the 6-part bar into three 2-part bars symbolized splitting the hypothetical fractional whole into an 8-part bar where each part was split into two parts. So his splitting operation was a splitting operation for connected numbers. Nathan also split a bar that he was told was two-sevenths of another bar into two parts and iterated the part to establish the other bar. This way of operating was also sufficient to impute the splitting operation into unit parts to him and corroborates the inference that his splitting operation also functioned in the case of splitting into multiple parts.

The Common Partitioning Scheme and Finding the Sum of Two Fractions

By the end of his third grade, Nathan had transformed his reversible partitive fraction scheme into a reversible iterative fraction scheme for connected numbers. We distinguished his iterative fraction scheme for connected numbers from the iterative fraction scheme that the other children constructed and used in their fourth and fifth grades based on the differences in the splitting operations of the two schemes. The common partitioning scheme that Nathan constructed was also based on the splitting operation for connected numbers and on his coordination of the elements of two number sequences (Protocols III and IV). In Protocol III, to find a common partition of two bars of the same size, where one was partitioned into three parts and the other into five parts, Nathan counted by three and by five until reaching fifteen, which he regarded as the number of parts needed. His activity might be regarded as a harbinger of finding the least common multiple of the denominators of two fractions, which is a common method used when attempting to "teach" children to find the least common denominator when finding the sum of two fractions. If such a method is to be based on the result of productive mathematical activity of children, then certainly the generalized number sequence must be available.

But Nathan's common partitioning scheme that he constructed to find a common partition of two given unit fractions was not immediately applicable when finding the sum of two given unit fractions (cf. Protocols V and VI). On the basis of his recursive partitioning operations, he could easily establish four-twelfths as commensurate to one-third, but based on Protocol V, he was yet to realize that the four-twelfths and one-third were the *same size*. That is, he was yet to establish them as identical parts of the fractional whole. Previously, one could be transformed into the other. But he seemed yet to take them as input for operating as indicated by his question in Protocol V, "Do the one-third pieces have to be the same size?" To establish them as the same size, he reasoned that, "Because one-third is one of three equal pieces, and four of twelve pieces will match, so any other number has to match." To reason in this way involves mentally splitting a bar into three equal parts and coordinating that with mentally splitting the three equal parts into equal parts while focusing attention on the three parts as component parts of a unit containing them. Doing this implies that Nathan viewed the three parts as units from the vantage point of the unit containing them, which means that the operations Nathan used to take one-third and four-twelfths as input were splitting operations for connected numbers. So, equal fractions differ from commensurate fractions in that the former are identical parts of a fractional whole whereas the latter are related by the activity of the unit commensurate fraction scheme, where one is the situation of the scheme and the other the result of the scheme. This difference was manifest in both Nathan's and Arthur's construction of a plurality of fractions equal to one-third when they were in the fourth grade (cf. Protocol VII)

Nathan did reconstruct his common partitioning scheme as a scheme for finding equal fractions when finding the sum of two unit fractions (cf. Protocol VI), but there still seemed to be a lacuna in his reasoning that involved embedding multiple partitions in the same bar. He used his counting strategy to find 36 as the common partition for a 1/9-bar and a 1/6-bar in the task following Protocol V, and then made a 36/36-bar to use to generate fractions equal to the two given fractions. To find these equal fractions, he copied a 4/36-bar to form the 1/9-bar but visually compared the 1/6-bar to the 36/36-bar. This was a surprise as we had expected Nathan simply to copy a 6/36-bar from the 36/36-bar as he did for the 4/36-bar. In retrospect, establishing multiple partitions in the same bar would seem to involve a recursive use of the splitting operation for connected numbers.

The Fractional Composition Scheme

Before exiting his third grade, Nathan did construct a unit fraction composition scheme. But given that he had also constructed the generalized number sequence, it was a surprise that he was yet to construct fractional distributive reasoning. It would seem that when he found one-fourth of a nine-sevenths bar that he engaged in distributive reasoning (cf. Fig. 9.7). However, he could not find how much the part he made was of the original bar, which entails distributive reasoning. The situation changed when Nathan and Arthur were in their fourth grade (cf. Fig. 9.10). To find one of four parts of a 3-part stick, they partitioned each part of the stick into four equal subparts and pulled three of these subparts out from one of the three sticks. Distributive reasoning seemed implicit in their actions but they did not explicitly express it. The three parts were parts of a 7-part stick but the children, after naming the three subparts, they made as three-fourths of a seventh of the stick, could not find what fractional part of the whole stick that these three subparts comprised. That is, they did not engage in recursive partitioning operations to find into how many parts the 7-part stick would be partitioned based on partitioning each of the three parts into four parts each. This was a surprise, but the children tried to iterate the 3-stick made from the three subparts enough times to make a stick congruent to the 7-part stick, which indicated that they did regard the subparts of the 3-stick also as subparts of the original 7-part stick.

So, the goal that was needed to engage in recursive partitioning operations had been established and, in Protocol VIII, all it took was an appropriate question by the teacher to provoke these operations in Arthur. Recursive partitioning operations were easily provoked because they were based on his units-coordinating scheme, where the modification in that scheme was to insert partitioning into five parts into each of nine sticks. Distributive partitioning operations were not so easily provoked because the antecedent operations on which they were based were the operations involved in sharing three sticks among four people and distributive reasoning was only implicit in those operations, whereas the units-coordinating scheme was a well-established scheme that was within Arthur's awareness. Arthur did make distributive partitioning operations explicit in the context of sharing five-elevenths of a stick among seven people (cf. Protocol X). He had solved only two tasks (cf. Protocols VIII and IX) prior to the emergence of distributive partitioning operations in Protocol X, so it can be said that he made them explicit in the context of solving what to him constituted problems.

There were three crucial steps in Arthur's construction of the fraction composition scheme in Protocol X. The first was making distributive partitioning operations explicit, and the second was taking the results of distributive partitioning as input for recursive partitioning. The third step was assimilating the results of what was a sharing task that involved three levels of units using his iterative fraction scheme for connected numbers. In Protocol X, he had found the fractional part of an 11-part stick comprised by one-seventh of five-elevenths of the stick, but he had not conceived of the situation as one-seventh of five-elevenths. Rather, he solved a sharing task – share five of 11 parts of a stick equally among seven people. How much of the whole stick does one person get? To reconstitute the sharing task into a fraction composition task entailed a change in goals from finding equal shares of a share of equal shares to finding a fraction of a fraction of a fractional whole. In the latter case, the operations involved in the sharing task would become operations of a reorganized iterative fraction scheme for connected numbers that we call a fraction composition scheme.

The focus was on Arthur's solving activity in the construction of the fraction composition scheme, but Nathan had been involved in the constructive activity as well. In Protocol XI, he worked out the number of small parts in the whole bar in that case of sharing four of nine pizza's equally among seven people and said that one person would get four sixty-thirds. But he was yet to interiorize recursive partitioning operations at three levels of units, which was an internal constraint that he experienced in constructing the fraction composition scheme.

Nathan, however, could engage in recursive partitioning, so in the third year of the teaching episode, we focused on a sequence of several tasks like those we presented to Joe and Melissa [cf. Chapter VIII, *On the Possible Construction of a Scheme of Recursive Partitioning Operations*] to investigate Nathan's construction of a fractional composition scheme. As the investigation progressed, it became clear that not only had Nathan constructed a fractional composition scheme that was on a par with Arthur's scheme, but both children constructed a scheme of recursive partitioning operations as well. The analysis of the construction of this scheme has been reported previously (Olive, 1999). In the activity of successive "thirding" and "fourthing" of a unit bar, the two boys realized that the order of their sequences would not change the end result but different intermediate fractions would be produced. They also constructed a strong correspondence between the operation of taking a unit fraction of something and dividing by the whole number reciprocal of that fraction.

Chapter 10
The Partitioning and Fraction Schemes

Leslie P. Steffe

As stated at the beginning of the first chapter, the basic hypothesis that guided our work is that *children's fraction schemes can emerge as accommodations in their numerical counting* schemes. We explained our way of understanding this hypothesis as follows. The child constructs the new schemes by operating on novel material in situations that are not a part of the situations of the preceding schemes. The child uses operations of the preceding schemes in ways that are novel with respect to the situations of the schemes as well as operations that may not be a part of the operations of the preceding schemes. The new schemes that are produced solve situations that the preceding schemes did not solve, and they also serve purposes the preceding schemes did not serve. But the new schemes do not supersede the preceding schemes because they do not solve all of the situations the preceding schemes solved. They might solve situations similar to those solved by the preceding schemes in the context of the new situations, but the preceding schemes are still needed to solve their situations. Still, the new schemes can be regarded as reorganizations of the preceding schemes because operations of the preceding schemes emerge in a new organization and serve different purposes.

In an articulation of the reorganization hypothesis in Chap. 4, I argued that the operations that produce continuous quantity and the operations that produce discrete quantity are unifying quantitative operations and that, without such unifying operations, the reorganization hypothesis would not be viable. As a result of the case studies, there are various schemes that can be used in evaluating the hypothesis. I begin with the partitioning schemes because these schemes serve in the construction of fraction schemes of all kinds.

The Partitioning Schemes

The Equipartitioning Scheme

In Chap. 4, I identified the equipartitioning scheme as the fourth level of fragmenting. At the beginning of the teaching experiment and at the time of conceptualizing the five levels of fragmenting presented in Chap. 4, equipartitioning was solely a hypothesis based on conceptual analysis. At the beginning of Chap. 5 (cf. Protocol II), Jason's way of marking off one of four parts of a stick and justifying that the marked part was one of four equal parts afforded the opportunity to use the equipartitioning scheme to explain Jason's ways of operating. The equipartitioning scheme was indeed the first confirmation of the reorganization hypothesis because Jason used his concept of four in two novel ways. The first was to partition the stick into four parts by using his concept of four as a template for partitioning. Rather than using the unit items of his concept as unitizing operations, which he would do in the case of assimilating a situation that he construed as a discrete situation using his concept of four, he used his unitizing operations to mark off four units on an unmarked stick. Mentally marking off four more or less equal length units on a stick corroborates the hypothesis that both simultaneity and sequentiality are involved in using a number concept in partitioning because there are at least implicit comparisons of the parts in an evaluation of their equality. An even stronger corroboration of the hypothesis resides in Jason's iteration of the marked part he broke off from the rest of the stick to produce a stick that he used as a term of comparison with the original stick in an evaluation of whether the part was one of four equal parts.

I referred to the result of Jason's partitioning and iterating activity as a connected number, four. In Chap. 4, I developed the scenario that children construct segmented but connected numbers in an analysis of Piaget's et al. (1960) extensive quantity, and their and Hunting's and Sharpley's (1991) subdivision tasks. Although there was no attempt to replicate these tasks with the children in the teaching experiment, my analysis points to the presence of at least records of such subdividing activity and length units in the unit items of Jason's concept of four. That is, my assumption is that Jason's concept of four already contained records of segmented but connected equal length units as well as discrete units, and that he could use his number concept to establish either a connected number or a discrete number in his experiential field.

The Simultaneous Partitioning Scheme

The third level of fragmenting that I developed in Chap. 4 anticipated Laura's simultaneous partitioning scheme because there was an a priori and unspoken necessity that she share the whole of the stick into ten equal parts (cf. Chap. 5, Protocols IX and X). Because Laura had constructed the ENS, I hypothesize that this certainty was made possible by the awareness of the copresence of the ten unit items comprising Laura's concept of ten and of the composite unit to which they belonged. In other words, I believe that Laura could move between her awareness of the copresence of the ten unit items and their unitary wholeness and, consequently, there seemed to be no question concerning whether the whole of the stick be partitioned. But there was little physical experimentation when partitioning a stick into ten or fewer parts, which suggests that she could organize the unit items of these numbers into experiential, linear patterns and use them as partitioning templates.

10 The Partitioning and Fraction Schemes

That Laura made evaluations of the size of the length units she drew or marked off in simultaneous partitioning is suggested by her equisegmenting activity in which she engaged in the case of making three equal sized parts of a stick (cf. Chap. 5, Protocol III) and in making seven equal sized parts of a stick (cf. Chap. 5, Protocol XI). The early observation of her equisegmenting activity turned out to be predictive of her lack of the construction of the disembedding operation in the case of connected numbers and, correspondingly, her lack of construction of the iterable length unit until well into her fifth grade. The lack of both of these operations severely constrained her construction of the partitive fraction scheme, a scheme that I consider as the first genuine fraction scheme. Still, given her uncannily accurate estimates of one of ten equal parts and one of eight equal parts of a stick along with her equisegmenting activity, the assumption that her numerical concepts were already constructed as connected numbers especially in the linear case is warranted.

The Splitting Scheme

Prior to the teaching experiment, I neither anticipated constructing the splitting operation nor was I aware of its implications for children's construction of fraction schemes. Primarily based on Confrey's (1994) definition, currently, splitting is almost universally equated with partitioning without iteration being involved. When working with the children in the teaching experiment, however, I found good reason to construct the splitting operation as involving a composition of partitioning and iterating (e.g. Chap. 7, Protocol IX). In retrospect, the argument that I advanced in Chap. 1 that neither simultaneity[1] nor sequencing is the more fundamental has far reaching consequences in children's construction of fraction schemes. The argument was central in the advancement of the reorganization hypothesis and its consequences are found at least in part in the splitting operation as well as in the equipartitioning scheme and the simultaneous partitioning scheme.

The distinction between the simultaneous partitioning scheme and the equipartitioning scheme resides in the operations that can be performed using the unit items contained in a unit of connected length units. Unlike children who have constructed only the simultaneous partitioning scheme, children who have constructed the equipartitioning scheme can disembed and then iterate a length unit that has been marked off on a segment as an estimate of one of, say, five equal connected length units to produce a sequence of five connected units that can be compared with the original segment to test whether it is one of five equal length units. The distinction between the equipartitioning scheme and the splitting scheme[2] resides in the levels of units that are available to the child prior to as well as in operating. In the former, the operations that produce two levels of units are available

---

[1] I use this term to mean an awareness of the copresence of several unit items.
[2] The splitting scheme contains the splitting operation as an assimilating operation.

as operations of assimilation whereas in the latter case, the operations that produce three levels of units are available as well (cf. Chap. 7, Protocols VI and IX).

In the case study of Patricia (Chap. 7), I found that the assimilating operations that produce three levels of units should be regarded as enabling operations in the construction of the splitting operation. In the case of the equipartitioning scheme, the child uses the operations implied by a number concept to operate on a continuous unit that is present in her visual field. Partitioning and iteration are sequentially enacted as operative activities of the scheme. In the case of the splitting operation, I explained in Chap. 5 (cf. Protocol XII) that for Jason to make a stick such that a given stick was five times longer than the stick to be made, he would need to posit a *hypothetical stick* such that repeating that stick five times would be the same length as the given stick. That is, he would need to not only posit a hypothetical stick, but also to posit the hypothetical stick as one of five equal parts of the given stick that could be iterated five times and see the results of iterating as constituting the given stick. In the splitting operation, the given stick is *mentally* split into five parts prior to sensory-motor action. What this means is that the partitioning and iterating operations of the equi-partitioning scheme that constituted its activity become assimilating operations of the splitting scheme and that the results of the equipartitioning scheme, such as the 4-stick that Jason produced in Protocol II of Chap. 5, become situations of the splitting scheme. So, heuristically, one can think of the operations of a splitting scheme as the entire equipartitioning scheme realized in one fell swoop. That is, establishing a situation of the splitting scheme – a split stick – implies the operations of partitioning the stick into equal sized parts and iterating one of these parts to produce the split stick.

The distinction I make between the splitting scheme and the splitting operation can be clarified by considering the first observation of Patricia's splitting operation (cf. Chap. 7, Protocol IX). In that Protocol, Patricia drew a short stick and said that the stick was ninety-nine times as long as the stick she was thinking of. Because it became clear that she was thinking of a very small stick that could be produced by splitting the given stick into ninety-nine parts, her comments are indicative of the splitting operation. Just prior to these comments, to make her teacher's stick after the teacher said that she was thinking of a stick such that a given stick was forty-eight times as long as her stick, Patricia partitioned the given stick into forty-eight equal parts and pulled one of the parts out of the forty-eight and asked, "Is that the stick?" So, the activity of her splitting scheme consisted of Patricia implementing her splitting operations. The result of her splitting scheme became clear after the teacher asked Patricia, "How did you know that was the stick I was thinking of?" Patricia replied, "Because you said this (pointing to the long stick) was forty-eight times as long as the stick so one of these must be one forty-eighth!" So, the result of her splitting scheme was a unit fraction that she established as a multiplicative relation between the stick she made and the given stick – the given stick was forty-eight times as long as the 1/48-stick she made and the stick she made was one forty-eighth as long as the given stick. I should emphasize, however, that the result of the splitting scheme does not need to be a unit fraction because Joe did not construct the stick he made as one-ninth of the given stick in Protocol VII of Chap. 7. Rather, he said, "This long

The Equipartitioning Scheme for Connected Numbers

Equipartitioning or splitting operations discussed above operate on an unpartitioned continuous unit and partitions it into individual units of equal size. In contrast, equipartitioning operations for connected numbers operate on a *split* continuous unit and partition it into equipartitioned units. I make a distinction between equipartitioning operations for connected numbers and the scheme the operations generate in a way that is quite analogous to the distinction between splitting operations and the scheme these operations generate. In Protocol V of Chap. 8, the teacher posed a question that provoked equipartitioning operations for connected numbers in Joe and Melissa when they were in fifth grade. The question was, given a 5/5-bar, "How many parts would you have to break this up into, the one-fifth, if each little part was one-eightieth of the unit bar?" Both Joe and Melissa interpreted the question to mean that there would be eighty parts in the whole of the 5/5-bar and attempted to find how many of the eighty eightieths would fit into each of the five-fifths. Only after several attempts to distribute the eighty eightieths unequally among the five-fifths did both children realize that each fifth would need to be partitioned into an equal number of little parts. So, in the process of assimilating the situation that was posed by the teacher, with the aid of the teacher, the children constructed equipartitioning operations and the situation that was implied by Melissa's question of *"How many eightieths can you get to fit into one-fifth?"*

In the case study, I inferred that the children's situation was a unit containing five connected units each of which contained an unknown numerosity of units. I also inferred that the concomitant goal the children constructed was to find into how many parts to partition each of the five connected units so that together there would be eighty one-eightieths in the 5/5-stick. Joe's equipartitioning scheme for connected numbers followed on from the equipartitioning operations and the goal he generated. The activity of Joe's scheme involved an implementation of the equipartitioning operations, but it involved more as well. As explained in the discussion of Protocol V, Chap. 8, Joe used multiplicative strategic reasoning, "I know five times twelve, so I did thirteen would be sixty-five, fourteen would be seventy, fifteen would be seventy-five, and sixteen would be eighty!" to find how many eightieths would fit into each one-fifth. There definitely was a sense that the result of operating, which was the number of eightieths in each fifth, when iterated five times, would produce eighty eightieths.

Although Melissa used her multiplication algorithm to find how many eightieths would go into each fifth in Protocol V, at the time of the teaching episode (1st of

December) I would not judge that she established an equipartitioning scheme for connected numbers, although she did engage in equipartitioning operations in the context of her assimilating activity with the aid of the teacher. But these operations were not operations she could independently use as assimilating operations. Still, like Joe, Melissa used her units-coordinating[3] scheme in an attempt to find how many of the eightieths would fit into one-fifth. Both children, then, used their whole number multiplying scheme in producing a result of their equipartitioning operations for connected numbers, which constitutes another corroboration of the reorganization hypothesis.

The Splitting Scheme for Connected Numbers

The equipartitioning scheme for connected numbers is based on using the operations that produce three levels of units as assimilating operations whereas the splitting scheme for connected numbers is based on using the operations that produce the generalized number sequence as assimilating operations. The operations of the generalized number sequence that are novel entail coordinating the basic units of two number sequences, such as the unit of three of a sequence of such units and a unit of four of a sequence of such units, prior to engaging in activity (cf. Chap. 9, *Nathan's Generalized Number Sequence*). This coordination opens new possibilities in coordinating two recursive partitionings that were manifest in Nathan finding a common partition of two bars of the same size, one partitioned into three parts and one into five parts, by iterating a unit of three and the unit of five until producing 15, as the comon partition (cf. Chap. 9, Protocol III). This activity of coordinating the two composite units was carried out for the purpose of producing a fractional unit of which both one-third and one-fifth were multiples. This goal was in turn predicated on Nathan's goal of recursively partitioning each part of the 3/3-bar into a sufficient number of subparts and each part of the 5/5-bar into a sufficient number of subparts so that the whole of both bars would be partitioned into an equal number of these subparts. In that these recursive partitionings were never enacted in Nathan's solution, I interpret them as splits of each part of the 3/3-bar and each part of the 5/5-bar, where the whole of each bar would be split into an equal number of subparts. My interpretation is based on Nathan's comment, "Ah, I found it!" after producing 15 by means of coordinating the two composite units, three and five. Nathan did not know the specific numerosity of the subparts of the fraction bars after he split them until he actually coordinated the two composite units, three and five. So, splitting operations for connected numbers simultaneously split each unit in a composite unit containing the units into an equal but unknown number of subunits. The sense of simultaneity is achieved because the splits of each unit are not actually enacted. Rather, because each unit in the composite unit containing them is an iterable unit, splitting one of the units implies splitting them all. In fact, all that is

---

[3] A units-coordinating scheme is a multiplying scheme for whole numbers (cf. Chap. 5, *Necessary Errors*, for an explanation).

needed is for the child to intend to split one of the units of the sequence. This opens the way for splitting operations for connected numbers to take an unpartitioned continuous unit (one as a connected number) and *split* it into an equipartitioned unit where each part of the split unit is again split into smaller parts.

Similar to the case where the whole of the equipartitioning scheme is contained in the first part of the splitting scheme that I explained above, the whole of the equipartitioning scheme for connected numbers is contained in the first part of the splitting scheme for connected numbers, but in a reorganized form. What this means is that the results of the equipartitioning scheme for connected numbers is reconstituted as a situation of the splitting scheme for connected numbers minus, perhaps, knowing the specific numerosity of the subparts of the units in the sequence of units produced, such as 16 as the number of subparts in the five units of the 5/5-stick that Joe produced in Protocol V of Chap. 8.

Depending on the situations used to infer the equipartitioning scheme for connected numbers or the splitting scheme for connected numbers, it may be difficult to make a behavioral distinction. To exemplify the difficulty in making a behavioral distinction, Nathan (Chap. 9, *Corroboration of the Splitting Operation for Connected Numbers*) broke off two-thirds of a 15-bar leaving a 5-part bar, and then reasoned that one-fifth of the 15-part bar had to be three parts. He resisted simply saying that one-fifth of the bar was one of the remaining five parts, and reasoned that it was a 3-part bar because three five times was fifteen (he actually counted by three to fifteen). So, the inference was made that Nathan could mentally split a 15-part bar when the whole of the bar was not in his visual field into five composite parts with three smaller parts in each of the five parts prior to any observable action and iterate the unit containing the three smaller parts five times to reconstruct the 15-part bar. This task was not presented to Joe while he was in his fifth grade, but given the situation where I inferred that he had constructed an equipartitioning scheme for connected numbers (Chap. 8, Protocol V), it is reasonable that the task where I inferred Nathan's splitting scheme for connected numbers would be also included in the situations of Joe's equipartitioning scheme for connected numbers. Even though it would be reasonable to expect Joe to solve a similar task when he was in his fifth grade that was solved by Nathan while he was in his third grade, Nathan's generativity in constructing the common partitioning that I discussed above as well as other fraction schemes while he was in his third grade and on into the next year indicates that there was a fundamental difference in the partitioning schemes for composite units of the two children.

The Distributive Partitioning Scheme

Based on a study by Lamon (1996) in the section, *Levels of Fragmenting* in Chap. 4, I categorized distributive partitioning as the most advanced partitioning operation. While they were in their fourth grade, neither Joe nor Patricia could share four slices of pizza among 6 friends during the first part of March, even though there was good reason to believe that both children could use the operations that produce three levels of units as assimilating opertions (Chap. 7, *A Lack of Distributive Partitioning*). Moreover, both childen had constructed recursive partitioning operations well before the first part of March. Recursive partitioning does involve distributing the partitioning operation across the parts of another partition as indicated in the case study of Nathan on the 24th of April while he was in third grade (cf. Chap. 9, *Multiplication of Fractions and Nested Fractions*). To make a 20-part bar, he partitioned a bar into five parts, pulled out one of the parts and partitioned this part into four parts, and then pulled out one of these parts and iterated it 20 times. To reason in this way, partitioning one of the five parts into four parts would need to imply partitioning each of the five parts into four parts, an implication that was based on the iterability of his connected number, four. Jason engaged in recursive partitioning as well when finding three-fourths of one-fourth at the beginning of his fifth grade (cf. Chap. 6, Protocol I) in a way that did not involve the iterability of his connected number, four. Jason seemed to distribute partitioning into four parts across each part of a 4/4-stick. Although Nathan's reasoning was more advanced than Jason's in that it was based on the iterability of a composite unit, Nathan was yet to construct distributive fractional reasoning (cf. Chap. 9, Fig. 9.8). He was also yet to share two identical bars equally among three mats by partitioning each bar into three parts and then distributing one part from each bar to each mat (cf. Chap. 9, Fig. 9.1). With his partner Drew, he did partition each of the two bars into six parts for a total of 12 parts, but these operations were based on his whole number knowledge and could not be judged as distributive partitioning. Consequently, there is a distinction between distributing partitioning operations across the parts of a partition in recursive partitioning and partitioning *n* items among *m* shares by partitioning each of the *n* items into *m* parts and distributing one part from each of the *n* items to the *m* shares, which involves distributive reasoning. Distributive partitioning operations proved to be even beyond Joe by the end of his fifth grade (cf. Chap. 8, Protocol XVII). Arthur (and eventually, Nathan) did, however, demonstrate distributive partitioning during his fourth grade, which was his first year in the teaching experiment (cf. Chap. 9, Protocol XI). I elaborate on Arthur's construction of this scheme in the section on *The Fraction Composition Scheme* below.

The Fraction Schemes

The Part-Whole Fraction Scheme[4]

Laura constructed her part-whole fraction scheme using her simultaneous partitioning scheme. The numerical concepts that Laura used in simultaneously partitioning segments were composite units containing specific numbers of connected but equal length units. Initially, when using connected number concepts in simultaneous

---

[4] A child, Jerry, had constructed only initial number sequence when he began the teaching experiment in his fourth grade. After working with him in the teaching experiment during his fourth and fifth grades, he was still to construct a part-whole fraction scheme in his fifth grade (cf. Biddlecomb 2002).

partitioning, Laura focused on the number of the length units and interpreted unit fraction number words such as "one-tenth" as referring to those numbers ("ten"). The accommodations she made in the simultaneous partitioning scheme that produced her part-whole fraction scheme occurred in the context of using partitioned sticks as stand-ins for quantitative items to be equally shared among the elements of a discrete quantity (cf. Chap. 5, Protocols XV and XVI). Laura could operate on the elements of the discrete quantities and disembed a numerical part of them and compare the numerical part back to the numerical whole without the discrete quantities being in Laura's perceptual field. Her operations on the numerical parts of the stick referred to her discrete quantitative operations and she used the parts as if they were elements of the discrete quantity. After this accommodation in her simultaneous partitioning scheme, Laura could disembed a numerical part of a partitioned stick and make a part-to-whole comparison between the part and the whole using appropriate fraction language. In retrospect, appealing to Laura's discrete quantitative concepts such as the number of children in her classroom was made necessary by her lack of her construction of an iterable length unit.[5] Her part-whole scheme was very resistant to our attempts to engender changes in the scheme, but that resistance had little to do with the part-whole fraction scheme per se. Rather, it had to do with the fact that she was yet to construct length units as iterable units.

The purpose or goal of a part-whole fraction scheme is to partition a segment into so many parts and establish a part-to-whole relation.[6] The relation between the part and the whole does not yet refer to the length of the pulled part nor is a unit fractional part an iterable unit. That is, a fraction like seven-tenths, say, refers to seven out of ten equal parts of a segment that has been partitioned into ten parts, but neither does it refer to the length of the seven parts nor does it refer to one-tenth iterated seven times to produce the seven parts. The seven parts can be said to be equivalent parts, but they are yet to be established as identical one to the other as they would be if one-tenth were an iterable fractional unit. The unit fractions like one-tenth are part-whole unit fractions and are yet to be construed as true unit fractions where the part-to-whole relation is also a whole-to-part relation.

The Partitive Fraction Scheme

Children who have constructed the partitive fraction scheme still make part-to-whole comparisons. In fact, part-to-whole comparisons are a critical aspect of the partitive fraction scheme as they are of any fraction scheme. The partitive fraction

---

[5] We restricted fractional parts of the 22 children in the classroom to involve only individual children to avoid reasoning with three levels of units (cf. Chap. 5, *Necessary Errors*).

[6] We used segments and their length in working with Laura, so I restrict the discussion to linear units. On the basis of findings of Hunting and Sharpley (1991) (Chap. 4), engendering the part-whole fraction scheme for other continuous units would be no easier than engendering it for linear units.

scheme does not replace the part-whole fraction scheme. Rather, the operations of the latter scheme are a part of the operations of the partitive fraction scheme because the scheme that is used in the construction of the partitive fraction scheme – the equipartitioning scheme – contains all of the operations of the simultaneous partitioning scheme as well as the disembedding and iterating operations as operations of the scheme.

The partitive fraction scheme is the first fraction scheme that I would refer to as a genuine fraction scheme (cf. Chap. V, *Jason's Partitive Unit Fraction Scheme*). The reason is that the equipartitioning scheme, whose operations serve as assimilating operations of the partitive scheme, satisfy six of the seven criteria of operational subdivision identified by Piaget et al. (1960, pp. 309–311). In Chap. 4, I commented that the sixth criteria of Piaget et al.'s. operational subdivision – the units produced by a partition can be taken as units to be subdivided further – would be a major issue investigated in the case studies. Recursive partitioning is the operation that is necessary to satisfy Piaget et al.'s sixth criteria for operational subdivision (cf. Chap. 5, Protocol I).

The partitive unit fraction scheme. The partitive *unit* fraction scheme is the scheme that emerges naturally from the equipartitioning scheme because the purpose of the latter scheme is to mark off one of several parts of a continuous unit and to verify that the part marked off is one of several equal parts of the continuous unit. When the part-to-whole meaning of "one-fifth," say, is established as how much one out of five equal parts of the continuous unit is of the continuous unit, this fraction language can be used to stand in for marking off one of five parts of the continuous unit and verifying that the part is indeed one of five equal parts by iterating the part five times. It is by this means that the equipartitioning scheme is reconstituted as a partitive unit fraction scheme.

Similar to the part-whole fraction scheme, how much a part is of a whole is based on a comparison between the part and the whole. So, the purpose of the partitive unit fraction scheme is to mark off a unit fractional part of the fractional whole, disembed the part, and iterate it to produce another partitioned continuous unit to compare with the fractional whole in a test of whether the part marked off is a unit fractional part. When the continuous unit is a segment, Jason taught us that when he iterated one-tenth ten times not only was ten-tenths ten little pieces and the whole stick, it was also the length of the whole stick [cf. Chap. 5, Protocol XII (Cont)]. So, fractional number words or numerals refer to length and length comparisons as well as to numerical comparisons.

Children who have constructed the partitive unit fraction scheme can also test which of a collection of segments is, say, one-fifth of a given segment by iterating a segment of the collection five times and comparing the resulting 5-part segment with the given segment until one is found that works. However, children who have constructed even the more general partitive scheme likely would not be able, say, given a segment that is said to be three-fifths of some other segment that is not in the visual field of the children, construct the other segment by partitioning the given segment into three parts, and iterating one of the parts five times or by copying two of the parts and affixing them to the given segment. To engage in these operations entails having already constructed three-fifths as a fractional number, which is beyond the scope of the partitive fraction scheme because it requires splitting.

The partitive fraction scheme can be constructed as a functional accommodation of the partitive unit fraction scheme by asking children to, say, given a sharable continuous unit partitioned into 24 parts, make the share for three people (cf. Chap. 5, Protocol XVII). Jason pulled out one part and iterated it three times to make the share for three people and called it "three-twenty-fourths" because it was three parts out of 24. The context of the problem was to share a birthday cake equally among 24 people, so "three-twenty-fourths" referred the amount of the birthday cake three people would get as well as to the length of the 3-part stick when compared with the 24-part stick. We assumed that the children could generate an image of the birthday cake and an image of the cake being cut into the 24 parts. So, drawing a stick in the TIMA: Sticks and using PARTS to partition it into 24 parts would constitute an enactment of the partitioning operations on an image of the cake and provide meaning for the children's actions when using the computer tool.

An additive scheme. It would seem as if the unit fraction, one-twenty-fourth, would be an iterable unit fraction and, consequently, as if the partitive fraction scheme was a multiplicative scheme. It is necessary to appeal to the history of the construction of the equipartitioning scheme and the partitive fraction scheme to understand why partitive unit fractions are not iterable units, even though they can be used in iterating. When placed in the context of the iterative fraction scheme, which is a multiplicative scheme, it is also possible to understand why partitive unit fractions are yet to be constructed as iterable units. When children have constructed the explicitly nested number sequence, their units of one are iterable units and number words like "seven" can refer to a singleton, an arithmetical unit, that can be iterated seven times to produce seven unit items. This is essentially why, when such children are asked to mark off one of seven equal parts of a continuous unit, they can focus on a part of the continuous unit as an instantiation of their iterable unit of one as a representative part and gauge where to place the mark in relation to the whole of the continuous unit. They can also transfer the marked unit along the continuous unit in a mental test of whether it is one of seven equal parts, which is, in effect, segmenting.

But they can do more because they can break off the marked unit or, as the case may be, make a replicate of the marked unit, and iterate it seven times along the stick in a test to find if the marked unit is one of seven equal units. This iterability of the marked part is also inherited from the iterable length unit. But these parts are parts of the partitioned continuous unit and their fractional meaning is constructed as a comparison of the parts to the whole of the partitioned continuous unit. The parts are still *parts of the partitioned continuous unit* and the partitioned continuous unit might be said to be the universe for the comparisons of the parts to the whole. The iterable unit of one can be used to give meaning to a number such as 45 in that it can be iterated 45 times to produce the extension of the number. But one-seventh cannot yet be iterated 45 times to produce an extension of the fraction forty-five sevenths. Fractions that extend beyond the universe, seven-sevenths, are yet to be constructed. Because one-seventh is produced by using the concept, seven, as a partitioning template, construction of a fraction connected number sequence that is parallel to the number sequence for units of one (the ENS) entails constructing the iterative fraction scheme.

Given that the partitive fraction scheme is an additive scheme, one might think that it is not an important scheme. But the assimilating operations of the partitioning scheme on which it is based, the equipartitioning scheme, produces two rather than three levels of units. Only children who construct the operations that produce three levels of units as assimilating operations can remove this internal constraint. When Jason entered his fifth grade, for example, the operations that produce three levels of units were available to him as assimilating operations. We had no indications of these operations throughout his fourth grade, so whatever might have occurred over the summer months contributed to this development.

While Jason was in his fourth grade, the partitive fraction scheme was the maximal scheme he constructed during that time, and it was resistant to our attempts to provoke major changes in it. Still, relative to his partitive fraction scheme, Jason engaged in self-initiated and independent mathematical activity in the context of interacting with his teacher and with Laura. He was always deeply engaged in solving activity regarding the tasks and situations in which he found himself and seemed to experience great satisfaction in producing solutions. Furthermore, he used his partitive fraction scheme (as did Joe and Melissa) in the construction of higher order fraction schemes once the operations that produce three levels of units became available to him; the unit fraction composition scheme and the unit commensurate fraction scheme. So, like the part-whole fraction scheme, the partitive fraction scheme is an indispensible fraction scheme that is used in the construction of higher-order fraction schemes as well as having its own situations and results.

Anticipatory iteration. When interacting with Joe in the teaching episodes, we stressed anticipatory iteration in his construction of the partitive fraction scheme (cf. Chap. 7, *Joe's Construction of a Partitive Fraction Scheme*).[7] The first task that we presented to Joe in the 7 December teaching episode, in which we asked him to find which of four given sticks was one-fifth of another given stick (cf. Chap. 7, Fig. 7.5), was not as demanding as the task we presented to Jason and Laura on the same day in the teaching experiment where I inferred Laura's simultaneous partitioning scheme (cf. Chap. 5, *Laura's Simultaneous Partitioning Scheme*). Joe made rapid progress, however, and in the next teaching episode held on the 10th of January, he produced six-fifths of a stick that was already marked into five equal parts. In the same teaching episode, he successfully estimated one-seventh of a stick and used this estimate to mark off all seven-sevenths on the original stick. He then iterated the 1/7-stick nine times at the request of the teacher and said that the resulting stick was nine-sevenths and explained why it would be nine-sevenths because each of the nine parts would be one-seventh (Chap. 7, Protocol VI). So, one might think that Joe's partitive scheme was a fleeting scheme and not worth mentioning. But that is not the case at all because four months later in the 5 April teaching

---

[7] See Tzur (2007) for a discussion of anticipatory iteration.

episode, Joe still believed that a fraction could not be greater than the fraction whole when he asked, "*How can a fraction be bigger than itself?*" (cf. Chap. 7, Protocol XIII). Joe had constructed the operations that produce an iterative fraction scheme on the 7th of December, but he was yet to construct such a scheme.

On the basis of Joe's progress, one might conclude that an iterative approach to teaching of fractions is superior to a partitioning approach and essentially replaces it. When taking into consideration that iterating was an anticipatory operation for Joe, however, it becomes necessary to explain the anticipatory aspect of the operation. In Fig. 7.5 of Chap. 7, Joe's choice of which stick was one-fifth of the long blue stick entailed a comparison of the chosen stick against the long blue stick and a partition of the blue stick engendered by the comparison. So, the anticipation involved in iterating one of the sticks five times in a test to find which one of them was one-fifth of the long blue stick was predicated on an image of either a 5-part stick or an image of a stick that could be partitioned into five parts.

Although we stressed iteration with Joe, I would not call our approach with Joe an iterative approach that excluded partitioning because partitioning was involved both as an assimilatory operation in constructing his fraction situations as well as in constructing the results of iterating. In fact, in Protocol V of Chap. 7, Joe commented that a 5-part stick was five-elevenths because "it's five and it's part of eleven." Another reason that I do not consider our approach to teaching Joe as excluding partitioning is that the operations that produce three levels of units were available to Joe in his constructive activity. These operations were manifest in Joe's construction of the splitting operation and the operations that produce an iterative fraction scheme soon after he constructed his partitive scheme.

The reversible partitive fraction scheme. Although Jason's (and Laura's) units-coordinating scheme was a reversible scheme while he was in his fourth grade, his partitive fraction scheme was not.[8] At the beginning of his fifth grade, however, he found how much three-fourths of one-fourth of a stick was in such a way that imputing reversibility to his partitive fraction scheme was compelling (cf. Chap. 5, Protocol I). To recap, his partner Laura had partitioned a stick into four parts, pulled one part out of the three parts, partitioned this part into four smaller parts and pulled three of these smaller parts out of the four parts. The teacher asked the children how much these three parts were of the whole. After thinking, Jason answered "three-sixteenths" and explained why that was the answer. The three smaller parts constituted the results of a partition of the original stick that had been only partially enacted. Based on his answer of three-sixteenths, I inferred that Jason regarded the three smaller parts as parts of a complete partitioning of the original stick. To regard the three smaller parts in that way, it would be necessary for him to take them as *input* for operating further and at least mentally project

---

[8]That his partitive fractional scheme was not reversible throughout his fourth grade is not documented in Chap. 5. But we worked extensively on this problem with him in teaching episodes toward the end of his fourth grade that are not analyzed in the case study.

them into the original stick that Laura started with. So, Jason took what constituted results of his partitive fraction scheme as a situation of the scheme and reversed the relation between the situation and the results of the original scheme. What before was a result was now situation and what before was a situation was now a possible result. This reversal of the relation between situation and result is sufficient to infer reversibility in the scheme, and the inference is solidly corroborated by his construction of the operations necessary to mentally complete the partition.

Both Joe and Patricia's partitive fraction schemes were reversible schemes as well (cf. Chap. 6, Protocols X, XI, XII, XVI). Consequently, I regard splitting as the operation that generates reversibility in the partitive fraction scheme. This is illustrated by Nathan's producing a whole bar given an unmarked bar that was said to be two-sevenths of the bar (cf. Chap. 9, *Nathan's generalized fraction language*). Nathan split the bar into two pars, pulled one out, and iterated the part seven times to produce the bar. To operate in this way, two-sevenths would need to be constituted as a fractional number so that it stood in a multiplicative relation to the unit fraction it contained, one-seventh. The difference in Nathan's task and Jason's task is that Jason was not told that the three parts were three-sixteenths of the whole bar. Still, Jason knew that one of the parts was a constitutive part of the bar and, if iterated a sufficient number of times, would complete a partition of the bar, which is simply the iterative aspect of the splitting operation. In contrast, Nathan used the splitting operation twice, once to split the bar into two equal parts to produce the hypothetical bar, one-seventh, with which two-sevenths stood in multiplicative relation, and the second time to iterate one of these parts to construct a hypothetical bar, which was seven times as long as the 1/7-bar.

Nathan had already constructed the splitting operation for connected numbers, so when he was presented a 6-part bar as three-eighths of a whole bar right after he was presented with the bar that was said to be two-sevenths of another bar, he operated analogously and split the 6-part bar into three units with two parts in each unit to produce one-eighth, disembedded one of the 2-part units and took it as one-eighth of a hypothetical fractional whole, and iterated it five times to complete eight-eighths. Nathan, of course, had constructed a more advanced form of splitting than the three other children, but this solution does solidly corroborate that claim that the splitting operation is that operation that generates reversibility in the partitive fraction scheme.

The Unit Fraction Composition Scheme

One of the surprises of the teaching experiment that we did not foresee is that finding one-fourth of three-fourths involves operations that are not a part of the scheme that the children constructed to find three-fourths of one-fourth. So, to distinguish these two schemes, we use "unit fraction composition scheme" to refer to the latter scheme and simply "fraction composition scheme" to refer to the former scheme. I have already discussed the reversible partitive fraction scheme that was involved in

Jason's construction of the unit fraction composition scheme, and noted that the splitting operation is the operation that generates reversibility in that scheme. But alone, the splitting operation is not sufficient to construct the unit fraction composition scheme. Recursive partitioning is also involved as the basic operation.

For a composition of two partitions to be judged as recursive, there must be good reason to believe that the child implements the two partitions in the service of a nonpartitioning goal. The importance of this judgment is that the child must *intentionally choose* to partition each of the parts of an original partition using the second partition. In Jason's case (cf. Chap. 5, *Jason's Fraction Compositional Scheme*), what this means is that he must intentionally choose to partition each of the remaining fourths of the original stick into fourths. This amounts to embedding recursive partitioning operations in the reversible partitive fraction scheme and the embedding produces the unit fraction composition scheme. Joe and Patricia both constructed the unit fraction composition scheme while they were in their fourth grade (cf. Chap. 7, Protocols X and XVI) and Nathan constructed it while he was in his third grade (cf. Chap. 9, *Multiplication of Fractions and Nested Fractions*) also by means of embedding their recursive partitioning operations in their reversible partitive factional scheme.

At the beginning of his fifth grade, Joe demonstrated that his unit fraction composition scheme was a reversible scheme (cf. Chap. 8, Protocols I and II). In the discussion of the unit fraction composition scheme immediately above, I did not emphasize that recursive partitioning is operationally equivalent to a reversible units-coordinating scheme for connected numbers at three levels of units. In the case study of Nathan on the 24th of April while he was in third grade (cf. Chap. 9, *Multiplication of Fractions and Nested Fractions*), he was asked to make a 20-part bar without using the number 20. He partitioned a bar into five parts, pulled out one of the five parts and partitioned it into four parts, and then pulled out one of these parts and iterated it 20 times. In this case, he was given a potential result of his units-coordinating scheme, the 20-part bar, and was asked to make the bar without using 20 to partition the bar. Upon choosing five as the first partition, he constructed the situation as a situation of his reversible units-coordinating scheme[9] that included the goal of finding into how many parts would each of the five parts need to be partitioned to produce 20 parts, which is to say that he recursively partitioned a prior partition.

Had Nathan not known that five times four is 20, he would have needed to enage in reasoning of the same kind as Joe engaged in when finding how many eightieths must be put into each of five parts so there would be an equal number in each part (cf. Chap. 8, Protocol V). Joe's situation was indeed a situation of his reversible units-coordinating scheme because he had been asked to find into how many parts he would need to partition one of five parts of a 5/5-stick so that the part was one-eightieth of the whole stick. On the basis of the question, he inferred that the whole

---

[9] In conventional langugage, he established one of the factors of a product, 20, and his goal was to find the other factor, which is conventionally regarded as a missing factor problem, or a dvision problem.

stick would be partitioned into 80 parts using his reversible partititive fraction scheme. So, he knew the result, 80, of his reversible units-coordinating scheme as well as the number of parts, five, and the goal was to find how many of the eightieths would go into each of the five parts. So, the reversibility of the unit fraction composition scheme is generated by the reversibility of the partitive fraction scheme, and the reversibility of the units-coordinating scheme for connected numbers. The reversible units-coordinating scheme becomes embedded in the reversible unit fraction composition scheme, so the latter scheme is truly a reorganization of the former scheme.

The Fraction Composition Scheme

In the above section, *The Distributive Partitioning Scheme*, I commented that while Joe and Patricia were in their fourth grade, they could not engage in distributive fractional reasoning, even though there was good reason to believe that they could use the operations that produce three levels of units as assimilating operations. Further, while Nathan was in the last part of his third grade, even though he had constructed the generalized number sequence, he was yet to construct distributive fractional reasoning. Distributive fractional reasoning proved to be even beyond Joe by the end of his fifth grade (cf. Chap. 8, Protocol XVII).

Distributive fractional reasoning emerges when distributive sharing is made explicit. A task that Nathan and Arthur solved in a teaching episode held on the 8th of April while they were in their fourth grade illustrates distributive sharing (cf. Chap. 9, *Construction of a Fraction Composition Scheme*). To solve the following task:

> A pizza is cut into seven slices. Three friends each get one slice. A fourth friend joins them and they want to share the three slices equally among the four of them. How much of one whole pizza does each friend get?

The children partitioned a stick in TIMA: Sticks into seven parts, pulled three of the parts out of the seven, partitioned each of the three parts into four smaller parts and pulled three of these smaller parts out for the share of one friend. This solution exemplifies distributive sharing because they distributed partitioning into four parts across each of the three parts and used three of these 12 parts as the share of one friend (cf. Fig. 9.10). Coordination of the units four and three was also involved in the distributive sharing because the connected number four was inserted into each unit of the connected number three by means of partitioning. But units-coordinating activity occurred only after the children formed the goal to share the three parts among four friends. The operation of units-coordination did not engender the sharing goal; rather, it was the other way around. The sharing goal evoked the children's reversible units-coordinating scheme.

To appreciate that the children engaged in reversible units-coordinating operations, consider Protocol V of Chap. 8 where both Joe and Melissa generated the goal of finding how many of eighty eightieths go into each of five parts of a 5/5-stick if an equal number is placed into five parts.[10] This is a goal of a reversible units-coordinating scheme for connected numbers. Analogously, embedded in the pizza task, Arthur and Nathan generated the goal of finding how much of three slices of a pizza does each of four people get if each person gets an equal amount of pizza. Of course, three is not a multiple of four, so the units-coordinating operations that Joe used to find 16 as the number of eightieths that go into each of five parts were relevant for Arthur and Nathan in the pizza task only *after* they formed a goal that *initiated* cutting each of the three slices of pizza into four parts to produce 12 smaller parts. So, distributive sharing replaces making estimates for how many eightieths might fit into each of five parts and using the units-coordinating scheme to test the estimates. When distributive sharing is reconstituted so that distributive reasoning is made explicit, distributive reasoning becomes embedded in the inverse units-coordinating scheme and the distributive partitioning scheme emerges as a reorganization of the inverse units-coordinating scheme.

A task that might clarify what I mean by the distributive partitioning scheme is where a child forms a goal of sharing three different-sized pizzas equally among five people. If the child partitions each pizza into five parts, distributes one part from each of the three pizzas to each of the five people, and if the child understands that the share of one person can be replicated five times to produce the whole of the three pizzas, this would constitute an enactment of the distributive partitioning scheme. If the three pizzas were considered as identical, then the child would also know that three-fifths of one pizza is identical to one-fifth of all of the pizza.

Had distributive partitioning been explicit for Arthur and Nathan in the pizza task, then the children would have known prior to engaging in activity that the share of one friend could be made by partitioning each of the three pizzas into four parts each and taking one part for each friend. Further, they would have known that one-fourth of the whole of the three parts together could be found by finding one-fourth of each part and joining the three parts together, which is distributive reasoning. In that case, when Arthur said that the three parts were three-fourths of one-seventh, he would have also known that it is equal to one-fourth of three sevenths. On the basis of Arthur's language and actions in Protocol X, we did infer that Arthur made distributive sharing explicit and corroborated that inference in Protocol XI.

In the pizza problem, it was surprising at the time that neither Arthur nor Nathan completed the split of the 7-part stick into 28 parts because finding three-fourths of one-seventh was a situation of their unit fraction composition scheme. So, in the teaching experiment, our efforts were devoted to organizing tasks and questions in such a way that recursive partitioning would be evoked and the results of the distributive partitioning scheme would be used as material of the operation. That we were successful in at least Arthur's case (cf. Chap. 9, Protocols X and XI) indicates that Arthur was in the process of transforming his splitting scheme for connected numbers that he used in assimilating the situations of Protocols X and XI as well as the Pizza situation into a fraction composition scheme.

---

[10] This situation can be considered as what is known as a partitive division situation.

There is an important distinction between the splitting scheme for connected numbers and the equipartitioning scheme for connected numbers that I have not made explicit. The reason that children like Nathan can coordinate the composite units of two sequences of composite units is that they are aware of each of the two sequences they are coordinating *as units*. What this means is that they are aware of the units that contain the sequences. Metaphorically, they are "outside" of the units that contain the sequences and can hold the units at a distance and contemplate operating using their elements. In the case of the equipartitioning scheme, the sequence is contained in a unit, but the child is not aware of the unit containing the sequence. They can operate inside of the unit containing the sequence, as it were, and operate with and on the elements of the sequence. So they are yet unable to coordinate the composite units of two sequences of composite units.

So, when a child like Nathan or Arthur mentally splits an unmarked segment into seven parts where each of these seven parts are further split into four parts, the child takes the segment *as if* it were already split in this way without actually implementing the splitting operations in the same way that a child who has constructed the equipartitioning scheme would do. But further, the child can take any one of the seven parts as if it were disembedded from the seven parts, take this part as if it were already split into four parts, and then take this split part as if it were already iterated seven times to produce the whole of the seven units; all of which are only *results* of the equipartitioning scheme for composite units.

In constructing the fraction composition scheme, the accommodation that was needed was explained as embedding the distributive partitioning scheme and the recursive partitioning scheme in the scheme. What we did not mention is the crucial ability of Arthur to take a unit of units of units as input for further operating in making the accommodations. In Protocol X, after Arthur partitioned the stick into 11 parts, pulled out five, partitioned one of these five parts into seven mini-parts, and pulled one of these mini-parts out of the 7-part stick, in response to a question, "What were you going to do?" after he sat thinking, he said,

> I don't know...everybody would get... If you divide each piece into seven pieces then everybody would get a piece from each piece, so five pieces. "If there are seven in one-eleventh then there would be seventy-seven in the whole and that would be five seventy-sevenths."

Reflection on the results of operating is implicit in his comments, and those results were all produced by means of reasoning. That is, "If you divide each piece into seven pieces everybody would get a piece from each piece, so five pieces" referred to the results of hypothetically using his distributive partitioning scheme to share the five pieces equally among seven people.[11] What this means is that the whole scheme had now been constituted as an operative concept or, if you will, as a program of operations that contained its possible situations and results. Further, the comment, "If there are seven in one-eleventh then there would be 77 in the whole and that would be five seventy-sevenths" referred to the results of hypothetically using

---

[11] I consider reasoning as the hypothetical enactment of one or more schemes.

his recursive partitioning scheme to complete the partition of the partition of the stick that he originally partitioned into 11 parts. That he so quickly interiorized the operations of his distributive partitioning scheme and his recursive partitioning scheme and used them in hypothetical reasoning corroborates the assumption that he was aware of the unit that contained his units of units because it enabled him to monitor using his schemes, which is the operation that interiorizes a scheme. Further, Arthur coordinated the unit containing five parts each of which consisted of seven hypothetical miniparts and the unit of 11 units that originally contained these five parts by using one of parts that belonged to both; "If there are seven in one-eleventh then..."

The upshot of this finding itself is quite surprising. That is, children do not construct the fraction composition scheme as a functional accommodation of the unit fraction composition scheme.[12] Rather, the fraction composition scheme is primarily constructed as an accommodation of children's splitting scheme for connected numbers, where the accommodation involves embedding the distributive partitioning scheme and the recursive partitioning scheme into the splitting scheme. But the accommodation involves more because after Arthur shared five-elevenths of a pizza among seven people (cf. Chap. 9, Protocol X) upon being asked he did not know what one-seventh of five-elevenths was. Another step was needed because he had to construe the results of his partitioning activity as a fraction situation. To accomplish this, he assimilated the results of his sharing activity using his iterative fraction scheme for connected numbers in the construction of the fraction composition scheme. The accommodation in the iterative fraction scheme for connected numbers that this assimilation produced can perhaps be best understood in terms that Arthur could now use his iterative fraction scheme for connected numbers to find, say, one-seventh of five-elevenths rather than simply find a proper (or improper) fraction of a fractional whole. He could now find fractions of fractions by means of reasoning without appealing to any paper-and-pencil computational procedure such as "multiply numerators and denominators" or variants thereof. Nathan, too, was on the verge of constructing a fraction composition scheme.

The Iterative Fraction Scheme

The splitting operations that Patricia produced in the 3rd of March teaching episode that I discussed above served in her construction of the iterative fraction scheme [cf. Chap. 7, Protocol XII (First Cont)]. We interpreted her explanation for why an 8/7-stick was indeed eight-sevenths ["The eight parts were the same as the little pieces in the 7-stick, so its eight-sevenths."] as indicating that one-seventh was freed from the fractional whole, seven-sevenths, of which it was a part. When the fractional

---

[12] The fraction composition scheme is a stage above the unit fraction composition scheme that Jason, Joe, and Patricia constructed.

whole is split into seven parts, this serves to free the parts from the fractional whole because the parts have been already conceived of as hypothetical parts apart from the fractional whole as well as contained in the fractional whole such that the fractional whole is seven times as long as any one of the hypothetical parts. So, the eight units of the eight-sevenths retain their fractional meaning relative to the fractional whole and eight-sevenths is conceived of as eight of one-seventh, or as one-seventh eight times.

The operations that produce three levels of units are involved in producing the composite unit containing the fractional whole. That is, eight-sevenths is produced as a composite unit containing two other units, one containing seventh-sevenths and the other containing the additional unit fraction. Of course, the child can integrate these two units together into one composite unit containing eight unit fractions each of which is one-seventh of the fractional whole. So, when Patricia said, "The eight parts were the same as the little pieces in the 7-stick, so its eight-sevenths," we interpreted this as meaning that the sevenths were freed from their fractional whole and integrated into a unit containing them. The status of her fractional numbers is illustrated in Protocol XIII of Chap. 7 when she constructed a stick that was ten times as long as a 14/99-stick. She used the calculator and explained why the stick would be a 140/99-stick by saying, "Because you always have the same little stick you started off with!"

There were other corroborations of the iterative fraction scheme in Patricia's case as well. Perhaps the most compelling corroboration in the case of both Patricia and Joe occurred in the 28 April teaching episode where both children produced a fraction connected number sequence for twelfths. When the children produced the sequence, <1/12, 2/12, ..., 12/12, 13/12, 14/12, 15/12, 16/12, 17/12...>, Patricia said "Infinity-twelfths!" after Joe said "seventeen-twelfths," indicating that she abstracted her number sequence for one when producing the sequence of fractions. Joe, too, used his number sequence to propel him forward in saying the fraction number words in sequence. So, a fraction connected number sequence is a constructive generalization of the number sequence for one, and it has major implications in children's mathematical education that is on a par with children's number sequences.

Melissa's iterative fraction scheme. Melissa produced an iterative fraction scheme, although she was yet to use the operations that produce three levels of units as assimilating operations (cf. Melissa in Chap. 8, Protocol VI). Her construction of the iterative fraction scheme seemed to be an anomaly especially because she was yet to construct the commensurate fraction scheme and the unit fraction composition scheme. It was indeed a surprise, but in the retrospective analysis of the video records of the teaching episodes, it became apparent that she produced three levels of units in the context of operating with connected numbers and, as a result, she also produced the splitting operation. Even though both operations were specific to the situations in which she constructed them, Melissa seemed to be in transition from using the operations that produce two levels of units as assimilating operations to those that produce three levels of units.

The Iterative Fraction Scheme: Connected Numbers [IFS: CN]. The iterative fraction scheme that Nathan constructed transcended the iterative fraction scheme constructed by the other children in the teaching experiment (cf. Chap. 9, *Construction of the Splitting Operation for Connected Numbers*). The assimilating operations of the iterative fraction scheme constructed by Joe et al. consisted of (1) the splitting operation where the unit to be split was an unpartitioned continuous unit and the split was into individual units, (2) the disembedding and iterating operations, and (3) the operations that produce three levels of units used in the context of the other operations. What was novel in the scheme that Nathan constructed was that the first splitting operation had been transformed into a splitting operation for connected numbers. Because the IFS: CN was a reorganization of Nathan's partitive fraction scheme, we referred to the former scheme rather than to the latter scheme in Nathan's and Authur's case study as their general fraction scheme.

The Unit Commensurate Fraction Scheme

Similar to his construction of a unit fraction composition scheme (Chap. 6, Protocol V), Jason transformed his partitive fraction scheme into a unit commensurate fraction scheme while he was in his fifth grade. As a result of explaining why three-fifteenths is one-fifth and four thirty-sixths is one-ninth in previous protocols, he became able to reason that five parts were one-third of a 15-part stick because "there are three groups, and one is fit into the other three!" [cf. Chap. 6, Protocol VII (Cont)]. His language indicates that he had established three units with five parts in each unit within the 15-part stick using equipartitioning operations. His language also indicates that he then "fit" the disembedded unit back into the three units within the 15-part stick, which means that he iterated it along the three units while inserting it into each one using units-coordinating operations. The activity of the unit commensurate fraction scheme, then, transforms a 5-part stick out of a 15-part stick into a 1-part stick out of a 3-part stick.

There is an additional operation in the activity that involves unitizing the 5-part stick into a singleton unit to be used in the units-coordination and unitizing the three 5-part sticks in the 15-part stick into three singleton units. I model that operation as collapsing the five subunits into a single unit in the process of unitizing, an operation that is made possible by his iterable units of one. His comment "there are three groups" referred to the three singleton units within the 15-part stick and the "one" in his comment "one is fit into the other three" referred to the singleton unit he made using the 5-part stick he pulled out and the phrase "is fit into the other three" referred to inserting this singleton unit into each of the three others.

I use the phrase "commensurate fractions" then, to refer to a situation of the unit commensurate fraction scheme (to three-fifteenths) and to a corresponding result produced by the scheme (to one-fifth). Jason had to establish one-fifth as commensurate to three-fifteenths as a result of the activity of his unit commensurate fraction scheme, and the result was not available to him prior to actually operating. He could also start with a unit fraction, like one-third, and by means of using his equipartitioning operations, establish a commensurate such as five-fifteenths (cf. Chap. 6, Protocols X and the continuations). But this was a creative act that was not available to him prior to operating and producing five-fifteenths was a result of operating. He was also able to produce a sequence of fractions each commensurate to one-third, but he did not abstract producing a plurality of such fractions like he did for the fraction one-half (cf. Chap. 6, Protocol IX). In the case of one-half, he apparently abstracted the operations that he used to make fractions commensurate to one-half and realized that he could use them to make any fraction that he wanted to make as long as making the fraction entailed using some specific number. Had he been able to operate analogously in the case of other unit fractions, I would have imputed an equal fraction scheme to him.

The Equal Fraction Scheme

While Nathan and Arthur were in their fourth grade, they established an equal fraction scheme (cf. Chap. 9, *Generating Equal Fractions*). The primary difference in an equal fraction scheme and a unit commensurate fraction scheme is that the assimilating operations of the former scheme are splitting operations for connected numbers. The presence of the splitting operation opened new possibilities for Arthur and Nathan. In Protocol XIV of Chap. 9, given an unmarked 3/8-stick, the task was to partition the stick into as many parts as desired under the constraint that a 1/8-stick could still be pulled out using the particular partition. After Nathan partitioned the unmarked 3/8-stick into 15 parts, Arthur found how many small parts there would be in the original stick by reasoning, "How many is that? Fifteen? And plus fifteen is thirty, and then, if you had to take two out of that [pointing to the 15-part 3/8-stick] it would be ten, and that would make forty." In this way, Arthur produced fifteen-fortieths as equal to three-eights. Other than the 3/8-stick, there was no material in Arthurs' perceptual field he could use to aid in his reasoning, which alone is an indication of the splitting operation for connected numbers. To reason as he did entails splitting the 15-part 3/8-stick into three parts after he produced six-eights as containing 30 parts so he could find how many more parts needed to be added to 30 parts to produce eight-eights. Equal fractions differ from commensurate fractions in that the former are identical parts of a fractional whole whereas the latter are related by the activity of the unit commensurate fraction scheme, where one is the situation of the scheme and the other the result of the scheme.

Both Arthur and Nathan produced a plurality of fractions equal to a basic fraction (a fraction in lowest terms) in thought by operating with the basic fraction one-third with only the most minimal material in their sensory field (cf. Chap. 9, *Generating Equal Fractions*). This was in contrast to Jason who could produce such a plurality for only one-half. For example, Nathan commented that, "All we've got to do is go up…This would be fifteenths (pointing to the one-third partitioned into five parts), the next one is eighteenths, then twenty-firsts, twenty-fourths." The teacher asked Arthur what the next two would be. Arthur responded with twenty-sevenths and thirtieths. He was asked: "How many thirtieths would make one third?" He responded "Ten…because ten times three is thirty." Given any number of their number sequences, they could, in principle, use it to generate a fraction equal to one-third if the number was a multiple of three.

But when asked to make a basic fraction for twenty-three sixty-ninths just after they had generated it by symbolic means, both children regarded it as a new problem that was not related to how they just generated it (cf. Chap. 9, *Working on a Symbolic Level*). There was also no indication that they could judge whether any two fractions belonged to the same sequence. For example, given thirteen-thirty-ninths and eighteen-fifty-firsts, there was no indication that they could judge the equality of the two fractions. For these reasons, we could not infer that the children's relation of the equality of two fractions was also a relation of equivalence of the two fractions.

School Mathematics vs. "School Mathematics"

That we could not infer that Arthur and Nathan's relation of the equality of two fractions was an equivalence relation calls into serious question the practice of considering the concept of equivalent fractions as a central concept in the teaching of fractions in the elementary school. Both students did develop a concept of equal fractions, but that concept was yet to evolve into an equivalence relation. Both children could generate any fraction starting with, say, one-third by correlating their number sequence with the operations they used to generate an equal fraction. That is, given any particular number of their number sequence, they could produce a fraction equal to one-third. This provided them with a sense of the extension of what later might become an equivalence class of fractions, but currently it is better to think of it as an unbounded plurality of fractions. Equivalence classes of fractions should not be taken as a given in the teaching of fractions. Rather, it is a *construction* that is quite nontrivial even for children as advanced as Arthur and Nathan.

The example of equal fractions is but one of a host of examples that I could choose to discuss the difference between school mathematics and "school mathematics." Fraction multiplication is another example that is ready-at-hand. In this case, the children who constructed the unit fraction composition scheme could *reason* to find, say, two-thirds of one-seventh, but they had no idea that they could simply multiply the numerators and the denominators to produce two-twenty-firsts as the answer. But they could use recursive partitioning to *reason* that one-third of one-seventh is one-twenty-first, so two-thirds of one-seventh is two-twenty-firsts. The issue this raises is not the hoary dichotomy between understanding and computational algorithms and basic facts. Rather, the issue goes much deeper, and it concerns how "school mathematics" should be understood.

Historically, school mathematics has been conceived of in terms of first-order models only (cf. Chap. 1). Second-order models in mathematics education consist of the models that the observer may construct of the observed person's knowing. In the main, little or no consideration has been given or is being given to second-order models like the fraction composition scheme or to how these second-order models might be used in mathematics teaching. If a teacher has constructed the fraction composition scheme as a second-order model, this model can then be regarded as constituting *first-order mathematical knowledge of the teacher* that she can use in interacting harmoniously with her students and in thought experiments in formulating *mathematics for the students* she is teaching.

For example, after Arthur and Nathan's fraction composition schemes had become practiced, the teacher decided that it was time for the children to construct the computational algorithm for finding the product of two fractions. I choose this example to illustrate that the computational algorithm should be constructed as a curtailment of the activity of the fraction composition scheme. The following scenario occurred during the 22nd of February of Arthur and Nathan's fifth Grade, so it is not reported in their case study in Chap. 9. To begin, the teacher asked the two children to show her "two-thirds times four-fifths" using TIMA: Bars. Unexpectedly, Nathan said out loud "two times four is eight and three times five is fifteen, so it has to be eight-fifteenths." So, the teacher asked him to show that using TIMA: Bars. To the teacher's surprise, Nathan created a bar, copied it, and partitioned the bar into three-thirds and the copy into five-fifths. He then pulled out two-thirds from the three-thirds bar and four-fifths from the five-fifths bar and lined these two pieces up at the bottom of the screen and said that he would put "X" between them if he could!

Nathan's actions were a complete surprise to the teacher because the algorithm for multiplying fractions that he had learned in his classroom and the fraction composition scheme that he had been using in TIMA: Bars were unrelated. Still, the teacher knew that the fraction composition schemes of the two children were compatible, so she pressed Arthur to verify that the result would, indeed, be eight-fifteenths. Arthur partitioned the 4/5-bar horizontally into thirds and pulled out two of these thirds. He then superimposed this cross-partitioned piece on the original five-fifths bar and explained that there were eight small pieces in his piece and fifteen of them in the whole bar. But he still had not curtailed the activity of his fraction composition scheme so that its activity and its results could be symbolized by "two times four is eight and three times five is fifteen, so it has to be eight-fifteenths" because he didn't explain what he would do if he used TIMA: Bars.

Being chagrined by Nathan's performance and encouraged by Arthur's use of his scheme, the teacher asked the children to show her three-fourths of nine-sixteenths in an attempt to establish a situation in which Arthur could curtail the activity of his scheme and Nathan could engage in the activity of his scheme and perhaps curtail the activity. But Nathan again used his classroom algorithm to obtain twenty-seven sixty-fourths. Arthur agreed with Nathan's answer, but when asked to show the result using TIMA: Bars both children appeared stumped. Arthur's agreement with Nathan apparently oriented him to use his computational algorithm as well because, in an attempt to find a way to show the result using TIMA: Bars, they resorted to implementing their equal fraction scheme in an attempt to find the sum of the two fractions. It is only after the teacher reminded them that she wanted three-fourths of nine-sixteenths that Arthur curtailed the activity of his scheme.

A:	Duah! Just like we did the other time! We cut it into four the other way ("it" refers to nine of the sixteen vertical parts) and take out three lines. (Arthur carries out his actions, creating a piece with twenty-seven parts.)
N:	See! Twenty-seven sixty-fourths. I told you!
T:	I see "twenty-seven" but I don't see "sixty-four!"
N:	(Partitions the 16-part *unit bar* horizontally into four parts, pulls one piece out and measures it obtaining "1/64" in the NUMBER box.) There!
A:	You can just do four times sixteen gives you sixty-four (pointing to the four rows and sixteen columns on the unit bar).
T:	What makes more sense to you, the numbers or the model?
N:	The numbers!
A:	The model!
N:	The numbers are easier for me. All you have to do is multiply them.
T:	I didn't ask which was easier, I asked which makes more sense.
A & N:	The model!
T:	Why?
A:	Well, it makes more sense with the model because you can show anybody on the model, but people who can't figure it with numbers have no idea!

As the teacher hoped, Nathan justified his computational result using Arthur's conceptual results. At that point, Arthur curtailed the activity of his scheme by means of establishing the number of elements in the two arrays as the product of the number of rows and the number of columns. He established a *child generated* computational algorithm by means of relating the results of using the scheme (three times nine and four times sixteen) to the situation of the scheme (three-fourths of nine-sixteenths). In the future, if upon encountering a phrase like "find three-fourths of nine-sixteenths," he used his computational algorithm to produce the spoken (or written) result that would constitute a curtailment of the activity of his scheme. There was a lot of work left for them in modifying their child-generated computational algorithm, but that "work" can be regarded as functional accommodations in their fractional composition scheme. Constructing and modifying child-generated computational algorithms constitute mathematics of children and the involved creative and productive activity is the result of children engaging in mathematical activity that is problematic in nature.

As important as it may be, we did not focus on children's construction of computational algorithms during the teaching experiment. Rather, our focus was on constructing models of children's fraction schemes that they can use in the construction of such algorithms when such a goal might be appropriate. When fractions are regarded as schemes that are functioning reliably and effectively that have to be constructed and modified by children in on-going mathematical activity, emphasis is on the creative and productive thinking of children. This emphasis, we believe, orients us appropriately in the mathematics education of children. We can see no reason for the learning of fractions to be not only very hard, but also a dismal failure (Davis et al. 1993) if the teaching of fractions is in harmony with children's fraction schemes. When the teaching of fractions is in harmony with such conventional knowledge such as equivalence classes of fractions, our research has demonstrated why the learning of fractions is not only very hard, but also a dismal failure. In that case, children are being asked to learn the first-order knowledge of competent mathematical adults that is far removed from the schemes of action and operation that we have explained in this book. But the situation is exacerbated by the realization that the schemes of action and operation that we have explained are not a result of spontaneous development such as the seriation scheme that was explained in Chap. 2. Rather, fraction schemes are constructed by children in the context of interactive mathematical activity in or as a result of the activity. When the emphasis is on teaching children conventional school mathematics, whether children construct these fraction schemes is very problematic. What the children do construct is all too often exemplified by Nathan's computational algorithm.[13]

Our overriding goal is to develop a "school mathematics" that is based on second-order models; that is, on knowledge that is constructed through social processes and that replaces school mathematics with "school mathematics." The distinction between the two is similar to Maturana's (1988) distinction between objectivity with parentheses and objectivity without parentheses. Objectivity without parenthesis refers to an explanatory path that "necessarily leads the observer to require a single domain of reality – a universe, a transcendental referent – as the ultimate source of validation for the explanations that he or she accepts" (p. 29). This comment fits well with the usual notion of school mathematics where conventional mathematical knowledge is the transcendental referent. The alternative, objectivity with parentheses, implies that knowledge is a process in the domain of explanation, which is basic in our concept of "school mathematics."

When focusing on the productive mathematical thinking of children, listening to children is paramount to bring forth their current conceptual operations (Confrey 1991). Ackermann (1995) made the case for listening to children as well as it can be made.

> As a teacher learns to appreciate her students' views for their own sake, and to understand the deeply organic nature of cognitive development, she can no longer impose outside standards to cover "wrong" answers. She comes to realize that her teaching is not "heard" the way she anticipated, and that the children's views of the world are more robust than she thought (p. 342).

"Listening" is itself a constructive activity, and what we "hear" is essentially established using our current concepts and operations. When mathematics teachers construct fraction schemes as a dynamic organization of second-order mathematical concepts and operations in their mental life, not only does it enable them to "hear" their students, but it also enables them to interact harmoniously with their students and to hypothesize pathways along which they can guide their students' mathematical activity. Teachers are the chief mathematicians with respect to the mathematics of children, and they are by necessity integrally involved in the production, modification, refinement, and elaboration of whatever "school mathematics" transpires in their classroom.

---

[13] Also see Chap. 7, *Protocol XX* for another example of the violence that computational algorithms can do to children's reasoning.

Chapter 11
Continuing Research on Students' Fraction Schemes

Anderson Norton and Amy J. Hackenberg

Directly or indirectly, *The Fractions Project* has launched several research programs in the area of students' operational development. Research has not been restricted to fractions, but has branched out to proportional reasoning (e.g., Nabors 2003), multiplicative reasoning in general (e.g., Thompson and Saldanha 2003), and the development of early algebra concepts (e.g., Hackenberg accepted). This chapter summarizes current findings and future directions from the growing nexus of related articles and projects, which can be roughly divided into four categories. First, there is an abundance of research on students' part-whole fraction schemes, much of which preceded *The Fractions Project*. The reorganization hypothesis contributes to such research by demonstrating how part-whole fraction schemes are based in part on students' whole number concepts and operations.

Second, several researchers have noted the limitations of part-whole conceptions and have advocated for greater curricular and instructional focus on more advanced conceptions of fractions (Mack 2001; Olive and Vomvoridi 2006; Saenz-Ludlow 1994; Streefland 1991). *The Fractions Project* has elucidated these limitations while articulating how advancement can be realized through the construction of key schemes and operations that transcend part-whole conceptions. In particular – and deserving of its own (third) category – research on fraction schemes has highlighted the necessity and power of the splitting operation in students' development of the more advanced fraction schemes, such as the reversible partitive fraction scheme and the iterative fraction scheme.

Finally, and more recently, researchers have used results from *The Fractions Project* to demonstrate how advanced fraction schemes can contribute to students' development toward algebraic reasoning. Although this research is in its infancy, one of the main findings so far is that the more advanced fraction schemes are critical in the construction of proportional reasoning (Nabors 2003), reciprocal reasoning (Hackenberg, accepted), and in solving basic linear equations of the form $ax = b$ (Tunc-Pekkan 2008).

Research on Part-Whole Conceptions of Fractions

The reorganization hypothesis has roots in work by McClellan and Dewey (1895), who argued, "the psychological process by which number is formed is first to last essentially a process of 'fractioning' – making a whole into equal parts and remaking the whole from the parts" (p. 138). We see this in Steffe's (2002) work, as he has described numerical operations that become reorganized as vital components of fraction schemes, such as unitizing, partitioning, disembedding, and iterating. Working with Steffe, and building on the ideas of McClellan and Dewey, Hunting (1983) carefully examined the progress of a 9-year-old student named Alan, from whole-number concepts toward the development of fraction concepts. Hunting identified partitioning (for which "fractioning" might be an euphemism) as the key operation in Alan's development of a part-whole conception for fractions.

Before elaborating on Hunting's findings, we briefly comment on a subtle distinction between fraction schemes and fraction concepts, which was alluded to in previous chapters. We consider fraction concepts as fraction schemes whose results are available prior to engaging in the activity of the scheme. This implies that the activity of the scheme has been interiorized and that the child can engage in operating in the absence of material in the child's perceptual field. Tzur (2007) has made a similar distinction in terms of *participatory* and *anticipatory* schemes. In those terms, concepts are anticipatory schemes. Although we cannot elaborate further here, Tzur's study empirically demonstrated the negative consequences of classroom instruction that does not support students' development from the participatory stage of scheme construction to the anticipatory stage.

In a fraction concept, the operations of the fraction scheme are contained in the first part of the scheme (the recognition template, or "trigger"), which enables the scheme to become anticipatory; that is, the scheme can be activated prior to its enactment in the sense of a resonating tuning fork (Steffe, 2002), with no need for carrying out a sequence of mental actions to establish meaning for a particular situation or numeral. In the case of the part-whole fraction scheme, part-whole conceptions can be inferred once the part-whole fraction scheme is symbolized by any given fraction word or numeral. A child who has developed a part-whole conception of fractions immediately understands "¾," say, as three parts disembedded from a whole that has been partitioned into four equal parts. However, as we have pointed out, a part-whole conception of fractions is a bit of a misnomer because the *partitive fraction scheme* is the first genuine fraction scheme.

Hunting (1983) found that Alan was able to develop a part-whole conception of fractions by applying his knowledge of numerical units to situations involving partitioning and sharing parts. Thus, he demonstrated the utility of partitioning operations and coordinating units at two levels in the construction of early fraction knowledge. However, Hunting was surprised to find that, although Alan seemed to understand one-fourth and one-eighth as one of four and eight equal parts, respectively, Alan did not understand that one-eighth was less than one-fourth (the "inverse order relationship," also addressed in Tzur 2007). Subsequent research, which we discuss in the next section, has indicated that such understanding requires the iterating operation and partitive conceptions of fractions.

Several researchers have affirmed the value of engaging students in situations involving sharing and partitioning, in support of students' construction of part-whole concepts (Behr et al. 1984; Empson 1999; Kieren 1988; Mack 2001, Saenz-Ludlow 1995). *The Fractions Project* has provided a theoretical basis to support such findings by identifying the role of the partitioning operation in early fraction concepts, and by explaining the construction of the partitioning operation in terms of the construction of composite wholes. Subsequent research by psychologists unfamiliar with *The Fractions Project* has independently affirmed its main theoretical underpinning – namely, the reorganization hypothesis. Working with three 7-year-old students using nonverbal whole-number and fractions tasks, Mix et al. (1999) came to a conclusion that contradicted earlier work by researchers who had advanced an interference hypothesis:

> There were striking parallels between the development of whole-number and fraction calculation. This is inconsistent with the hypothesis that early representations of quantity promote learning about whole numbers but interfere with learning about fractions. (p. 164)

At least in the mathematics education research community, it is now commonly accepted that numerical operations, such as partitioning and disembedding, constitute students' development of fraction concepts – when students have constructed these operations in continuous contexts, such as with connected numbers. This is clearly illustrated in recent work, even by researchers unaffiliated with *The Fractions Project*. In particular, Mack (2001) implicitly relied on the reorganization hypothesis in her study of six fifth-grade students, examining the development of fraction multiplication. Mack found that, indeed, students' informal knowledge of partitioning contributed to their construction of fraction concepts.

On the one hand, findings such as Mack's and Hunting's underscore the foundational importance of part-whole reasoning in developing fraction conceptions. On the other hand, to construct "genuine" fractions, students need to transcend part-whole conceptions. In fact, in the very same work cited above, Mack (2001) found that "students' informal knowledge of partitioning did not fully reflect the complexities underlying multiplication of fractions" (p. 291). The problem is confounded when we recognize that – as Streefland noted in 1991 – "teaching efforts have focused almost exclusively on the part-whole construct of a fraction" (p. 191).

The singular focus of curricula and instruction on part-whole concepts has contributed to students' difficulties in working with fractions operations and even algebraic reasoning. For example, in working with a student named Tim, Olive and Vomvoridi (2006) found that restriction to part-whole concepts hindered his ability to meaningfully engage in classroom activities that implicitly required more advanced conceptions. "Sparse conceptual structures limit students' understanding; once these conceptual structures had been modified and enriched, Tim was able to function within the context of classroom instruction" (p. 44). However, educators must recognize that they cannot change students' structures at will. Laura's case study (Chaps. 5 and 6) exemplifies this fact: Despite persistent efforts to provoke accommodations to her part-whole fraction scheme, Laura did not construct a partitive fraction scheme for over a year.

Transcending Part-Whole Conceptions

Olive (1999) and Steffe (2002) have demonstrated that numerical schemes contribute to the construction of fraction schemes, even beyond initial constructions such as the part-whole fraction scheme. Earlier work by Saenz-Ludlow (1994, 1995) elucidated those contributions by establishing explicit links between the numerical and fraction conceptions of two third-grade students named Michael and Anna. In this section, we build on the previous one by describing how students like Michael used numerical schemes to construct partitive fraction schemes. At the same time, we share findings on ways in which partitive reasoning transcends part-whole reasoning.

Saenz-Ludlow began her teaching experiment with the hypothesis that "Michael's well-grounded conceptualization of natural-number units would facilitate the generation of fractional-number units" (1994, p. 63). In fact, Michael seemed to reorganize two key numerical operations – coordinating two levels of units and iterating – to construct a partitive fraction scheme for composite units. As Michael demonstrated, the new scheme transcended the power of his previously constructed part-whole fraction scheme. For example, consider Michael's response to the following task:

T: If I give you forty-fiftieths of 1,000 dollars, how much money will I give you?
M: (After some thinking.) Eight hundred dollars.
T: Why?
M: (Quickly.) Because one-fiftieth is 20 dollars and five 20s is 100, so five, ten, fifteen, twenty, twenty-five, thirty, thirty-five, forty (Keeping track of the counting of fives with his fingers and finally showing eight fingers.); that is 800.

Michael's ability to anticipate the value of forty-fiftieths before actually double-counting fives and hundreds on his fingers indicates that, in fact, he had interiorized three levels of units, at least for whole numbers. He was able to consider the given fraction (forty-fiftieths) as a quantity relative in size to the given whole (1,000). Moreover, the units he was iterating (100's) were each composed of five-fiftieths, which provides indication of a composite unit fraction. His overall way of operating resembles the partitive fraction scheme for connected numbers that Nathan constructed (cf. Chap. 9). It enabled Michael to perform such tasks, whereas, in using his part-whole fraction scheme, he would have been restricted to interpreting forty-fiftieths as 40 parts out of 50 equal parts within a referent whole.

Saenz-Ludlow (1994) alluded to partitive reasoning when she advocated student conceptions of fractions as quantities. Such conceptions enable comparisons of size between part and whole, or even part and part, through the iteration of units. However, the operations of the partitive fraction scheme remain constrained within the referent whole. Both of Jason and Laura (Chap. 5) experienced the necessary errors that result from this way of operating. Even late in their fourth grade, 9/8 became "nine-ninths" or "eight-ninths" or "one-eighth plus one, where the "eighth" referred to 8/8 (p. 406). So, "conceptualizing improper fractions is not a simple extension of iterating a unit fraction within the whole" (Tzur, 1999 p. 409). Thus, there are at least two developmental hurdles with regard to conceptualizing fractions: moving from part-whole to partitive conceptions, and moving from partitive conceptions of proper fractions to iterative conceptions of proper and improper fractions. Subsequent research has indicated the critical role splitting plays in clearing the latter hurdle.

The Splitting Operation

Several researchers have independently adopted the term "split" from their students (Confrey 1994; Empson 1999; Olive and Steffe 2002; Saenz-Ludlow 2004). Confrey was first in promoting use of the term in research, especially with regard to her splitting hypothesis. Her splitting hypothesis posits that children develop a multiplicative operation – splitting – in parallel with additive operations. According to Confrey (1994), splitting applies to actions of "sharing, folding, dividing symmetrically, and magnifying" (p. 292). "In its most primitive form, splitting can be defined as an action of creating simultaneously multiple versions of an original, an action often represented by a tree diagram... a one-to-many action" (p. 292).

According to Steffe (Chaps. 1 and 10), the splitting operation is the composition of partitioning and iterating, in which partitioning and iterating are understood as inverse operations. For example, a student with a splitting operation can solve tasks like the following: "The bar shown below is three times as big as your bar. Draw your bar" (see bar and student response in Fig. 11.1). Finding an appropriate solution requires more than sharing (or any other act of partitioning), and even more than sequentially applying acts of partitioning and iterating; the student must anticipate that she can use partitioning to resolve a situation that is iterative in nature. Namely, by partitioning "my" bar into three parts, the student obtained a part that could be iterated three times to reproduce the whole bar.

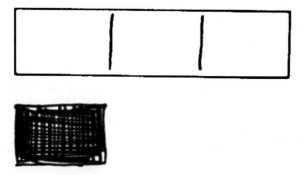

Fig. 11.1. Task response providing indication of a splitting operation.

As exemplified in Fig. 11.1, splitting involves partitioning, but it involves more. It supersedes even the levels of fragmenting identified in Chap. 10: simultaneous partitioning and equi-partitioning. As such, Confrey and Steffe's definitions for splitting contain similarities, but differ in one key regard: Confrey's definition makes no mention of iterating. In fact, Confrey (1994) intentionally juxtaposed splitting with iterating, which she viewed as contributing to repeated addition rather than the multiplicative reasoning that splitting supports. If she did not take exception to the inclusion of iterating operations, Confrey's splitting might include Steffe's splitting, as well as equipartitioning and simultaneous partitioning.

Splitting, as defined by Steffe, is especially powerful, as illustrated in the following case. During a semester-long teaching experiment with three pairs of sixth-grade students, Norton (2008) worked with a student name Josh who had constructed a splitting operation, but no genuine fraction schemes. That is to say, he could solve tasks like the one illustrated in Fig. 11.1 and he had developed a part-whole conception for fractions, but he had not yet constructed a partitive unit fraction scheme. Among the three pairs of students, only one other student, Hillary, had constructed a splitting operation (Norton and D'Ambrosio 2008). Relative to their peers, both students made impressive advancements in their constructions of fraction schemes, but we focus on Josh.

At the beginning of the teaching experiment, Josh was unable to unambiguously use fractional language. For example, when shown a 7/7-bar and asked what amount would remain if two-sevenths were removed, Josh answered, "5 pieces." When pressed for a fraction name, he could not decide between "five-sevenths," "fifty-sevenths," and "seven-fifths." Furthermore, when presented with a stick that had been partitioned in half, with the left half partitioned in half again, Josh thought the leftmost piece would be "one third." These responses indicate that Josh had not yet constructed a partitive unit fraction scheme. However, toward the end of the teaching experiment, Josh began estimating fractional sizes for proper fractions. Using the computer fractions software, TIMA: Sticks, Josh's partner produced an unpartitioned 2/9-stick along with its unpartitioned whole. When asked what the stick would measure, Josh lined up four copies of the fraction stick along the whole stick and estimated, "two-ninths." His estimate indicated that he had constructed a general partitive fraction scheme.

Throughout the teaching experiment, Josh formed conjectures that involved novel uses of his splitting operation. These conjectures seemed to support his construction of fraction schemes, including a partitive fraction scheme and a commensurate fraction scheme. Norton (2008) hypothesized that the splitting operation was particularly powerful in supporting his constructions because it composed two operations critical to the construction of fraction schemes: partitioning and iterating. In fact, studies cited in the previous two sections (e.g., Mack 2001) have illustrated the critical roles of those operations. Their composition then, provides powerful opportunities for growth, including the construction of more advanced fraction schemes.

In all of the fractions teaching experiments cited here, no student constructed an iterative fraction scheme – or a reversible partitive fractional scheme – without first constructing splitting. We have seen examples of this phenomenon from students mentioned in previous chapters, as well as Hillary, whose splitting operation supported her construction of a reversible partitive fractional scheme. In addition, all four of the students in Hackenberg's (2007) 8-month teaching experiment fit that pattern of development. All four students began sixth grade with splitting operations; all four constructed reversible partitive fraction schemes; and two of the students constructed iterative fraction schemes.

Consider the following exchange between the teacher–researcher and one of the pairs in Hackenberg's study, Carlos and Michael. The teacher–researcher asked Carlos to produce fourteen-thirteenths. Carlos began by partitioning a copy of the whole stick into 14 parts. Seeing this, Michael exclaimed, "No – no – no! You made fourteenths – (looks at Carlos) yours is thirteenths (gives a little laugh)." Carlos responded by asking the teacher–researcher, "didn't you say fourteen-thirteenths?" (Hackenberg 2007, p. 39). Carlos eventually produced the 14/13 by appending an extra piece to a 13/13 stick, but the exchange indicates Carlos's struggles in producing improper fractions. But then, he had much more success in solving tasks like the following: "Tanya has $16, which is 4/5 of what David has; how much does David have?" (p. 45).

Notice that the latter task requires a way of operating that is in reverse of the task Saenz-Ludlow (2004) posed to the third grade student named Michael (illustrated in the previous section). In particular, it requires a reversible partitive fraction scheme, which Carlos seemed to have constructed. However, Carlos had not yet constructed the kind of operating that his partner, Michael, used to solve the task involving 14/13. Namely, Michael had constructed an iterative fraction scheme. Both ways of operating require splitting because the students had to use partitioning in service of an iterative goal: Producing 14/13 required Carlos to partition the whole into 13 parts so that one of them could be iterated 14 times to produce the improper fraction; producing David's amount of money required Carlos to partition the given 4/5 part into four parts so that one of them could be iterated five times to produce the unknown whole. However, only Michael could readily solve the former task, and Hackenberg (2007) attributes this difference to the interiorization of three levels of units. A student would need to posit three levels of units prior to activity to purposefully produce a bar containing 14 thirteenths, with the understanding that the whole is produced from 13 iterations of any one of those thirteenths.

Findings from Norton's and Hackenberg's teaching experiments have challenged previous hypotheses about the origins and nature of splitting. Steffe (2004) had hypothesized that the splitting operation is based on the reversible partitive fraction scheme: "I presently consider the splitting operation to be the result of a developmental metamorphic accommodation of the reversible partitive fractional scheme" (p. 161). He revised this hypothesis after considering the case of Josh (Norton 2008) who had constructed a splitting operation even before constructing a partitive unit fraction scheme. We now understand that – to the contrary of the initial hypothesis – splitting is required for the construction of the reversible partitive fraction scheme.

Revising Steffe's (2004) hypothesis about the origin of the splitting operation begs the question: Where does splitting "come from" in students' constructive itineraries?

Confrey (1994) attributed the origins of splitting to abstractions from fair sharing activities, which makes sense given that for her splitting is based on making equal partitions. Saenz-Ludlow (1994) demonstrated how students transform Confrey's split from whole number to fractions contexts. However, in Steffe's splitting, which includes iterating as well as partitioning, fair sharing activities alone would likely be insufficient as an origin. Although Josh's case offers an exception, it seems that students' experiences in partitive fraction situations support their construction of splitting. After all, the partitive fraction scheme involves the sequential use of partitioning and iterating. Applying those operations as part of one scheme could plausibly contribute to their eventual composition as a single operation: splitting.

This view aligns with Steffe's revised splitting hypothesis expressed in Chap. 10: Construction of splitting results from interiorization of the equipartitioning scheme – a necessary prequel to the reversible partitive fraction scheme. In fact, the partitive unit fraction scheme is also a derivative of the equipartitioning scheme (cf. Chap. 10), so it makes sense that most students construct partitive unit fraction schemes prior to their construction of splitting (Norton and Wilkins, in press). The revised hypothesis also aligns with Hackenberg's (2007) findings regarding units coordination. Namely, students operating with a partitive fraction scheme should have constructed the two levels of interiorized units required to construct a splitting operation.

Steffe had also previously hypothesized that, "upon the emergence of the splitting operation," the partitive fraction scheme would be reorganized as an iterative fraction scheme (2002, p. 299). He revised this hypothesis in light of Hackenberg's (2007) teaching experiment and one of its key findings:

> Although the splitting operation still seems to be instrumental in the construction of an iterative fraction scheme, it does not appear to be sufficient for it… Students can construct the splitting operation without also interiorizing the coordination of three levels of units, and this interiorized coordination appears to be necessary for constructing improper fractions, and therefore the improper fraction scheme. (p. 46)

Steffe revised his hypothesis to its present form: If the child's operations that produce three levels of units become assimilating operations of the partitive fraction scheme, then the partitive fraction scheme can be used in the construction of the iterative fraction scheme. In other words, the partitive fraction scheme requires two levels of interiorized units, but if it, furthermore, includes a structure for producing three levels of units, the splitting operation might indeed be used to reorganize the partitive fraction scheme into an iterative fraction scheme. In fact, Joe and Patricia's case studies (cf. Chap. 10) illustrate such constructions. Further, based on Melissa's case study, Steffe claims that children construct the splitting operation using three levels of units in activity, though the operations that produce these units are not necessarily assimilating operations.

Students' Development Toward Algebraic Reasoning

In the past decade, researchers have begun to work on the problem of how students' construction of fraction schemes and operations may support students' construction of algebraic reasoning. This research focus is part of a larger effort to understand how to help students base their construction of algebraic reasoning on robust quantitative reasoning (Kaput 2008; Smith and Thompson 2008; Thompson 1993). One thrust of this effort is to understand how students can develop significant, conceptually coherent quantitative reasoning that would actually warrant generating and using powerful symbolic tools of algebra (Smith and Thompson 2008). Researchers who study students' construction of fraction schemes have made some progress in this area, as we will outline below. In contrast, little research based in scheme theoretic approaches has as of yet made significant progress on how students construct algebraic symbol systems (cf. Tillema 2007; Tunc-Pekkan 2008).

Before discussing the research that has been done in this area, we give a brief outline of how we characterize algebraic reasoning. From a very broad perspective, Kaput (2008) posited that algebra has two core aspects: (A) systematically symbolizing generalizations of regularities and constraints, and (B) engaging in syntactically guided reasoning on generalizations expressed in conventional symbol systems. He envisioned these core aspects as embodied in three strands: (1) algebra as the study of structures and systems abstracted from computations and relations, including algebra as generalized arithmetic and quantitative reasoning; (2) algebra as the study of functions, relations, and joint variation; and (3) algebra as the application of a cluster of modeling languages. Much of the research on children's early algebraic reasoning focuses on Kaput's core aspect A and strand 1 (e.g., Carpenter et al. 2003; Carraher et al. 2006; Dougherty, 2004; Knuth et al. 2006). We do so as well, but, as Tunc-Pekkan (2008) has pointed out, we do not take for granted the quantitative operations that may be required to build algebraic reasoning – in fact, we aim to specify them in our work with students.

We also aim to specify how quantitative operations may be reorganized to produce algebraic operations, a potential extension of the reorganization hypothesis of *The Fractions Project* (Hackenberg 2006; Tunc-Pekkan 2008). One possible "bridge" from quantitative fraction schemes (with the partitive fraction scheme being the first of these) to algebraic reasoning lies in students' construction of ratios and proportional reasoning. Nabors (2003) investigated this arena in her teaching experiment. She worked with seventh grade students to help them construct fraction schemes prior to investigating how they constructed schemes to solve problems involving ratios and proportions and rates, such as the following:

Money Exchange Problem. "In England, pounds are used rather than dollars. Four US dollars can be exchanged for three British pounds. How many pounds would you get in exchange for 28 US dollars? (adapted from Kaput and West 1994)" (p. 136).

Nabors hypothesized that the construction of what we have called more advanced fraction schemes (such as a reversible partitive fraction scheme, an iterative fraction scheme, and a reversible iterative fraction scheme) would be sufficient for students to reason with unit ratios to solve problems like the Money Exchange Problem (see Kaput and West's third level of proportional reasoning, 1994). This hypothesis was not confirmed – Nabors found that the fraction schemes were likely necessary, but not sufficient, for students' construction and use of unit ratios (cf. Davis 2003). Although the students in her study made progress in solving problems like the Money Exchange Problem and other problems involving rates, they used "build-up" strategies, both additive and multiplicative in nature (cf. Kaput and West's first and second levels of proportional reasoning, 1994).

In particular, Nabors (2003) found that students who had constructed a units coordinating scheme for composite units – in which composite units were iterating units – could solve problems like the Money Exchange Problem by repeatedly coordinating iterations of two composite units (in this case, units of four and units of three). Nabors agreed with Kaput and West (1994) that this solution strategy is primarily additive in nature, and her description of it indicates that students who have constructed the ENS can engage in it. In contrast, to solve the problem by anticipating that twenty-eight is some number of composite units of four, using division to determine that number, and then iterating the composite unit of three that number of times required that composite units were iterable units for the students. In other words, Nabors indicates that students had to be aware of the operation of iterating composite units prior to iterating them (p. 139). In essence, this finding implies that the operations that produce the GNS are needed for solving problems involving ratios and proportions with this "more advanced" build-up strategy. Even though some students in her study appeared to have constructed these operations, they did not produce solutions involving unit ratios (e.g., in which students determine that three-fourths of a pound corresponds to 1 dollar, and so three-fourths of 28 will yield the number of pounds that correspond to 28 dollars). Nabors did not hypothesize what operations are necessary to construct such solutions, except for noting that the interiorization of three levels of units is likely necessary.

Finally, Nabors (2003) found that students in her study could use some standard notational forms to solve problems involving ratios and proportions, but she could not claim that doing so meant they were engaging in reasoning beyond the two kinds of solutions discussed above. In fact, she notes that her study was an initial foray into this area, and that future research should investigate how students construct "numerical and algebraic representations of their reasoning processes" (p. 177) in these situations (cf. Kaput and West's fourth level of proportional reasoning, 1994).

Hackenberg's (2005, accepted) research is similar to Nabors' research in that she aimed to understand how students construct schemes and operations that underlie another traditional "component" of beginning algebra: the construction and solution of basic linear equations of the form $ax=b$. In her teaching experiment, she investigated how students reverse their quantitative reasoning with fractions to solve problems that can be solved with a basic linear equation of the form $ax=b$. A central finding of her study was the interiorization of three levels of units (i.e., the operations that produce the GNS) was critical for the construction of schemes to solve problems like this one:

Candy Bar Problem. That collection of 7 inch-long candy bars [7 identical rectangles] is 3/5 of another collection. Could you make the other collection of bars and find its total length?

To solve this problem, one student, Michael, modified his splitting operation to include the units-coordinating activity of his multiplying scheme. That is, Michael had constructed a reversible iterative fraction scheme and he used it to assimilate this problem: He aimed to split the known quantity into three equal parts, each of which would be one-fifth of the unknown quantity. However, he had no immediate way of operating to use to split seven units into three equal parts – the *seven* seemed to be at the "heart" of his perturbation in solving the problem. He eliminated this perturbation by splitting each of the 7 in. into a number of mini-parts (three) that would create a total number of mini-parts (21) that he *could* split into three equal parts (each containing 7 mini-parts). Hackenberg proposed that Michael could operate in this way because he could flexibly switch between two three-levels-of-units structures. That is, he conceived of the collection as a unit of seven units each containing three units, and then he could reorganize (in thought) the 21 mini-parts into a unit of three units each containing seven units. This way of reasoning is based on the splitting scheme for composite units in which the distributive partitioning scheme is embedded.

However, Hackenberg (2005, accepted) also found that the interiorization of three levels of units was not sufficient to provoke or explain the construction of reciprocal reasoning – although it seems to be necessary for it. In particular, Michael did not reason reciprocally to solve problems like the Candy Bar Problem, but another student in the study, Deborah, at least began to do so. Hackenberg hypothesized that Deborah had abstracted a fraction as a multiplicative concept, i.e., as a program of operations that included those of Deborah's iterative fraction scheme and reversible iterative fraction scheme. As Tunc-Pekkan (2008) has identified, how Deborah produced this abstraction was not clear. A related limitation of Hackenberg's study was that she and her student-participants did not engage in operating explicitly on unknowns, an important characteristic of algebraic reasoning. In the context of solving problems like the Candy Bar Problem, reasoning reciprocally would facilitate operating on the unknown quantity.

Tunc-Pekkan (2008) conducted a teaching experiment specifically to investigate students' construction of reciprocal reasoning in stating and solving equations of the form $ax=b$ where a and b are both fractional numbers. She differentiated between reversible reasoning and inverse reasoning in this context. On the basis of her analysis of one of the two pairs of eighth grade students with whom she worked, she hypothesized that constructing an inverse relationship between two quantities requires (1) conceptualizing *both* quantities independently (rather than solely that the known can be used to make the unknown); (2) constructing explicit equivalencies between fractional parts of the known and unknown quantities; and (3) using operations such as disembedding and iterating parts of the known quantity to create the unknown quantity (i.e., using multiplicative reasoning to construct the unknown quantity).

Tunc-Pekkan's findings indicate that the construction of *measurement units* were critical for producing and operating with equivalency; constructing only identity relationships between parts of quantities, which is possible when composite units are iterable, was insufficient to construct standard measurement units as independent quantities. The construction of measurement units involves the coordination of sequences of units, and so surpasses the operations that produce the GNS (alone). This finding is important because it indicates that "numerical aspects" of reasoning with quantities must be included in what we call "quantitative reasoning" for it to be powerful enough to be a basis for algebraic reasoning, something that some advocates of quantitative reasoning as a basis for algebra have downplayed or ignored (cf. Smith and Thompson 2008).

In addition, Tunc-Pekkan (2008) found that the students' construction of a *symbolic* fraction multiplication scheme was critical for students' construction of reciprocal reasoning. By symbolic she meant that students need to construct "more" than an anticipatory scheme in which they can find (make) a fraction composition – i.e., in which they construct a new quantity (the composition) as a result of operating on known quantities. In addition, students need to be able to construct the measurement of those quantities using what she called *recursive distributive partitioning operations*. Constructing these operations involves constructing partitioning and iterating as inverse operations, as well as distributive partitioning.[1] Her conclusions are interesting in light of the central role that Steffe gives to splitting in the construction of fraction schemes: Constructing the splitting operation may be the first step in the construction of an awareness of partitioning and iterating as inverse operations (cf. Chap. 10). In this way, construction of the splitting operation is important in the development of algebraic reasoning.

A central message of Tunc-Pekkan's (2008) research is that the power of algebraic thinking comes from *not* being dependent on quantities produced through operating but from being able to think of and interpret quantitative situations in terms of measurement units. More needs to be understood regarding how students construct measurement units and recursive distributive partitioning operations. However, together the work we have reviewed in this section suggests that researchers have made progress in understanding two hallmarks of algebraic reasoning: how students build conceptual structures and operate on them further, and how students learn to operate explicitly on unknown quantities.

---

[1] For Tunc Pekkan (2008), distributive partitioning is the operation that a student might use to find, say, 1/7 of 3 in. The student might partition each of the 3 in. into seven equal parts, disembed three parts (e.g., one part from each of the three inches), and unite them together to make 3/7 of 1 in. Recursive distributive partitioning involves, further, being able to engage in distributive partitioning of parts of quantities that are not perceptually present in service of taking a fractional amount of a quantity. For example, consider taking 1/7 of 3/5 of a liter, when only the 3/5-liter is present in the student's visual field. If a student uses distributive partitioning but also applies it to the two fifths of the liter that are not present, to conclude that the result is 3/35 of a liter, then the student has used recursive distributive partitioning.

References

Ackermann E (1995) Construction and transference of meaning through form. In: Steffe LP, Gale J (eds) Constructivism in education. Erlbaum, Hillsdale, NJ, pp 341–354

Behr MJ, Wachsmuth I, Post TR, Lesh R (1984) Order and equivalence of rational numbers: A clinical teaching experiment. J Res Math Educ 15:323–341

Biddlecomb BD (2002) Numerical knowledge as enabling and constraining fraction knowledge: an example of the reorganization hypothesis. J Math Behav 21:167–190

Booth L (1981) Child methods in secondary school mathematics. Educ Stud Math 12:29–41

Brouwer LEJ (1913) Intuitionism and formalism. Bull Am Math Soc 20:81–96

Brownell WA (1935) Psychological considerations in the learning and teaching of arithmetic. In: Reeve WD (ed) Teaching of arithmetic: the tenth yearbook of the National Council of Teachers of Mathematics. Teachers College, Columbia University, New York

Carpenter TP, Franke ML, Levi LW (2003) Thinking mathematically: integrating arithmetic and algebra in elementary school. Heinemann, Portsmouth, NH

Carraher DW, Schliemann AD, Brizuela BM, Earnest D (2006) Arithmetic and algebra in early mathematics instruction. J Res Math Educ 37:87–115

Ceccato S (1974) In the garden of choices. In: Smock CD, von Glasersfeld E (eds) Epistemology and education. Follow Through Publications, Athens, GA, pp 125–142

Confrey J (1988) Multiplication and splitting: their role in understanding exponential functions. In: Behr MJ, Lacampagne CB, Wheeler MM (eds) Proceedings of the tenth annual meeting of the North American chapter of the international group for the psychology of mathematics education. Northern Illinois University, DeKalb, IL, pp 250–259

Confrey J (1991) Learning to listen: a student's understanding of powers of ten. In: von Glasersfeld E (ed) Radical constructivism in mathematics education. Kluwer, Dordrecht, The Netherlands, pp 111–138

Confrey J (1994) Splitting, similarity, and rate of change: a new approach to multiplication and exponential functions. In: Harel G, Confrey J (eds) The development of multiplicative reasoning in the learning of mathematics. State University of New York Press, Albany, NY, pp 291–330

Cooper RG Jr (1990) The role of mathematical transformations and practice in mathematical development. In: Steffe LP (ed) Epistemological foundations of mathematical experience. Springer, New York, pp 102–123

Curcio FR, Bezuk NS (1994) Understanding rational numbers and proportions. Curriculum and evaluation standards for school mathematics: Addenda series, grades 5–8. National Council of Teachers of Mathematics, Reston, VA

Davis G (2003) Teaching and classroom experiments dealing with fractions and proportional reasoning. J Math Behav 22:107–111

Davis G, Hunting RP, Pearn C (1993) What might a fraction mean to a child and how would a teacher know? J Math Behav 12:63–76

Davydov VV (1975) The psychological characteristics of the "prenumerical period" of mathematics instruction. In: Steffe LP (ed) Children's capacity for learning mathematics. School Mathematics Study Group, Stanford

Dörfler W (1996) Is the metaphor of mental object appropriate for a theory of learning mathematics? In: Steffe LP, Nesher P, Cobb P, Goldin GA, Greer B (eds) Theories of mathematical learning. Erlbaum, Mahwah, NJ, pp 467–476

Dougherty B (2003) Voyaging from theory to practice in learning: measure up. Proceedings of the twenty-seventh international group for the psychology of mathematics education conference held jointly with the twenty-fifth North American chapter of the international group for the psychology of mathematics education conference, Vol 1. University of Hawai'i, Honolulu, HI, pp 18–23

Dougherty B (2004) Early algebra: perspectives and assumptions. For the learning of mathematics 24(3):28–30

Empson SB (1999) Equal sharing and shared meaning: the development of fraction concepts in a first-grade classroom. Cognit Instruct 17:283–342

Erlwanger SH (1973) Benny's concept of rules and answers in IPI mathematics. J Children's Math Behav 1:7–26

Foxman DD, Cresswell MJ, Ward M, Badger ME, Tuson JA, Bloomfield BA (1980) Mathematical development: primary survey report no. 1. Her Majesty's Stationery Office, London

Freudenthal H (1983) Didactical phenomenology of mathematical structures. D. Reidel, Dordrecht, The Netherlands

Gelman R, Gallistel CR (1978) The child's understanding of number. Harvard University Press, Cambridge

Ginsburg H (1977) Children's arithmetic: the learning process. D. Van Nostrand, New York

Guthrie ER (1942) Conditioning: a theory of learning in terms of stimulus, response, and association. In: Henry NB (ed) The Psychology of Learning. University of Chicago Press, Chicago, pp 17–60

Hackenberg AJ (2005). Construction of algebraic reasoning and mathematical caring relations. Unpublished doctoral dissertation, The University of Georgia, Athens

Hackenberg AJ (2007) Units coordination and the construction of improper fractions: a revision of the splitting hypothesis. J Math Behav 26:27–47

Hackenberg AJ (accepted) Students' reversible multiplicative reasoning with fractions. Cognition and Instruction.

Hart KM (1983) I know what I believe, do I believe what I know? J Res Math Educ 2:119–125

Hoffman P (1987) The man who loves only numbers. The Atlantic Monthly 260:60–74

Hunting RP (1983) Alan: a case study of knowledge of units and performance with fractions. J Res Math Educ 14:182–197

Hunting RP, Sharpley CF (1991) Pre-fraction concepts of preschoolers. In: Hunting RP, Davis G (eds) Early fraction learning. Springer, New York, pp 9–26

Inhelder B, Piaget J (1964) The early growth of logic in the child. W. W. Norton, New York

Kagan VF (1963) Essays on geometry. Moscow University, Moscow

Kaput JJ (2008) What is algebra? What is algebraic reasoning? In: Kaput JJ, Carraher DW, Blanton ML (eds) Algebra in the early grades. Erlbaum, New York, pp 5–17

Kaput JJ, West MM (1994) Missing-value proportional reasoning problems: factors affecting informal reasoning patterns. In: Confrey J, Harel G (eds) The development of multiplicative reasoning in the learning of mathematics. State University of New York Press, Albany, NY, pp 235–287

Kerslake D (1986) Fractions: children's errors and strategies. NFER-Nelson, Windsor, England

Kieren TE (1988) Personal knowledge of rational numbers: its intuitive and formal development. In: Behr MJ, Hiebert J (eds) Number concepts and operations in the middle grades. National Council of Teachers of Mathematics, Reston, VA, pp 162–181

Kieren T (1993) Rational and fractional numbers: From quotient fields to recursive reasoning. In: Carpenter T, Fennema E, Romberg T (eds) Rational numbers: an integration of research. Erlbaum, Hillsdale, NJ, pp 49–84

Kieren T (1994) Multiple views of multiplicative structures. In: Harel G, Confrey J (eds) The development of multiplicative reasoning in the learning of mathematics. State University of New York Press, Albany, NY, pp 387–397

Knuth EJ, Stephens AC, McNeil NM, Alibali MW (2006) Does understanding the equal sign matter? Evidence from solving equations. J Res Math Educ 37:297–312

Lamon SJ (1996) The development of unitizing: its role in children's partitioning strategies. J Res Math Educ 27:170–193

Long C, DeTemple DW (1996) Mathematical reasoning for elementary teachers. Harper Collins, New York

Mack NK (2001) Building on informal knowledge through instruction in a complex content domain: partitioning, units, and understanding multiplication of fractions. J Res Math Educ 32:267–296

Maturana H (1988) Reality: a search for a compelling argument. Irish J Psychol 9(3):25–82

McLellan J, Dewey J (1895) The psychology of number and its applications to methods of teaching arithmetic. D. Appleton, New York

Menninger K (1969) Number words and symbols: a cultural history of numbers. The Massachusetts Institute of Technology Press, Cambridge

Mix KS, Levine SC, Huttenlocher J (1999) Early fraction calculation capability. Dev Psychol 35:164–174

Nabors WK (2003) From fractions to proportional reasoning: a cognitive schemes of operation approach. J Math Behav 22:133–179

Nik Pa NA (1987) Children's fractional schemes. Unpublished doctoral dissertation, The University of Georgia, Athens

Norton A (2008) Josh's operational conjectures: abductions of a splitting operation and the construction of new fractional schemes. J Res Math Educ 39:401–430

Norton A, D'Ambrosio BS (2008) ZPC and ZPD: zones of teaching and learning. J Res Math Educ 39:220–246

Norton A, Wilkins J (2009) A quantitative analysis of children's splitting operations and fractional schemes. J Math Behav 28:150–161

Olive J (1999) From fractions to rational numbers of arithmetic: a reorganization hypothesis. Math Think Learn 1:279–314

Olive J (2002) The construction of commensurate fractions. In: Cockburn A (ed) Proceedings of the twenty-sixth conference of the international group for the psychology of mathematics education, vol 4. University of East Anglia, Norwich, England, pp 1–8

Olive J, Steffe LP (2002) The construction of an iterative fractional scheme: the case of Joe. J Math Behav 20:413–437

Olive J, Vomvoridi E (2006) Making sense of instruction on fractions when a student lacks necessary fractional schemes: the case of Tim. J Math Behav 25:18–45

Piaget J (1937) The construction of reality in the child. Basic Books, New York

Piaget J (1964) Development and learning. In: Ripple RE, Rockcastle VN (eds) Piaget rediscovered: a report of a conference on cognitive studies and curriculum development. Cornell University Press, Ithaca, NY, pp 7–19

Piaget J (1970) Genetic epistemology. Columbia University Press, New York

Piaget J (1980) Opening the debate. In: Piattelli-Palmarini M (ed) Language and learning: the debate between Jean Piaget and Noam Chomsky. Harvard University Press, Cambridge, pp 23–34

Piaget J, Szeminska A (1952) Child's conception of number. Routledge, London

Piaget J, Inhelder B, Szeminska A (1960) The child's conception of geometry. Basic Books, New York

Post TR, Cramer KA, Behr M, Lesh R, Harel G (1993) Curriculum implications of research on the learning, teaching, and assessing of rational number concepts. In: Carpenter T, Fennema E, Romberg T (eds) Rational numbers: an integration of research. Erlbaum, Hillsdale, NJ, pp 327–362

Renshaw P (1992) The psychology of learning and small group work. In classroom oral language: reader. Deakin University Press, Geelong, Victoria, Australia

Reys BJ (1991) Developing number sense in the middle grades. Curriculum and evaluations standards for school mathematics: Addenda series, grades 5–8. Reston. National Council of Teachers of Mathematics, VA

Sáenz-Ludlow A (1994) Michael's fraction schemes. J Res Math Educ 25:50–85

Sáenz-Ludlow A (1995) Ann's fraction schemes. Educ Stud Math 28:101–132

Sáenz-Ludlow A (2004) Metaphor and numerical diagrams in the arithmetical activity of a fourth-grade class. J Res Math Educ, 35:34–56.

Schwartz J (1988) Intensive quantity and referent-transforming operations. In: Hiebert J, Behr M (eds) Number concepts and operations in the middle grades. National Council of Teachers of Mathematics, Reston, VA, pp 41–52

Smith J (1987, April). What is fraction conceptual knowledge? Paper presented at the Annual Meeting of the American Educational Research Association, Washington, DC

Smith JP, Thompson PW (2008) Quantitative reasoning and the development of algebraic reasoning. In: Kaput JJ, Carraher DW, Blanton ML (eds) Algebra in the early grades. Erlbaum, New York, pp 95–132

Steffe LP (1988) Children's construction of number sequences and multiplying schemes. In: Hiebert J, Behr M (eds) Number concepts and operations in the middle grades. Erlbaum, Hillsdale, NJ, pp 119–140

Steffe LP (1991) Operations that generate quantity. Learn Indiv Differ 3:61–82

Steffe LP (1992) Schemes of action and operation involving composite units. Learn Indiv Differ 4:259–309

Steffe LP (1994a) Children's construction of meanings for arithmetical words. In: Tirosh D (ed) Implicit and explicit knowledge: an educational approach. Ablex, New Jersey, pp 131–169

Steffe LP (1994b) Children's multiplying schemes. In: Harel G, Confrey J (eds) Multiplicative reasoning in the learning of mathematics. State University of New York Press, Albany, NY, pp 3–39

Steffe LP (1996) Social-cultural approaches in early childhood mathematics education: a discussion. In: Mansfield H, Pateman NA, Bednarz N (eds) Mathematics for tomorrow's young children: international perspectives on curriculum. Kluwer, Boston, pp 79–99

Steffe LP (2002) A new hypothesis concerning children's fractional knowledge. J Math Behav 20:267–307

Steffe LP (2004) On the construction of learning trajectories of children: the case of commensurate fractions. Math Think Learn 6:129–162

Steffe LP, Cobb P (with von Glasersfeld E) (1988) Construction of arithmetical meanings and strategies. Springer, New York

Steffe LP, Thompson PW (2000) Teaching experiment methodology: underlying principles and essential elements. In: Kelly AE, Lesh RA (eds) Research design in mathematics and science education. Erlbaum, Mahwah, NJ, pp 267–307

Steffe LP, Wiegel H (1996) On the nature of a model of mathematical learning. In: Steffe LP, Nesher P, Cobb P, Goldin GA, Greer B (eds) Theories of mathematical learning. Erlbaum, Mahwah, NJ, pp 477–498

Steffe LP, von Glasersfeld E, Richards J, Cobb P (1983) Children's counting types: philosophy, theory, and application. Praeger Scientific, New York

Stolzenberg G (1984) Can inquiry into the foundations of mathematics tell us anything interesting about mind? In: Watzlawick P (ed) The invented reality: how do we know what we believe we know?. W. W. Norton, New York, pp 257–309

Streefland L (ed) (1991) Fractions in realistic mathematics education: a paradigm of developmental research. Kluwer, Boston, MA

Thompson PW (1982) A theoretical framework for understanding young children's concepts of numeration. Unpublished doctoral dissertation, The University of Georgia, Athens.

Thompson PW (1993) Quantitative reasoning, complexity, and additive structures. Educ Stud Math 25:165–208

Thompson PW (1994) The concept of speed and its relationship to concepts of rate. In: Harel G, Confery J (eds) The development of multiplicative reasoning in the learning of mathematics. State University of New York Press, Albany, pp 179–234

Thompson PW, Saldanha LA (2003) Fractions and multiplicative reasoning. In: Kilpatrick J, Martin WG, Schifter D (eds) A research companion to principles and standards for school mathematics. National Council of Teachers of Mathematics, Reston, VA, pp 95–113

Tillema E (2007) Students' construction of algebraic symbol systems. Unpublished doctoral dissertation, The University of Georgia, Athens

References

Tunc-Pekkan Z (2008) Modeling grade eight students' construction of fraction multiplying schemes and algebraic operations. Unpublished doctoral dissertation, The University of Georgia, Athens

Tzur R (1999) An integrated study of children's construction of improper fractions and the teacher's role in promoting that learning. J Res Math Educ 30:390–416

Tzur R (2007) Fine grain assessment of students' mathematical understanding: participatory and anticipatory stages in learning a new mathematical conception. Educ Stud Math 66:273–291

von Glasersfeld E (1980) The concept of equilibration in a constructivist theory of knowledge. In: Benseler F, Hejl PM, Kock WK (eds) Autopoiesis, communication, and society. Campus, Frankfurt, Germany, pp 75–85

von Glasersfeld E (1981) An attentional model for the conceptual construction of units and number. J Res Math Educ 12:83–94

von Glasersfeld E (1991) Abstraction, re-presentation, and reflection: An interpretation of experience and Piaget's approach. In: Steffe LP (ed) Epistemological foundations of mathematical experience. Springer, New York, pp 45–67

von Glasersfeld E (1995a) Radical constructivism: a way of knowing and learning. Falmer Press, Washington, DC

von Glasersfeld E (1995b) Review of the book learning mathematics: constructivist and interactionist theories of mathematical development. Int Rev Math Educ 27:120–123

Washburne C (1930) The grade placement of arithmetic topics. In: Whipple GM (ed) Report of the society's committee on arithmetic (29th Yearbook of the National Society for the Study of Education. University of Chicago Press, Chicago, pp 641–670

Watzlawick P (1984) The fly and the fly-bottle. In: Watzlawick P (ed) The invented reality: how do we know what we believe we know?. W. W. Norton, New York, pp 249–256

Index

A
Accommodations, 1
Algebraic reasoning, 343
   construction of ratios and proportional reasoning, 349
   core aspects, 349
   reciprocal reasoning, 351
   solving problems of ratios and proportions, 350
   symbolic fraction multiplication scheme, 352
   using linear equations, 350
Anticipatory scheme, 24–25
Arithmetical numerosity, awareness of, 43

B
Brouwer, L.E.J., 7

C
Children's fraction knowledge, perspectives
   difficulties in learning fractions, 13–14
   first-order mathematical knowledge, 16
   fractions as schemes, 18–25
   interference hypothesis, 2–4
   invention and construction, 15
   mathematics for children, 17–18
   mathematics of children, 16–17
   mathematics of living, 25
   second-order mathematical knowledge, 16
   separation hypothesis, 5–6
   trap, 14
Children's mathematics, 16–17
Commensurate fraction scheme, 284–286
   commensurate one third with four-twelfths, 131–133
   fractions commensurate with one-half, 138–141
   fractions commensurate with one-third, 142–147
   fractions commensurate with two-thirds, 147–148
   further investigations, 136–138
   hypothesis testing, 241–246
   independent mathematical activity and, 163–164
   to nonunit fractions, 216
   renaming unit fractions of 24-part stick, 215
   social interaction and, 133–136
Common partitioning scheme construction
   for fifths and sevenths, 291
   for thirds and fifths, 290
Composite unit fractions, 129–138
   case study, discussion, 217–223
   conflation of, 172–174
   multiplying schemes in, 17–18
      engendering construction, 86–87
      finding fractional parts of a 24-stick, 87–89
      necessary errors, 90–92
      operating on three-level of units, 89–90
      units-coordinating schemes, children's use in, 83–86
   partitioning and disembedding operations, 176–180
      one-fourth of 27-stick, estimates of, 177–178
      one-seventh of mystery stick, finding, 179
      sticks as fraction of 24-stick, estimates of, 175–176
Conceptions of fractions, 342–344
Confrey, J., 5–6, 9–11, 49, 345–346, 348
Connected number sequence, 80–83, 118–119
Constructivist learning, 15
Cross-partitioning, 308–309

D
Davis, G., 13
Davydov, V.V., 49
Dewey, J., 4, 342
Discrete units, 27–29
Disembedding operations, 176–180
Distributive partitioning scheme, 321–322, 331
Dörfler, W., 46

E
Equal fractions, 298
  commensurate fractions and, 312
  plurality of fractions, generating, 299–301
  working on symbolic level, 301–302
Equal fraction scheme, 336–337
Equi-partitioning operations for connected numbers
  case study, discussion, 266–268
  child-generated fraction adding scheme, 260–263
  commensurate fraction scheme, hypothesis testing, 241–246
  fractional connected number sequence construction, 236–241
  inferences, 235–236
  initial fraction scheme, Melissa's, 225
    recursive partitioning, contraindications, 227–228
    reversibility of unit fraction composition scheme, 228–231
  iterative fraction scheme, 268
    child-generated *vs.* procedural scheme for adding fractions, 275
    fraction multiplication, children's meaning of, 273–274
    interiorization of operations, 269–271
    recursive partitioning operations, 271–273
  levels of units in re-presentation
    fractions of fractional parts, 247–251
    partitioning by drawing, 251–253
    partitioning operations, test of accommodation, 254–260
  unit fraction adding scheme, 263–266
  units coordinating scheme, reorganization, 231–236
Equi-partitioning schemes, 66–67, 315–316
  breaking stick into two equal parts, 75–76
  composite units as templates, 76–78
  for connected numbers, 319–320
  splitting scheme and, 317–318
  vs. equi-segmenting schemes, 78–79
Equi-segmenting schemes, 78–79, 93

Explicitly nested connected number sequence, 82, 95, 119
Explicitly nested number sequence (ENS), 41–42, 45–47
Extensive quantitative comparisons, 55–56

F
Figurative counting schemes, 32–35
  figurative unit items, attentional pattern, 33
  visualized image of actions, 33
Figurative length, 50–51, 54
Figurative lots, 34
Figurative numerosity, awareness of, 42
Figurative plurality, 34, 53–55
First-order mathematical knowledge, 16, 18
Fraction adding scheme, 260–263
Fractional connected number sequences
  Melissa's construction of, 236–241
    generalizing assimilation, 239
    iterative fraction scheme, 237–238
    relation between fractional part and fractional whole, 238
  one-half and one-third of twelve-twelfths, finding, 212
  three-twelfths into one-fourth, transforming, 213–214
  for twelfths, 211
Fraction composition scheme, 303, 313–314, 330–333
  case study, discussion
    common partitioning scheme and sum of two fractions, 311–313
    reversible partitive fraction scheme, 310–311
  commensurate fractions, 284–286
  common partitioning scheme construction, 289–291
  equal fractions, 298
    plurality of fractions, generating, 299–301
    working on symbolic level, 301–302
  five-elevenths among seven people, sharing, 306–307
  four-fifths of one-ninth, finding, 305–306
  four-ninths among five people, sharing, 304–305
  generalized number sequence, 278–279
  hypothesis testing using TIMA: Bars, 307–310
  language of fractions
    generalized fraction language, 281–282
    improper fraction language, 282–284

partitioning and sharing situations, 279–281
multiplication of fractions and nested fractions, 295–298
renaming fractions, 288–289
splitting operation for connected numbers, corroboration, 286–288
two-thirds of one-seventh, finding, 306
unit fractions with unlike denominators, construction strategies, 291–295
Fraction language
generalized fraction language, 281–282
improper fraction language, 282–284
partitioning and sharing situations, 279–281
Fraction multiplication, 273–274
Fractions as schemes
ethnomathematical knowledge, 18
learning as accommodation, 21
scheme, parts, 20
scheme, structure, 22–24
sharing situation, 19, 22
sucking scheme, 21–22
Fractions beyond fractional whole, 99–204
iterative fraction scheme, 200
two times as long as four seventh, finding, 199–200
Fraction schemes, 1, 6
equal fraction scheme, 336–337
fraction composition scheme, 330–333
fraction concepts and, 342
iterative fraction scheme, 333–335
partitive fraction scheme, 323–328
additive scheme, 325–326
anticipatory iteration, 326–327
partitive unit fraction scheme, 324–325
reversible partitive fraction scheme, 327–328
part-whole fraction scheme, 322–323
research on
algebraic reasoning, students development toward, 348–352
part-whole conceptions of fractions, 342–344
splitting operations, 345–348
transcending part-whole conceptions, 344–345
unit commensurate fraction scheme, 335–336
unit fraction composition scheme, 328–330
Fractions of fractional parts, 247–251
Fragmenting, templates for, 57–58
attentional patterns
fragmenting a rope, 60–61

fragmenting circular cake, 59–60
subdividing circular cake, 61–64
experiential basis, 58–59
levels of, 68–70
number sequences and subdividing a line
intuitive subdivision, 65
lack of subdivision, 64–65
measuring without unit iteration, 65–66
measuring with unit iteration, 66–67
Freudenthal, H., 18

G
Gallistel, C.R., 59
Gelman, R., 59
Generalized fraction language, 281–282
Generalized number sequence, 43–45, 278–279, 291
Generated goal, 22–23
Gross quantitative comparisons, 52
Grouping fractions, 13

H
Hackenberg, A.J., 341–352
Hunting, R.P., 4, 60, 61, 66, 69, 342

I
Improper fractions, 282–284
Joe's production of, 185–188
operating with, 207–208
Inhelder, B., 29–31
Initial fraction scheme, Melissa's, 225
recursive partitioning, contraindications, 227–228
reversibility of unit fraction composition scheme, 228–231
Initial number sequence, 35–38, 45
Intensive quantitative comparisons, 52–53
Interference hypothesis
children's quantitative operations, 3
fraction knowing, 3
inverse relation between the number and size of the parts, 4
quantitative operations, 3
rational number knowing, 2
Iterative fraction scheme, 116–117, 180, 200, 268, 333–335
case study, discussion, 217–223
child-generated *vs.* procedural scheme for adding fractions, 275
construction of, 221–223

Iterative fraction scheme (*cont.*)
    fraction multiplication, children's meaning of, 273–274
    interiorization of operations, 269–271
    operating with improper fractions, 207–208
    recursive partitioning operations, 271–273
    twelve slices make 12/8 of pizza, 205

K
Kagan, V.F., 49
Kaput, J.J., 349–350
Kerslake, D., 13
Kieren, T., 2–3, 18–19, 74

L
Lamon, S.J., 69–70

M
Mack, N.K., 343
Mathematical development, children's, 13, 23, 25
Mathematics for children, 17–18
McLellan, J., 4, 342
Menninger, K., 7–8
Mix, K.S., 343
Multiplication of fractions and nested fractions, 295–298

N
Nabors, W.K., 349–350
Nik Pa, N.A., 3–4
Nongrouping fractions, 13
Norton, A., 341–352
Notational system, 256–260
Numerical counting schemes, operation
    awareness of numerosity, 42–43
    discrete units, complexes of, 27–29
    explicitly nested number sequence, 41–42
    figurative counting schemes, recognition templates, 32–35
    generalized number sequence, 43–45
    number sequence, 27
    numerical patterns and initial number sequence, 35–38
    perceptual counting schemes, recognition templates, 29–32
    principal operations, overview, 45–47
    recursive unitizing, 27
    tacitly nested number sequence, 38–40

Numerical finger patterns, 39
Numerical patterns, 36
Numerosity, awareness of, 42–43

O
Olive, J., 277–314, 341, 344–345

P
Partitioning and iterating
    disuniting composite unit structure, 68
    fragmenting levels, 68–70
    simultaneous co-occurrence of unit items, 67
Partitioning operations, 176–180
    by drawing, 251–253
    test of accommodation, 254–260
Partitioning schemes
    distributive partitioning scheme, 321–322
    equi-partitioning scheme, 315–316, 319–320
    partitive fraction scheme, 323
        additive scheme, 325–326
        anticipatory iteration, 326–327
        partitive unit fraction scheme, 324–325
        reversible partitive fraction scheme, 327–328
    part-whole fraction scheme, 322–323
    simultaneous partitioning scheme, 316–317
    splitting scheme, 317–321
Partitive fraction schemes, 114, 117, 121, 323
    additive scheme, 325–326
    anticipatory iteration, 326–327
    partitive unit fraction scheme, 324–325
    reversible partitive fraction scheme, 327–328
Partitive schemes
    connected number sequence, 80–83
    equi-partitioning schemes, 75–78
    fractional meaning for multiple parts of stick, 110–112
    fractional numbers, continued absence, 113–117
    independent use of parts, Laura's, 102–107
    partitive unit fraction scheme, Jason's, 100–102
    part-whole fraction scheme, Laura's, 107–110
    segmenting
        dual emergence of quantitative operations, 80
        equi-segmenting *vs.* equi-partitioning schemes, 78–79

simultaneous partitioning scheme, 92–98
splitting operations, lack of, 98–100
Partitive unit fraction scheme, 100–102, 110, 116, 180–185, 191, 206, 324–325
  case study, discussion, 218–219
  distributive reasoning, lack of, 191–193
  iterative fraction concept, 181
  numerosity and length, 184
  two-thirds and one whole one, 181–182
Part-whole conceptions of fractions, 342–344
Part-whole fraction scheme, 107–110, 114, 119–121, 322–323
Perceptual counting schemes
  collection of perceptual items, 29–30
  perceptual lots, 30–32
Perceptual density, 52
Perceptual length, 50–51, 54
Perceptual lots, 30–32
Perceptual plurality, 31–32
Perceptual unit item, attentional pattern, 31
Piaget, J., 19–20, 22, 24, 29–31, 52, 54–55, 57, 59–61, 63–65, 69, 72, 75, 80, 118–119, 122
Plurality of fractions, 299–301
Post, T.R., 2, 4

R

Reciprocal reasoning, 351
Recognition templates
  of figurative counting schemes, 32–35
  of perceptual counting schemes, 29–32
Recursion, 9–11
Recursive distributive partitioning operations, 352
Recursive numerical scheme, 41
Recursive partitioning, 151–153, 188
  accommodation in, 256–260
  case study, discussion, 220–221, 271–273
  in re-presentation, 250–251
  splitting operations and, 162
  unit fraction composition scheme and, 169
Recursive splitting operations, 300
Reorganization hypothesis
  construction of new schemes, 1
  counting problems, 1
  fragmenting levels, 68–70
  gross, intensive, and extensive quantity, 51–52
  partitioning and iterating, 67–68
  perceptual and figurative length, 50–51
  quantity, defined, 49
  templates for fragmenting, composite structures as, 57–58

experiential basis, 58–59
using attentional patterns, 59–64
Reversible partitive fraction scheme, 125, 196–197, 327–328
  for connected numbers, 281
  fractions beyond fractional whole, 199–204
  reversibility in Joe's, 197
  reversibility in Patricia's, 198
Reversible unit fraction composition scheme
  finding 1/4-stick out from 3/3-stick, 264–266
  sequences of fractions, 263–264
Reversible units-coordinating scheme, 229

S

Saenz-Ludlow, 344, 347–348
Scheme(s), 1, 3–6, 12
  anticipatory, 24–25
  definition, 19
  instrument of interaction, 19
  parts of, 20
  seriation, 24–25
  structure, 22–24
  sucking scheme, 21–22
School mathematics, 337–340
Second-order mathematical knowledge, 16
Sensory-motor units, attentional pattern, 28–29
Separation hypothesis
  simultaneity and sequentiality, 6–9
  splitting conjecture, 5
Sequentiality, 6–9
Seriation scheme, 24–25
Sharpley, C.F., 60, 61, 66, 69
Simultaneity, 6–9
  distribution and, 10–11
Simultaneous partitioning scheme, 92–98, 316–317
Smith, J., 14
Splitting, 5, 9–11, 73–74
Splitting operations, 121–123, 233
  for connected numbers, 286–288
  corroboration and contraindication, 188–191
  emergence of Patricia's, 193–195
  fraction schemes, research on, 345–348
  independent mathematical activity and, 163
  iterative fraction scheme and, 219–220
  recursive partitioning and, 220–221
  unit fraction composition scheme and, 162
Splitting scheme, 317–319
  for connected numbers, 320–321
  equi-partitioning scheme and, 317–318

Stolzenberg, G., 14–16
Streefland, L, 2
Sucking scheme, 21–22
Symbolic fraction multiplication scheme, 352
Szeminska, A., 52, 54–55

T
Tacitly nested number sequence (TNS), 38–40, 45–47, 66, 71
Thompson, P.W., 34
Tunc-Pekkan, Z., 349, 351–352

U
Ulrich, C., 225–275
Unit commensurate fraction scheme, 335–336
Unit fraction adding scheme, 263–266
 finding 1/4-stick out from 3/3-stick, 264–266
 reversible unit fraction composition scheme, 263–264
Unit fraction composition scheme, 124–129, 153, 229–231, 328–330
 apparent recursive partitioning, Laura's, 128–129
 composition of two fractions, 153–155
 constraint in children's
  one-half of one-third, finding, 209
  one-half of three-sevenths, finding, 210
 construction of, 148–151, 153–156
 emergence of Joe's, 195–197
 Jason's scheme
  corroboration, 126–127
  recursive partitioning, 126
  reversible partitive fraction scheme, 125
 Laura's construction of
  apparent construction of recursive partitioning, 169
  engendering accommodation, 165–167
  feedback system, 166
  feed-forward system, 166
  metamorphic accommodation, 167–168
  partitive fraction scheme, 168
 partitioning operations, progress in, 157–161
 reversibility of Joe's, 228
 splitting operations and, 162–163
Unit fraction of a connected number, 174–176
Unit fractions with unlike denominators, construction strategies, 291–295
 invariability of one-third of a fixed quantity, 292–293
 numerical relations between two different fractions, 294
Units-coordinating scheme
 anticipatory, 226–227
 children's use in, 83–86
 for improper fractions, 114–116
 reorganization in Melissa's, 231–236
  conjecture, 233
  fractions commensurate with one-fifth, 231–232

V
Vomvoridi, E., 341
von Glasersfeld, E., 11, 19–20, 28, 30, 37

W
Washburne, C., 13–14
Watzlawick, P., 14
West, M.M., 350
Whole numbers, 13, 15

CPSIA information can be obtained at www.ICGtesting.com
Printed in the USA
LVOW080313151111

255007LV00008B/22/P